Neuroscience Year

Neuroscience Year

Supplement 3 to the Encyclopedia of Neuroscience

Edited by

Barry Smith and George Adelman

With 71 Figures

A *Pro Scientia Viva* Title

Birkhäuser

Boston · Basel · Berlin

First Printing 1993

Library of Congress Cataloging-in-Publication Data

Neuroscience year : supplement 3 to the Encyclopedia of neuroscience /
   edited by Barry Smith and George Adelman.
        p.   cm.
      "A Pro scientia viva title."
      Includes bibliographical references and index.
      ISBN 0-8176-3592-0 (alk. paper).—ISBN 3-7643-3592-0 (alk.
   paper)
      1. Neurosciences. 2. Neurobiology. I. Smith, Barry H.
   II. Adelman, George, 1926–   . III. Encyclopedia of neuroscience.
      [DNLM: 1. Neurosciences. WL 13 E56 1987 Suppl.3]
   RC334.E53   1987, Suppl. 3
   612.8′03—dc20
   DNLM/DLC
   for Library of Congress   92-49552
                                                          CIP

Printed on acid-free paper.

© 1993 Birkhäuser Boston.

Birkhäuser 

Typeset by Compset, Inc.
Printed and bound by Edwards Brothers, Ann Arbor, MI
Printed in the United States of America

9 8 7 6 5 4 3 2 1

ISBN 0-8176-3592-0
ISBN 3-7643-3592-0

Preface

This is the third and final supplement to the 1987 edition of the *Encyclopedia of Neuroscience*. As with the earlier two supplements, *Neuroscience Year 1* and *Neuroscience Year 2*, the categories of entries in this volume include (1) developing topics in neuroscience that were not covered as separate entries in the *Encyclopedia of Neuroscience* or were only briefly covered in other related articles (for example, "AIDS and the Brain," given separate entry treatment in *Neuroscience Year 1*, is only briefly covered in the article, "Neurovirology," in the *Encyclopedia*); (2) newly emerging topics that were not covered in the *Encyclopedia* and that are associated with possible research breakthroughs (for example, "Atrial Natriuretic Factor and Brain"); and (3) updates of topics in which there have been significant developments (for example, "Alzheimer's Disease").

The *Encyclopedia of Neuroscience* is a two-volume work containing over 700 articles by authorities in the field and encompassing all of neuroscience. We have defined neuroscience broadly as an interdisciplinary science covering all the fields involved in the investigation of how the brain and nervous system mediate behavior, including the mental and emotional behavior of humans. The *Encyclopedia* surveys the physical-chemical-electrical substrates, the anatomical-biological structures, the physiological-neurological transductive mechanisms, and the behavioral-psychological outcome of all this interactive processing—the full spectrum of body and mind activities, from vegetative functions and motor activities to higher brain function. Clinical-medical neuroscience (including psychiatry, as well as neurology and neuropharmacology) is an essential aspect of *Encyclopedia* coverage as are the various and many fields of basic research, from molecular genetic neuroscience and membrane biophysics to the upper reaches of behavioral psychology and mathematical-theoretical neurobiology.

The contributors to *Neuroscience Year,* as in the *Encyclopedia,* are professional neuroscientists who were invited to write on the basis of their expertise and their contributions to the field. The articles are written for a wide audience—professional neuroscientists and their students as well as a broad range of professionals and students from other scientific areas who want brief, ready-reference access to the field. General readers, interested in and concerned about research progress in brain-mind problems and issues, will also find these articles interesting and useful. Some of the articles, because of the nature of the subject, are fairly technical; but even the most technical include introductory statements that place the topic in the context of the broader field.

The topics to be covered and the authors to write the articles were selected on the basis of recommendations by Scientific Advisory Board members as well as our ongoing review of the current literature of neuroscience. Our efforts were supported by library searches using electronic search techniques. All articles have been peer reviewed and carefully edited.

Neuroscience has made enormous research advances, in the past two decades especially. Powerful new tools—imaging devices, computing systems, molecular genetic analyses—are being used on a regular basis to solve problems that only a few years ago were not even approachable. Because of this rapid growth, neuroscience seems to have become compartmentalized into a mosaic of subspecialties, each with its own identity. But for practical as well as conceptual (heuristic) reasons, neuroscience also remains a unified, interdisciplinary or multilevel science brought together by its singular, even Promethean, goal: the understanding of how the brain-mind works. Our purpose with this ongoing neuroscience research information project—comprising the *Encyclopedia of Neuroscience* and its supplement volumes in the *Neuroscience Years* series—is to further this goal of a unified neuroscience and to assure that the explosive new developments that are the hallmarks of this field are made available to as wide an audience as possible. We expect that this reference source will be especially useful in this "Decade of the Brain."

The Neuroscience Research Information Database, using the *Encyclopedia of Neuroscience* as its structural base, continues to be an important goal. It will keep all entries of the *Encyclopedia* and its supplements readily available for updating and revisions. With its planned tie-in to other data bases in biology, psychology, and the various health sciences, our data base will allow users, expert and newcomer alike, ready information access to the world of basic and clinical neuroscience.

Entries in this volume appear in traditional encyclopedia alphabetical arrangement. For those who may find a subject guide to the topics useful, we have also included a listing of all the papers in a subject array. This grouping of the entries will allow the reader to appraise new developments in the four broad areas of neuroscience: molecular and cellular neuroscience, cell assemblies, brain and behavior, and clinical-medical.

Finally, we wish to acknowledge with sincere thanks the editorial support and assistance of Jennifer De Pasquale of The Health Foundation and James Doran of Birkhäuser Boston.

George Adelman
Barry H. Smith

Contents

* Updated from original *Encyclopedia of Neuroscience*

Contents by Subject

Contributors

György Ádám Department of Comparative Physiology, Eötvös Loránd University of Budapest, Budapest 1088, Hungary
Viscerosensory Functions

Garrett E. Alexander Department of Neurology, Emory University School of Medicine, Atlanta, Georgia 30322, USA
Basal Ganglia, Anatomy and Circuitry

Julie K. Andersen Neurology, Molecular Neurogenetics Unit, Massachusetts General Hospital and Neuroscience Program, Harvard Medical School, Charlestown, Massachusetts 02129, USA
Herpes Simplex Virus and Its Use in Neuroscience Research

Paul Bach-y-Rita Department of Rehabilitation Medicine, University of Wisconsin Medical School, Madison, Wisconsin 53705, USA
Volume (Extrasynaptic) Transmission

Anthony S. Basile Laboratory of Neuroscience, National Institute of Diabetes, Digestive and Liver Diseases, National Institutes of Health, Bethesda, Maryland 20892, USA
Hepatic Encephalopathy

Andrew M. Batchelor Department of Physiology, University of Liverpool, Liverpool L69 3BX, England
Cerebellum, Messenger Molecules

Michel Baudry Neuroscience Program, University of Southern California, Los Angeles, California 90089-2520, USA
Polyamines in the Nervous System

Laurence Edward Becker Department of Pathology (Neuropathology) and Pediatrics, University of Toronto and The Hospital for Sick Children, Toronto, Ontario, Canada M56 1X8
Sudden Infant Death Syndrome (SIDS): A Neural Perspective

Robert E. Becker Department of Psychiatry and SIU Alzheimer Center, Southern Illinois University School of Medicine, Springfield, Illinois 62794-9230, USA
Alzheimer's Disease, Pharmacologic Therapy

Yehezkel Ben-Ari Laboratoire de Neurobiologie et Physiopathologie du Developement, INSERM U-29, Paris 75014, France
Glucose-Sensitive Potassium Channels

Hagai Bergman Department of Physiology, The Hebrew University, Hadassah Medical School, Jerusalem 91010, Israel
Parkinsonism, Effects of Lesions of the Subthalamic Nucleus

Ilene L. Bernstein Department of Psychology, University of Washington, Seattle, Washington 98105, USA
Cachexia and Tumor Necrosis Factor

Jan Krzysztof Blusztajn Department of Pathology and Psychiatry, Boston University School of Medicine, Boston, Massachusetts 02118, USA
Signal Transduction in Neurons by Phospholipid Breakdown Products

Daniel F.B. Bossut Department of Physiology, University of North Carolina at Chapel Hill, Chapel Hill, North Carolina 27599, USA
Pain Receptors, Peripheral, and Chronic Pain

Mark R. Brann Department of Psychiatry, College of Medicine, University of Vermont, Burlington, VT 05405-0068 USA
G-Proteins and Neuronal Signal Transduction

Xandra O. Breakefield Neurology, Molecular Neurogenetics Unit, Massachusetts General Hospital and Neuroscience Program, Harvard Medical School, Charlestown, Massachusetts 02129, USA
Herpes Simplex Virus and Its Use in Neuroscience Research

Robert H. Brown, Jr. Day Neuromuscular Research Laboratory, Massachusetts General Hospital East and Harvard Medical School, Charlestown, Massachusetts 02129, USA
Hyperkalemic Periodic Paralysis

W. Ted Brown Department of Human Genetics, New York State Institute for Basic Research in Developmental Disabilities, Staten Island, New York 10314, USA
Fragile X Syndrome

Theodore H. Bullock Department of Neurosciences, University of California, San Diego, La Jolla, California 92093-0201, USA
Induced Rhythms, Oscillations in the Brain

Michel J. Calache Department of Psychiatry and SIU Alzheimer Center, Southern Illinois University School of Medicine, Springfield, Illinois 62794-9230, USA
Alzheimer's Disease, Pharmacologic Therapy

Dennis S. Charney Yale University School of Medicine, West Haven Veterans Affaris Medical Center, West Haven, Connecticut 06519, USA
Panic Disorder, Psychobiology
Post-Traumatic Stress Disorder (PTSD), Psychobiology

Jill Clayton-Smith Department of Clinical Genetics, St. Mary's Hospital, Manchester M13 0JH, England
Angelman Syndrome

David P. Corey Harvard Medical School and Massachusetts General Hospital, Howard Hughes Medical Institute, Boston, Massachusetts 02114, USA
Hair Cells, Sensory Transduction

Valérie Crépel Laboratoire de Neurobiologie et Physiopathologie du Developement, INSERM U-29, Paris 75014, France
Glucose-Sensitive Potassium Channels

Michael D. Crutcher Department of Neurology, Emory University School of Medicine, Atlanta, Georgia 30322, USA
Basal Ganglia, Anatomy and Circuitry

Lourdes J. Cruz University of the Philippines, Diliman, Quezon City, Philippines
Conus Venom Neuropeptides

Antonio R. Damasio Division of Behavioral Neurology and Cognitive Neuroscience, University of Iowa College of Medicine, Iowa City, Iowa 52242, USA
Prosopagnosia

Mahlon R. DeLong Department of Neurology, Emory University School of Medicine, Atlanta, Georgia 30322, USA

Parkinsonism, Effects of Lesions of the Subthalamic Nucleus
Basal Ganglia, Models of Function: Normal and Disease

Paul Ekman   Department of Psychiatry and Langley Porter Psychiatric Institute, University of California, San Francisco, San Francisco, California 94143, USA
Facial Expression and Emotion

Rodger J. Elble   Division of Neurology, Department of Medicine, Center for Alzheimer Disease and Related Disorders, Southern Illinois University School of Medicine, Springfield, Illinois 62794-9230, USA
Tremor

Joseph G. Feghali   Department of Otolaryngology, Albert Einstein College of Medicine, Montefiore Medical Center, Bronx, New York 10467, USA
Vertigo

Bertrand Fontaine   Service de Neurologie et de Neuropsychologie and INSERM U134, Hôpital de la Salpêtrière, Paris 75013, France
Hyperkalemic Periodic Paralysis

Fred H. Gage   Department of Neurosciences, University of California, San Diego, La Jolla, California 92093, USA
Brain Grafts, Genetic Engineering

John Garthwaite   Department of Physiology, University of Liverpool, Liverpool L69 3BX, England
Cerebellum, Messenger Molecules

Anne A. Gershon   Department of Pediatrics, Division of Pediatric Infectious Disease, Columbia University, New York, New York 10032, USA
Zoster and Postherpetic Neuralgia

Joseph A. Ghika   Department of Neurology, Centre Hospitalier, Universitaire Vaudois (CHUV), Lausanne, Switzerland
Wilson's Disease

Ezio Giacobini   Department of Pharmacology, Southern Illinois University School of Medicine, Springfield, Illinois 62794-9230, USA
Alzheimer's Disease, Pharmacologic Therapy

Sid Gilman   Department of Neurology, The University of Michigan Medical Center, Ann Arbor, Michigan 48109-0316, USA
Olivopontocerebellar Atrophy

Andrew W. Goddard   Department of Psychiatry, Yale University School of Medicine, Clinical Neuroscience Research Unit, New Haven, Connecticut 06519, USA
Panic Disorder, Psychobiology

James F. Gusella   Molecular Neurogenetics Laboratory, Massachusetts General Hospital East and Harvard Medical School, Charlestown, Massachusetts 02129, USA
Hyperkalemic Periodic Paralysis

Jeffrey C. Hall   Department of Biology, Brandeis University, Waltham, Massachusetts 02254, USA
Drosophila, *Genes and Behavior*

Richard Hawkes   Department of Anatomy, Neuroscience Research Group, University of Calgary Health Science Center, Calgary, Alberta, Canada T2N 4NI
Zebrins: Compartment Markers in the Cerebellum

Michael J. Ignatius   Department of Molecular and Cell Biology, Division of Neuroscience, University of California, Berkeley, Berkeley, California 94720, USA
Extracellular Matrix, Laminin and Integrins

Thomas R. Insel   Laboratory of Neurophysiology, National Institute of Mental Health, Poolesville, Maryland 20837, USA
Obsessive-Compulsive Disorder, Neurobiology of

Masao Ito   Frontier Research Program—RIKEN, Wako Saitama 351 01, Japan
Long-term Depression (LTD)

Edmund C. Jenkins   Department of Cytogenetics, New York State Institute for Basic Research in Developmental Disabilities, Staten Island, New York 10314, USA
Fragile X Syndrome

Alan E. Kazdin   Department of Psychology, Yale University, New Haven, Connecticut 06520, USA
Depression, Childhood, and Treatment

Louise S. Kiessling   Departments of Pediatrics and Family Medicine, Brown University School of Medicine and Memorial Hospital of Rhode Island, Pawtucket, Rhode Island 02860, USA
Attention Deficit Hyperactivity Disorder

Hideo Kimura   The Salk Institute, San Diego, California 92138, USA
Glial Growth Factors

Joan H.M. Knoll   Division of Genetics, Children's Hospital, Harvard Medical School, Boston, Massachusetts 02115, USA
Prader-Willi Syndrome

Martin Koltzenburg   Neurologische Universitäts-Klinik, D-8700 Wurzburg, Germany
Nociceptors, Novel Classes

Stephen M. Kosslyn   Department of Psychology, Harvard University, Cambridge, Massachusetts 02138, USA
Mental Imagery

John H. Krystal   Yale University School of Medicine, West Haven Veterans Affairs Medical Center, West Haven, Connecticut 06516, USA
Post-Traumatic Stress Disorder (PTSD), Psychobiology

Louis M. Kunkel   Howard Hughes Medical Institute, Children's Hospital Medical Center and Harvard Medical School, Boston, Massachusetts 02115, USA
Dystrophin and Duchenne Muscular Dystrophy

David J. Kwiatkowski   Division of Experimental Medicine, Brigham and Women's Hospital, Harvard Medical School, Boston, Massachusetts 02115, USA
Gelsolin

Marc Lalande   Division of Genetics, Children's Hospital, Harvard Medical School and Howard Hughes Medical Institute, Boston, Massachusetts 02115, USA
Prader-Willi Syndrome

Irwin B. Levitan   Department of Biochemistry and Center for Complex Systems, Brandeis University, Waltham, Massachusetts 02254, USA
Protein Phosphorylation and Neuronal Modulation

Benjamin Libet   Department of Physiology, University of California School of Medicine, San Francisco, California 94143, USA
Conscious Experience, Neural Basis

Hart G.W. Lidov   Departments of Pathology and Neurology, Children's Hospital Medical Center, Boston, Massachusetts 02115, USA
Dystrophin and Duchenne Muscular Dystrophy

Lisa A. Matsuda   Department of Psychiatry and Behavioral Sciences, Medical University of South Carolina, Charleston, South Carolina 29425, USA
Marijuana Receptor Gene, Cloning and Characterization

Uel Jackson McMahan Department of Neurobiology, Stanford University School of Medicine, Stanford, California 94305, USA
Agrin

Stephen McMahon Department of Physiology, United Medical and Dental Schools, St. Thomas' Campus, London SE1 7EH, England
Nociceptors, Novel Classes

Thomas H. Milhorat Department of Neurosurgery, State University of New York, Health Science Center at Brooklyn, Brooklyn, New York 11203, USA
Pseudotumor Cerebri

J.P. Mohr Stroke Unit, Neurological Institute, New York, New York 10032, USA
Transcranial Sonography

P. Read Montague Computational Neurobiology Laboratory, Salk Institute, San Diego, California 92186-5800, USA
The NO Hypothesis

Dominique Muller Department of Pharmacology, Centre Médical Universitaire, 1211 Genève 4, Switzerland
Wilson's Disease

Ulrich Müller Institut für Humangenetik, Justus-Liebig-Universität Giessen, D-6300 Giessen, Germany
Dystonia-Parkinsonism Syndrome, X-linked (XDP)

Imad Najm Neuroscience Program, University of Southern California, Los Angeles, California 90089-2520, USA
Polyamines in the Nervous System

Baldomero M. Olivera Department of Biology, University of Utah and Marine Science Institute, Salt Lake City, Utah 84112, USA
Conus *Venom Neuropeptides*

Edward R. Perl Department of Physiology, University of North Carolina at Chapel Hill, Chapel Hill, North Carolina 27599, USA
Pain Receptors (Peripheral) and Chronic Pain

Stephen J. Peroutka Department of Neurology, Stanford University, Stanford, California 94305, USA
Serotonin (5-Hydroxytryptamine) Receptor Subtypes: Clinical Relevance

Jerome B. Posner Department of Neurology, Memorial Sloane-Kettering Cancer Center, New York, New York 10021, USA
Paraneoplastic Disorders

Jasodhara Ray Department of Neurosciences, University of California, San Diego, La Jolla, California 92093, USA
Brain Grafts, Genetic Engineering

Richard M. Restak Department of Neurology, Georgetown University Medical School, Washington, DC 20009, USA
Neuropsychiatry

Mary M. Robertson Academic Department of Psychiatry, University College and Middlesex Schools of Medicine (UCMSM), London W1N 8AA, England
Gilles de la Tourette Syndrome and Self-injurious Behavior

Stephen D. Roper Department of Anatomy and Neurobiology, Colorado State University, Fort Collins, Colorado 80523, USA, and The Rocky Mountain Taste and Smell Center, University of Colorado Health Science Center, Denver, Colorado 80262, USA
Taste Transduction

Erika L. Rosenberg Department of Psychiatry and Langley Porter Psychiatric Institute, University of California, San Francisco, San Francisco, California 94143, USA
Facial Expression and Emotion

Lisa M. Shin Department of Psychology, Harvard University, Cambridge, Massachusetts 02138, USA
Mental Imagery

David M. Simpson Clinical Neurophysiology Laboratory, Mount Sinai Medical Center, New York, New York 10029, USA
Bell's Palsy

Barbara E. Slack Department of Brain and Cognitive Sciences, Massachusetts Institute of Technology, Cambridge, Massachusetts 02139, USA
Signal Transduction in Neurons by Phospholipid Breakdown Products

Steven M. Southwick Yale University School of Medicine, West Haven Veterans Affairs Medical Center, West Haven, Connecticut 06516, USA
Post-Traumatic Stress Disorder (PTSD), Psychobiology

David Sternman Department of Neurology, St. Luke's Roosevelt Hospital Center, New York, New York 10023, USA
Bell's Palsy

William H. Sweet Harvard Medical School, Massachusetts General Hospital, Boston, Massachusetts 02114, USA
Malignant Tumors of Brain and Head, Boron Neutron Capture Treatment

Daniel Tranel Division of Behavioral Neurology and Cognitive Neuroscience, University of Iowa College of Medicine, Iowa City, Iowa 52242, USA
Prosopagnosia

Bertrand Y. Tuan Division of Experimental Medicine, Brigham and Women's Hospital, Harvard Medical School, Boston, Massachusetts 02115, USA
Gelsolin

N. Venketasubramanian Stroke Unit, Neurological Institute, New York, New York 10032, USA
Transcranial Sonography

Joseph Wagstaff Division of Genetics, Children's Hospital, Harvard Medical School, Boston, Massachusetts 02115, USA
Prader-Willi Syndrome

Lawrence Weiskrantz Department of Experimental Psychology, University of Oxford, Oxford OX1 3PS, England
Blindsight, Residual Vision

Thomas Wichmann Department of Neurology, Emory University School of Medicine, Atlanta, Georgia 30322, USA
Parkinsonism, Effects of Lesions of the Subthalamic Nucleus
Basal Ganglia, Models of Function: Normal and Disease

Leonhard S. Wolfe Donner Laboratory of Experimental Neurochemistry, Montréal Neurological Institute, McGill University, Montréal, Québec, Canada H3A 2B4
Prostaglandins in the CNS: Arachidonic Acid

Clifford J. Woolf Department of Anatomy and Developmental Biology, University College London, London WC1E 6BT, England
Pain as Learning

Josef Zihl Max Planck Institute of Psychiatry, Clinical Institute, 8000 Munich 40, Germany
Reading Disorders, Nonaphasic, and Their Treatment

Sylvie Zini Laboratoire de Neurobiologie et Physiopathologie du Developement, INSERM U-29, Paris 75014, France
Glucose-Sensitive Potassium Channels

Agrin

Uel Jackson McMahan

The surface of skeletal muscle fibers at their neuromuscular junctions is characterized by aggregates of proteins. The aggregates include acetylcholine receptors (AChR) and acetylcholinesterase (AChE) which are essential for synaptic transmission. Agrin, a protein extracted from basal lamina fractions of the synapse-rich electric organ of the marine ray, induces cultured myotubes to form aggregates of AChR, AChE and other postsynaptic proteins. Several lines of evidence have led to the hypothesis that agrin in both embryonic and adult vertebrates is synthesized in the cell bodies of motor neurons in the CNS and is transported along their axons to muscle where it is released from the axons' terminals into the synaptic cleft. There it binds to components of the muscle fibers' basal lamina and to an agrin receptor in the muscle fibers' plasma membrane triggering the myofibers to aggregate AChRs and other proteins that compose the postsynaptic apparatus. Agrin is apparently in many neurons throughout the nervous system and, thus, it may also play a role in the formation of the postsynaptic apparatus at neuron-neuron synapses.

Activity

The electric organ of the ray is homologous to skeletal muscle and is richly innervated by motor neurons. It has been useful for purifying and characterizing other components of cholinergic synapses, such as AChR and AChE. There are 150 kDa and 95 kDa forms of active agrin in extracts of basal lamina fractions of the electric organ. They are proteolytic fragments of a larger protein cleaved either during extraction or during *in situ* posttranslational modification. When applied to cultured chick myotubes both forms cause a 3-20 fold increase in the number of AChR aggregates in the myotube plasma membrane. Agrin has little or no effect on myotube size, total number of AChRs on the myotube surface or the rate of AChR degradation. The AChR-aggregating activity is dose-dependent and due, at least in part, to the lateral migration of AChRs present in the plasma membrane at the time extracts are applied. The increase in number is first seen 2 hr after adding agrin to the medium and it is maximal by 24 hr. Agrin also causes chick myotubes to form aggregates of AChE (membrane and collagen-tail forms), butyrylcholinesterase, heparan sulfate proteoglycan and 43 kDa AChR-associated protein, all of which are aggregated in the postsynaptic apparatus of the chick neuromuscular junction. Each accumulates with a time course similar to that of the AChR aggregates and the aggregates tend to be coextensive.

Evidence that supports the agrin hypothesis

Agrin is synthesized in the cell bodies of motor neurons: Extracts of motor neuron-enriched fractions of the spinal cord are enriched in proteins that induce cultured myotubes to aggregate AChRs and AChE, and these proteins are immunoprecipitated by anti-agrin antibodies. mAbs against members of the agrin family stain the golgi apparatus in motor neurons. PCR experiments have revealed that spinal cord fractions enriched in motor neurons are enriched in transcripts that code for agrin, and *in situ* hybridizations using probes for agrin transcripts result in heavy labelling of motor neurons.

Agrin is transported along motor axons to muscle: If nerves are ligated material stained by anti-agrin mAbs accumulates on the cell body side of the ligature. Extracts of the cell body side of ligated nerves are enriched in proteins that induce cultured myotubes to form AChR aggregates and the active proteins are immunoprecipitated by anti-agrin mAbs.

Agrin is externalized by motor neuron axon terminals to trigger myofibers to aggregate AChRs at neuromuscular junctions: Antiserum that blocks the activity of chick agrin but not rat agrin inhibits the motor neuron-induced aggregation of AChRs at neuromuscular junctions in chick motor neuron-chick myotube cocultures but not in rat motor neuron-chick myotube cocultures.

Agrin binds to basal lamina in the synaptic cleft at the neurosmuscular junction: If adult muscles are damaged in ways that result in degeneration and phagocytosis of all muscle fibers and axon terminals but spare the muscle fibers' basal lamina sheaths, the new muscle fibers that develop inside the sheaths of the original muscle fibers aggregate AChRs and AChE at the former synaptic site on the sheaths even in the absence of innervation. Moreover, agrin antibodies stain the synaptic basal lamina and the antigenic proteins remain concentrated in the synaptic basal lamina for weeks after axons and myofibers have degenerated.

Structure

Agrin cDNAs have been isolated from ray, chick and rat gene libraries. Because the only agrin that has been purified is from ray, ray agrin cDNA was the first to be cloned. Northern blot analysis and sequence data have revealed that the ray cDNA codes for the carboxyl-terminal half of the protein. Ray cDNA has since been used to isolate homologous cDNAs from chick and rat libraries. These cDNAs also only code for the carboxyl-terminal half of their proteins, but full-length sequences have been obtained for chick and rat by overlapping cDNAs from primer extension libraries.

The proteins encoded by the full-length cDNAs for chick and rat are approximately 1900 amino acids long, with predicated molecular weights of approximately 200 kD. The mRNA size for both species is about 8kb. Two messages, of approximately 7.5 and 9 kb, are detected in northern blots of ray tissue; the two sizes may be due to variation in the size of the 5′ and/or 3′ untranslated regions. As illustrated in Fig. 1, agrin has a number of domains that show similarities

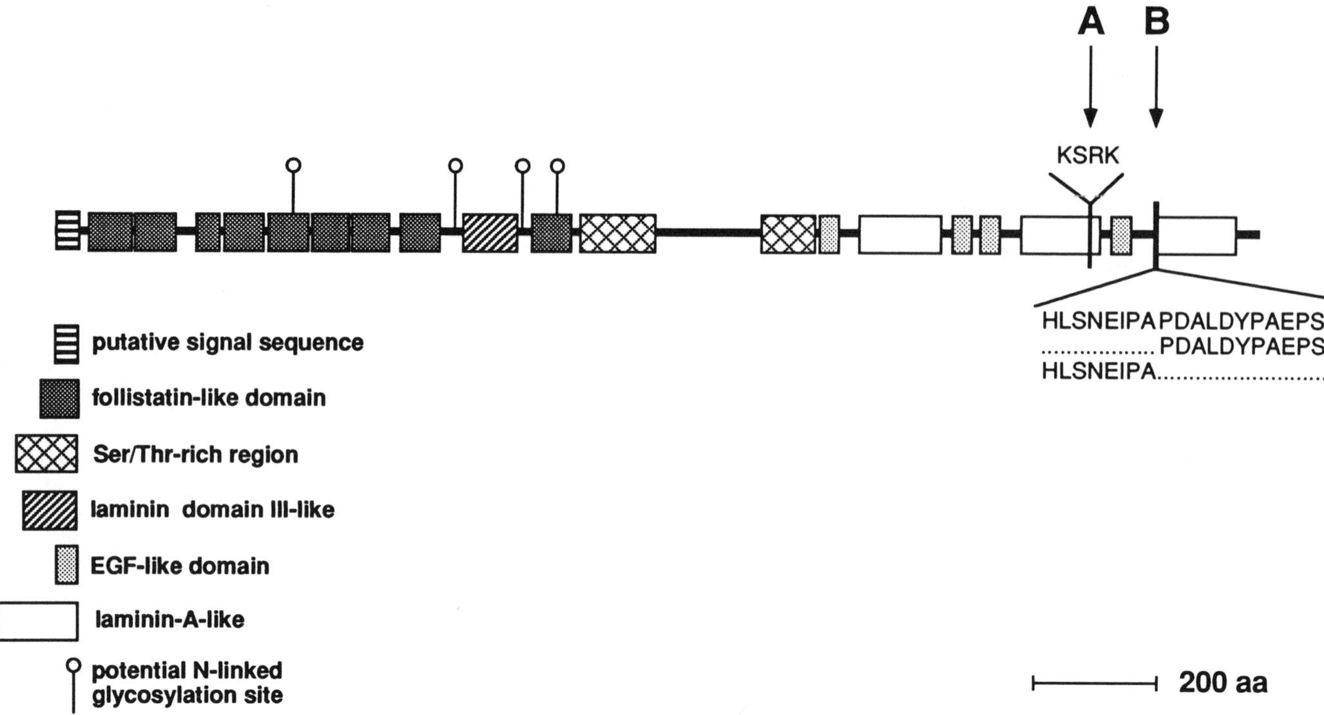

Figure 1. Structural map of chick agrin.

with regions of other extracellular proteins. The nature, number and polarized distribution of the different domains is consistent with idea of agrin being multifunctional, i.e. interacting with more than one protein.

Studies on ray agrin and its cDNA revealed that the carboxyl-terminal half of the full length protein is sufficient for aggregating AChRs, AChE and other postsynaptic proteins on the surface of cultured myotubes. Several isoforms of agrin result from alternative splicing of a common precursor mRNA in the region coding for the C-terminal half of the protein. The splicing occurs at two positions, designated A and B (Fig. 1). At position A there can be a 4 amino acid insert and at position B there can be inserts of 8, 11, or 19 (8 + 11) amino acids. The results of expression studies using chick and rat agrin cDNAs are consistent with the conclusion that only those isoforms having inserts at both positions A and B are active in aggregating AChR and AChE in chick myotube cultures. A rat isoform having the 4 amino acid insert at A but no insert at B has some AChR aggregating activity in rat and mouse myotube cultures but much less than isoforms having both the A and B inserts. Thus, the presence or absence of A and B inserts has a profound effect on agrin's function as an AChR/AChE aggregating protein.

Distribution

PCR and insitu hybridization experiment, provide strong evidence that motor neurons synthesize agrin isoforms having both the A and B inserts. Dissociated cells from embryonic days 5 and 6 spinal cords can be separated on a metrizamide gradient or by panning with a motor neuron-specific antibody into a fraction enriched in motor neurons and a fraction enriched in other cells including non-motor neurons. PCR shows that for either separation method the motor neuron-enriched fraction is enriched in agrin transcripts coding for the 19 or the 11 amino acid inserts at position B (B_{19} and B_{11} inserts). The enrichment for mRNA encoding isoforms having the B_{19} insert is much greater than that for transcripts encoding isoforms having the B_{11} insert. In situ hybridizations confirm the presence of transcripts encoding B_{19} and B_{11} isoforms in the spinal motor column (the region containing motor neurons) of embryonic day 5 and 6 chick spinal cords with the B_{19} transcript being in much greater abundance and highly concentrated in these regions. In situ hybridizations indicate that at least some motor neurons contain B_8 isoforms but the transcripts reach levels of detection at later embryonic stages than those at which the transcripts for the B_{19} and B_{11} isoforms are first observed. It has not yet been determined whether the different agrin isoforms, are in the same or different motor neurons. PCR on embryonic day 5 and 6 chick spinal cord and brain indicates that all agrin transcripts in the chick CNS that code for B inserts also code for A inserts. The finding that agrin having the A and B inserts, both of which are required for activity, is present in motor neurons in the spinal cord of embryonic days 5 or 6 chicks is consistent with the agrin hypothesis; it is at this time that the motor neurons begin to form neuromuscular junctions and induce the myotubes to aggregate AChRs and other postsynaptic proteins.

PCR experiments, indicate that spinal cord motor neurons in embryonic rat, as in embryonic chick, contain transcripts encoding agrin isoforms having both A and B inserts with those coding for B_{19} inserts more abundant than those coding for B_{11} or B_8 inserts. Motor neurons of the electric lobe of the ray, which provide the neuronal agrin to the electric organ from which active agrin has been purified, also contain agrin transcripts coding for isoforms having both the A and B inserts but most, if not all, of the transcripts code for B_8 inserts.

Agrin immunoreactivity is present in basal lamina in numerous non-neural tissues indicating that agrin is not specific to the nervous system. Several non-neuronal chick tissues including embryonic muscle, heart, lung, kidney and liver have been screened by PCR to determine the isoforms of they contain. The only agrin transcripts detected in these tissues coded for agrin isoforms lacking B inserts. These finding suggest that the only agrin having B inserts, and thus highly active in AChR aggregation on cultured chick myotubes, is synthesized in the nervous system.

Agrin antibodies stain throughout the central and peripheral nervous systems of embryos and adults. Among the most intensely stained structures are the basal lamina sheaths of capillary endothelial cells in the central nervous system and the basal lamina sheaths of Schwann cells in the peripheral nervous system. Insitu hybridizations on tissue sections of perpheral and central nervous system indicate that neurons other than motor neurons contains agrin transcripts as do at least some of the nervous systems' non neural cells. PCR using mRNA extracted from several different regions of the embronic chick brain which contain the mRNA of both neurons and non-motor neurons (endothelial cells and glia or Schwann cells) revealed that these regions had agrin isoform with and without B inserts. On the other hand, regions of the nervous system which contain mRNA of glial and endothelial cells but not neurons showed only isoforms lacking B inserts. Together these findings suggest that neurons other than motor neurons synthesize isoforms of agrin that have B inserts and that in most, if not all, cases where non-neuronal cells in the nervous system express agrin isoforms the isoforms lack B inserts. The finding indicating that neurons other than motor neurons express agrin isoforms having B inserts together with the observation that in chick, at least, all agrin isoforms having a B insert have an A insert, both of which are required for AChR aggregating activity, support the hypothesis that agrin directs the formation of postsynaptic protein aggregates at neuron-neuron synapses throughout the nervous system as it does at the neuromuscular junction.

Further reading

Ferns M, Hall ZW (1992): How many agrins does it take to make a synapse? *Cell* 70:1–3

McMahan, UJ (1990): The agrin hypothesis. *Cold Spring Harbor Symp. Quant.* Biol. 50:407–418.

McMahan UJ, Horton SE, Werle MJ, Honig LS, Kröger S, Ruegg MA, and Escher G (1992): Agrin isoforms and their role in synaptogenesis. *Curr Op Cell Biol* 4:896–874

Rupp F, Ozcelik T, Linial M, Peterson K, Francke U, Scheller R (1992): Structure and chromosomal localization of the mammalian agrin gene. *J Neurosci* 9:3535–3544

Alzheimer's Disease, Pharmacologic Therapy

Michel J. Calache, Ezio Giacobini, and Robert E. Becker

Alzheimer's disease (AD) is characterized by a slow progressive deterioration of memory and other higher cortical functions. These result in an impairment of thinking and judgment, and personality changes. In addition, depression, psychosis, and behavioral changes may occur during the course of the illness.

Effective therapy of AD must control the disease process as well as treat and prevent complications. Such a treatment approach requires the use of a multidisciplinary team that would implement the medical, behavioral, familial, and institutional interventions needed to care for AD patients and their families. The pharmacologic treatment that is discussed in this chapter represents only one tool in the management of AD.

Definitive treatment of AD

Effective treatment of the cognitive decline in Alzheimer's disease awaits an improved understanding of the disease process or identification of an efficacious drug. Unfortunately, the etiology and pathogenetic processes important to the clinical progression of AD are unknown. Theories address possible causes in heredity, amyloid accumulation, microtubular dysregulation, viral and immune etiologies, aluminum toxicity, increased oxygen radical concentrations, and nerve growth factor dysregulation. None of these hypotheses have been translated into effective treatment approaches. No efficacious drug has been found, although attempts have been made to correct deficiencies in the several neurotransmitter neuromodulator systems affected in AD. These include drugs that alter acetylcholine, norepinephrine, and serotonin systems. It is hoped that improved understanding of basic neurochemical alterations may lead to the development of alternative therapies that attenuate the expression of AD symptoms, and retard or halt the disease process.

The neurotransmitter approach. Cholinergic innervation to the medial temporal cortex and thalamic regions of the brain is reduced in AD. The cholinergic deficit consists of a reduction in (1) choline uptake, (b) choline acetyltransferase and acetylcholinesterase activity, and (c) acetylcholine (ACh) synthesis. Several strategies have been used to correct the cholinergic deficits in AD patients. These strategies are (a) precursor (choline) loading, (b) use of cholinesterase inhibitors (ChEI), and (c) administration of cholinergic muscarinic agonists. Both choline and lecithin are ACh precursors. Although four double-blind studies using choline or lecithin reported modest positive results, nine studies did not show any significant improvement.

Cholinesterase inhibitors (ChEI), on the other hand, prevent the breakdown of ACh in the synapse. This increases the availability of ACh at receptor sites. The main ChEI that have been used in AD research are physostigmine (Phy), tacrine, metrifonate, and heptyl-physostigmine. Physostigmine's half-life is only about 30 minutes. This limits its clinical usefulness. The results of single-dose or short-term trials with oral Phy were mainly negative. However, four out of five long-term studies using Phy for more than 2 months reported moderate improvement on neuropsychological testing, but not according to clinical or functional criteria. This could indicate that long-acting ChEI might lead to better results in the future. Tacrine, (tetrahydroaminoacridine, THA) is similar to Phy but with much longer duration of action. Both double-blind short-term and long-term (8 weeks and more) studies with THA reported small but positive results. The results of combination therapy of lecithin with either Phy or THA were similar to monotherapy with either drug. Improvement was not significantly different or impressive. Two other ChEI under investigation are metrifonate and heptylphysostigmine. Initial results with both of these drugs indicate a positive effect on memory in laboratory animals. Unfortunately many drugs shown to improve learning and memory performance in laboratory rats have not shown efficacy in humans. Metrifonate has been reported to improve cognitive function in one open clinical trial of AD patients.

Functional cholinergic neurons are necessary for ACh precursor and ChEI effect. Muscarinic agonists have the advantage that their effect is not dependent on the integrity of presynaptic cholinergic neurons. The main cholinergic muscarinic agonists used in AD patients are SE-86 and bethanechol. Positive results were obtained with SR-86. The reported improvement was on neuropsychological testing, but not clinically or functionally. The results with intracerebral injection of betanechol were less impressive. Unfortunately, the serious side effects associated with the use of these drugs limit their use. The side effects included nausea, dizziness, angina, bronchospasm, syncopy, confusion, seizure, and surgical morbidity such as infection and hematoma when intracerebral ventricular infusion is required.

Other neurotransmitters have also been implicated in AD. There is evidence of a loss of adrenergic neurons in the locus ceruleus as well as deficits in the serotonin system. A few clinical trials demonstrated a beneficial effect of L-tryptophan and a L-tryptophan tyrolase inhibitor in some patients. However, clinical trials using antidepressants to increase the availability of biogenic amines in nondepressed demented elderly failed to improve the cognitive and functional capacities of AD patients.

Other treatment approaches. Aluminum toxicity has been implicated in AD. In a two-year single-blind trial the chelating agent desferroxamine led to significant reduction in the rate of decline of daily living skills in AD patients. However, the treatment group did not show any evidence of improvement from a baseline level. The results of this study need to be replicated.

Among the neuropeptides and proteins believed to be altered in AD are somatostatin, substance P, and the nerve growth factor (NGF). These substances are currently under experimental use in AD patients. Other approaches include using drugs that would be effective in minimizing excitotoxin glutamate-induced injury; potassium channel blockers that delay repolarization and might increase evoked release of ACh; antioxidants such as tocopherol and the MAO-B inhibitor deprenyl to reduce free radical concentrations. Also, immunotherapy, depending on the nature of the immunologic deficit encountered in a particular patient, is currently being investigated. Four subsets of immunodeficiency have been found in AD: a subset with a defect (membrane fluidity) of a specific T lymphocyte, a subset with circulating antibodies to axon filaments, a third subset with antibodies to brain antigens, and a subset due to lymphocyte deficiencies of more than one biochemical factor.

Treatment of psychiatric complications

Depression. Depression occurs in up to 60% of cases of AD. The cause of depression in AD is unknown. Several factors may predispose demented patients to develop mood disturbance. These may include genetic factors that predispose an individual to depression, a history of premorbid depression, neuroanatomical deficits in the frontal cortex, or neurotransmitter changes in norepinephrine (NE) or 5-HT that occur in AD. Clinical experience has shown that depression in dementia is in many cases treatable and that frequently cognition improves secondary to mood improvement. Yet one double-blind study has indicated that the improvement in some depressed AD patients may not be related to the administration of antidepressant drugs.

The use of antidepressants with anticholinergic effects needs to be avoided in AD. Therefore, drugs such as trazodone, fluoxetine, and bupropion may be particularly useful because of their low side effects profile and low cardiac toxicity. As a rule, one third to one half the usual adult dose will be satisfactory for the elderly. Monoamine oxidase inhibitors are also useful. MAO activity has been found to be elevated in dementia, but it is not known if this is related to the appearance of symptoms. In case of resistance, electroconvulsive therapy (ECT) should be available.

Psychosis. Psychotic symptoms occur in a large majority of demented patients. The incidence of hallucinations varies from 3–49% and of delusions from 16–80%. The duration of these delusions may vary from a few days to several months or years. Neuroleptic treatment is indicated and frequently effective. Neuroleptics with low anticholinergic effects are recommended to avoid delirium. If indicated, long-acting neuroleptics can be used. These were shown to be effective and safe in the elderly. Side effects frequently include extrapyramidal syndromes. These may not be apparent until drug accumulation is reached following 1–2 weeks of treatment. For patients developing extrapyramidal side effects, it is recommended to decrease the dosage or switch to another class of neuroleptics rather than to add anticholinergic medication.

Behavioral disorders. About 50% of demented patients exhibit behavioral disturbances. These behavioral disturbances are not specific to AD. Similar behavioral disturbances are encountered in organic brain syndromes (e.g., head injury, mental retardation, and schizophrenia). Behavioral disturbances in dementia may or may not accompany secondary complications such as depression, psychosis, or delirium.

Certain behavior disturbances such as wandering and inappropriate voiding are frequently resistant to psychopharmacological interventions, and behavioral and environmental treatments are better approaches. Other behavioral disturbances, such as restlessness, tearfulness, and verbal outbursts and aggressivity, respond better to psychopharmacological intervention. In these conditions, treatment may include the judicial use of antipsychotics and antidepressants, or the discontinuation of drugs that may have precipitated the behavioral change.

Several antiaggressive agents are used in the treatment of behavioral disturbance in the demented elderly. These are the neuroleptics, carbamazepine, β-blockers, lithium, serotoninergic drugs, and benzodiazepines. Several placebo-controlled studies support the definite role of the neuroleptics in the treatment of agitation in the demented elderly. The majority of patients do respond quickly and do not require chronic treatment. However, there have been a few antipsychotic discontinuation studies that indicate that some agitated demented patients may need long-term antipsychotic treatment. Akathisia is a potentially important adverse effect. It should be identified and appropriately managed in order not to confuse it with agitation.

The patient who has mood lability, psychomotor hyperactivity, and occasional aggression may benefit from lithium or carbamazepine. The pharmacokinetics of lithium necessitate careful monitoring in the elderly patient. Smaller doses of lithium are needed because of its hydrophilic properties and its small volume of distribution. Also, there are reports that indicate that the elderly may have a higher incidence of neurotoxicity with presumed therapeutic doses of lithium. Five clinical trials demonstrated the effectiveness of carbamazepine as well as β-blockers in controlling agitation in a group of demented population resistant to neuroleptic treatment. The doses of propranolol used varied from 80 to 560 mg per day. Three clinical trials evaluated the effect of carbamazepine with and without a neuroleptic mediation. In all studies carbamazepine proved to be effective in controlling agitation in the elderly with blood levels from 8 to 12 μg/dL. Since it takes from 2 to 7 weeks for carbamazepine and 2 months for β-blockers to demonstrate antiaggressive effects, these drugs are more useful for the chronic resistant cases of agitation.

Benzodiazepines should be reserved for acute agitation, or for very brief courses of treatment to control restlessness at night. This class of medications has been known to be relatively safe. However, benzodiazepines are known to induce tolerance, rebound insomnia, anxiety, paradoxical rage, and impairment of memory. They are not the drugs of choice for long-term treatment of aggression, particularly in AD patients.

Serotoninergic abnormalities in AD include decrease in 5HT in frontal and temporal cortex, decrease in CSF 5-HIAA, decrease in cortical 5HT2 and 5HT1 receptors, decrease in 5HT uptake and release from presynaptic sites in biopsy specimens, and tangle formation in the dorsal raphe nucleus. Also, because of the implication of serotonin in aggression, depression, sleep, memory, and repetitive behavior, serotoninergic drugs are of particular interest in the treatment of AD. Several clinical reports indicate the usefulness of trazodone and buspirone in controlling sleep disturbances and aggression in the agitated demented patient. Buspirone is effective in controlling repetitive behavior (e.g., tapping) but has not been studied for this indication in the demented elderly. Clomipramine, fluoxetine, and sertriline are potent serotonin reuptake inhibitors. Clomipramine may not be the drug of choice for AD patients because of its anticholinergic properties. Fluoxetine is frequently associated with jitteriness and anorexia. It also has a very prolonged duration of action because of its slow metabolism, especially in the elderly. A newly released serotonin uptake inhibitor, sertraline, is more rapidly metabolized and may be more safely administered. Clinical trials are needed to compare the potential antiaggressive properties of these drugs.

Sleep

Sleep disorders in the demented elderly may be associated with delirium, depression, iatrogenic effects from medications, sleep apnea, or myoclonus. However, even when all these factors are taken into consideration, profound sleep disturbances and circadian rhythm abnormalities occur in dementia patients. These consist of loss of slow wave sleep and increased amount of nighttime wakefulness and, later, loss of REM sleep and the breakdown of the sleep-wake circadian rhythm, with significant amounts of sleep occurring during the day. Sleep disturbance needs to be treated only if disruptive to the milieu, or if the patient is exhibiting dangerous agitation with aggressiveness toward self or family. Sleep hygiene should be the first measure to recommend. This may include such measures as regular sleep habits, using the bed for sleep only, encouraging daily physical activity and ambulation, and avoiding stimulants such as caffeine. Naps should be discouraged. However, certain patients experiencing sundown syndrome with increasing confusion in the late evening do get better if they take short naps in the afternoon.

Once the diagnosis of the cause of sleep disturbance is established the treatment should be directed toward the identified cause. It is recommended not to add an additional medication to minimize the risk of complications. In case a specific treatment for insomnia is indicated, trazodone, diphenhydramine, and short-acting benzodiazepines are appropriate drugs to be used. Trazodone and diphenhydramine are relatively safe drugs, but tolerance may develop to the former. Benzodiazepines should be used briefly for the patient with transient episodes of insomnia. Short-acting benzodiazepines (e.g., triazolam) may be a better choice since they do not have a hangover effect and possess a low risk for accumulation. Benzodiazepines are highly lipophilic drugs and tend to accumulate in the adipose tissue of the elderly, who have an increased body fat to lean muscle tissue ratio. Benzodiazepines that are not oxidized but are directly eliminated as a glucuronide (e.g., lorazepam) have less tendency to accumulate than those that are oxidized (e.g., diazepam). Hepatic oxidation, but not glucuronidation of drugs, is impaired in the elderly. Thus, the clinician must consider the altered metabolism of drugs in the elderly in selecting and dosing drugs for patients with AD.

Further reading

Giacobini E, Becker R (1989): Advances in the therapy of Alzheimer's disease. In: *Familial Alzheimer's Disease: Molecular Genetics and Clinical Perspectives*. Miner GD, Richter RW, Blass JP, Valentine JL, Winters-Miner LA, eds. New York and Basel: Marcel Dekker

Lawlor BA (1990): Serotonin and Alzheimer's disease. *Psych Ann* 20 (10):567–569

Maletta GJ (1990): Pharmacologic treatment and management of the aggressive demented patient. *Psych Ann* 20 (8):446–455

Vitiello MV, Prinz PN (1989): Alzheimer's disease: Sleep and Sleep/wake patterns. *Clin Geriat Med* 5 (2):289–299

Angelman Syndrome

Jill Clayton-Smith

In 1965 Harry Angelman, an English pediatrician, reported three children with severe mental retardation, easily provoked laughter, a seizure disorder, and a similar pattern of facial features. They all had jerky, ataxic movements and he called them "puppet children" because of their resemblance to marionettes. In a further report by Bower and Jeavons the condition was referred to as "happy puppet syndrome" and this name continued in popular use over the next 20 years, although the condition was reported very infrequently during this time. In 1987 two cytogeneticists, Kaplan and Magenis, working independently described a deletion of chromosome 15 in the region 15q11–q13 in several patients with this syndrome. Following this the condition has been diagnosed and reported more frequently and is not as rare as previously thought, with an estimated incidence of 1 in 20,000. The term "happy puppet" is not popular with those involved in the care of these children and Angelman syndrome (AS) is now the preferred name.

The characteristic clinical features of Angelman syndrome (see Table 1) include microcephaly with a flattened occiput and a horizontal occipital groove. The mouth is wide and smiling and is held open and the upper lip is thin. The teeth are widely spaced and there may be bowing of the primary dentition due to persistent tongue thrusting. The mandible, although not large, is prominent and the chin is pointed. The midface is small and the eyes deep set. Although Angelman children share many features in common with each other and with other children who have chromosomal disorders, the facial dysmorphism is subtle and AS cannot always be diagnosed by looking at the face alone, without taking into consideration the characteristic behavioral and neurological features. The facial features evolve with age so that the infant with AS is not dysmorphic but becomes more so in later childhood. As the face elongates the jaw becomes more prominent and the mouth wider, perhaps as a result of the persistent smiling and tongue thrusting. In older patients the prominent chin is the most obvious feature. Hypopigmentation is a common finding in AS. Over half of the patients have blond hair and blue eyes and are significantly paler than other family members. Lack of pigmentation in the eye can

Table 1. Clinical Features Seen in Angelman Syndrome

OFC < 3rd percentile	25%
OFC < 50th percentile	98%
Brachycephaly	90%
Occipital groove	35%
Macrostomia	75%
Widely spaced teeth	60%
Blue eyes	88%
Blond hair	65%
Pointed chin or prognathism	95%
Tongue protrusion	70%
Scoliosis	10%
Ataxia	100%
Strabismus	40%
Feeding problems	75%
Seizures	80%
Delayed motor milestones	100%
Absent speech or < 3 words	98%

OFC = occipito-frontal circumference.

cause nystagmus and strabismus and in a few cases gives rise to reduced visual acuity.

The parents of an AS child usually notice developmental problems at around 6 months of age. Pregnancy and delivery are normal, with the children being on average 200 g lighter than their healthy siblings. Seventy-five percent have significant feeding problems during the first months and failure to establish breast feeding and severe gastroesophageal reflux are common. Motor milestones are delayed with average age at sitting being 12 months, crawling 22 months, and walking unaided 39 months. 80 % of AS children have a seizure disorder that may be very difficult to control. Seizures commonly begin around 18 to 24 months and all seizure types occur. They are precipitated in many cases by common triggers such as fever, teething, and tiredness, but in addition they often follow an episodic pattern with severe bouts of seizures lasting several weeks separated by fit-free periods of up to several months. The EEG in all cases is abnormal even when seizures are not occurring, and the bizarre changes have been well documented by Boyd et al. The severity of the seizures and the EEG changes are age related and become less frequent with time. Seizures decrease in frequency and may cease altogether over the age of 10 years.

AS patients have a characteristic gait with a wide base and stiff legs. The hands are upheld and there is a tendency to flap the hands, especially when excited. There is marked ataxia, and neurological examination reveals truncal hypotonia but increased tone in the limbs with brisk reflexes. Contractures may develop as a result of this. CT scans are normal or show mild cerebral atrophy. MRI scans have been performed on some patients and heterotopia is an occasional finding.

Speech is a specific problem area for AS children. In a study of 85 patients in the UK none had more than five words of speech and most had none or only a single word. Some can communicate using sign language or primitive gestures, and comprehension of receptive language is better than expressive language. Intellect is difficult to assess because of the lack of suitable scales with which to assess the capabilities of these nonverbal children, but is within the severe range of mental retardation.

Information regarding the neuropathology in Angelman syndrome is limited since it was described fairly recently in the pediatric literature and many adults with this condition remain undiagnosed. AS patients generally keep good health and there is only a single case report in the literature detailing autopsy findings. In this case a 21 year old patient who died from pneumonia had a small (910 g) brain with a normal gyral pattern and only mild cerebral atrophy. There was simplification of the dendritic tree within the cerebral cortex. The most striking feature was marked cerebellar atrophy with loss of Purkinje and granule cells and extensive gliosis. Neurochemical studies indicated that there was a decrease in gamma-aminobutyric acid (GABA) content within the cerebellar cortex and increased glutamate levels in the frontal and occipital cortex. This patient was on six different anticonvulsant preparations at the time of death, but the early onset of symptoms in AS may reflect the failure of the cerebellum to develop rather than degeneration related to medication. Further studies are needed to substantiate the findings in this isolated report, but the suggestion that reduction of GABA may be implicated in the neuropathology of AS is interesting in view of recent reports suggesting that one of the genes within the chromosome 15q11–13 region that is typically deleted in AS is a GABA receptor subunit.

Detailed high-resolution cytogenetic analyses have now

been performed on several hundred AS children, and in 60% of cases there is a *de novo* deletion of chromosome 15q11–13 that arises on the maternally derived chromosome. This is in contrast to the Prader-Willi syndrome (PWS), the other condition known to be associated with a 15q11–13 deletion, where the deletion arises on the paternal 15. The differing parental origins of the same deletion in the two conditions is considered to be due to genomic imprinting, a phenomenon whereby a gene or genes are expressed differently depending on the parent of origin. In a further 5% of AS patients there is a rearrangement of the chromosome 15, usually a translocation or inversion in association with a deletion of 15q11–13. The rearrangement may be carried in a balanced form by other family members. Several families have now been reported where there are first cousins with AS and PWS, both arising as a result of a *de novo* deletion when a familial rearrangement is inherited from the mother or father, respectively. The remaining 35% cases of AS have normal chromosomes on cytogenetic examination. None of the affected siblings with AS reported in the literature to date have had a cytogenetic deletion.

The 15q11–13 region has now been investigated extensively by molecular genetic techniques, using probes that span this region. A physical map of the region is now established, although it is possible that there is still a critical region within 15q11–13 for which probes are not available. Molecular studies have revealed that the deletions when present are usually large and that those in AS are of a similar size to those in PWS even though the latter appear smaller cytogenetically. There are a group of patients where there is no cytogenetic deletion, but there is a large molecular deletion. A further group do not have any deletions with the probes currently available. This last group includes all the sibling pairs with the exception of one family reported from Japan where the mother, the maternal grandfather, and three affected children all share a deletion with a single probe. This lends evidence to the hypothesis that familial Angelman syndrome is inherited as a dominant disorder, modified by genomic imprinting, and indicates that the AS and the PWS loci may well be separate. The mother inherited the deletion from her father but was not affected by PWS because the PWS locus was not involved. However, when she passed it on to her three children they all inherited the deletion at the AS locus from their mother and hence were all affected.

One of the most important aspects of Angelman research involves investigation of the mechanisms of genomic imprinting, since AS and PWS provide a perfect human model. Imprinting involves the modification of DNA, presumably at the time of development of the germ cells such that expression is altered dependent upon the parent of origin. When a gene is imprinted in this way it is necessary to have a copy of both the maternal and the paternal imprint for normal development because it is likely that for one stage of development one is reliant on the paternal copy and for another stage on the maternal copy. This has been proven in AS by the

demonstration of uniparental disomy in around 2% of sporadic patients. Both chromosome 15s are inherited from the father and there is no 15 from the mother. In effect, this has the same consequence as a maternal deletion. It is likely that uniparental disomy arises as a result of a trisomy 15 conceptus losing one chromosome 15 in order to remain viable as trisomy 15 is a common finding in spontaneous abortions. In one out of the three cases the maternal 15 will be lost and uniparental disomy will result.

The actual mechanism leading to imprinting of the germ line is unclear, but the imprint has to be erasable so that the germ cells for the next generation can be reimprinted depending on the parent of origin. It is thought that the process of methylation is related to imprinting, perhaps acting to lock in the imprint, and methylation studies performed on AS and PWS patients who have deletions or uniparental disomy have demonstrated distinct methylation differences for probes within 15q11–13 between the two conditions, both being different from the normal pattern.

It is likely that in the near future the gene or genes responsible for AS will be identified, and this may provide guidance as to which therapies are suitable, for example, for the seizure disorder of AS. Several different genetic mechanisms giving rise to AS have already been identified and were mentioned previously. There is a low risk of recurrence in families with a *de novo* deletion or uniparental disomy, providing the parents' chromosomes are normal, and in those with rearrangements recurrence risk depends on the individual rearrangement in question, and a prenatal test is often available. In those families where there is more than one affected child and where there is no deletion, the genetic mechanism has yet to be identified in order to offer accurate genetic counseling and a prenatal test if wished. These familial cases may be arising due to a point mutation at a critical locus and further work will help to clarify this.

Further reading

Angelman H (1965): Puppet children: A report on three cases. *Dev Med Child Neurol* 7:681–688

Boyd SG, Harden A, Patton MA (1988): The EEG in early diagnosis of the Angelman (happy puppet) syndrome. *Eur J Paediatr* 147:508–513

Jay V, Becker L, Chan FW, Thomas LP Snr (1991): Puppet-like syndrome of Angelman: A pathologic and neurochemical study. *Neurology* 41:416–422

Knoll JH, Nicholls RD, Magenis RE, Glatt K et al (1990): Angelman syndrome: Three molecular classes identified with chromosome 15q11–13 specific DNA markers. *Am J Hum Genet* 47:149–155

Malcolm S, Clayton-Smith J, Nichols M, Robb S, Webb T, Pembrey ME (1991): Uniparental disomy in Angelman syndrome. *Lancet* 337:694–697

Williams CA, Gray BA, Hendrickson JE, Stone JW, Cantu ES (1989): Incidence of 15q deletions in the Angelman syndrome: A survey of twelve affected persons. *Am J Med Genet* 32:339–345

Attention Deficit Hyperactivity Disorder

Louise S. Kiessling

Attention deficit hyperactivity disorder (ADHD) is the term used for the syndrome characterized by short attention span, impulsivity, distractibility, poor impulse control, and fidgety and sometimes hyperactive behavior. The official current description is based on the American Psychiatric Association's *Diagnostic and Statistical Manual III—Revised* (DSMIII-R) criteria.

Currently, DSMIII-R is being revised, but there is also a committee at work developing a primary-care–oriented manual of behavioral pathology, the *Diagnostic and Statistical Manual—Primary Care* (DSM-PC), scheduled for publication in 1994. This is particularly important in pediatric disorders because most children are treated and followed by primary care physicians or neurologists and relatively few by child psychiatrists.

Primary care physicians have long been dissatisfied with the way the criteria have been developed in DSMIII-R, particularly with regard to the occurrence of attention deficit disorder (ADD) without hyperactivity, which was essentially omitted from DSMIII-R. There are several recent studies in the pediatric literature reconfirming the presence of ADD without hyperactivity. From a neuroscience perspective these are important findings because the ADD without hyperactivity group must be studied separately from the group with hyperactivity to establish whether they differ neurophysiologically, and clinically they are important because the children with ADD without hyperactivity tend to be misdiagnosed and improperly treated.

The remainder of this article will address two areas where recent research has added to our knowledge even while raising still more questions about the basic physiology underlying ADHD. Recently, several researchers have been using the technology of radioactive tracers and positron emission tomography (PET) to study brain physiology.

One of these groups, led by Hans C. Lou in Denmark, has studied groups of children with ADHD, ADHD plus other deficits (ADHD+), and normal children and adolescents, using ^{133}xenon inhalation and PET to study regional blood flow. In one of these studies, 6 of the children had pure ADHD, 13 had ADHD+, and there were 9 controls. The studies were performed at rest with eyes open. Results showed that in the ADHD group, the striatal regions appeared hypoperfused, while the primary sensory and sensorimotor areas seemed relatively hyperperfused. Only the differences in the right striatum, the occipital lobe, and the left sensorimotor and primary auditory regions reached statistical significance when the ADHD and control children alone were compared. Relative hypoperfusion of the striatum suggests underfunctioning of the basal ganglia–thalamocortical circuits modulating the frontal lobe output through inhibition, while the hyperperfusion of the sensory areas suggests some release from subcortical (striatal) feedback inhibition. The combination could result in the increased responsiveness to multiple sensory stimuli and the decreased selective attention one sees in the ADHD child.

Of interest, 4 of the ADHD children and 9 from the ADHD+ group had the study repeated 30 to 60 minutes after oral administration of their daily dose of 10 to 30 mg of methylphenidate hydrochloride (Ritalin). This resulted in an increase in the regional cerebral blood flow to both striata,

with only the change in the left striatum reaching statistical significance. These changes paralleled clinical improvement.

Another research group studied adults who fit the criteria for childhood onset of ADD with hyperactivity according to the *Diagnostic and Statistical Manual of Mental Disorders* (third edition, not revised) as well as the Utah criteria for ADD in adulthood. These adults also had biologic children who met criteria for the diagnosis of ADD with hyperactivity. They underwent PET scanning using [^{18}F]fluoro-2-deoxy-D-glucose administered intravenously to measure cerebral glucose metabolism while they performed an auditory attention task. A control group of adults without ADD or other psychiatric disorders was also used.

Global cerebral glucose metabolism was just over 8% lower in the adults with hyperactivity than in the normal control adults. Regional glucose metabolism was diminished in 30 of 60 areas measured ($p < .05$) in the adults with hyperactivity, including both cortical and subcortical regions. When the data for specific regions were normalized (dividing the regional value by the global glucose values), four regions were found to have significantly lower metabolism in the patients than in the controls—primarily in the left premotor and somatosensory cortex. The regions with significantly diminished normalized glucose metabolism represent the left medial frontal, left anterior frontal, left posterior frontal, and Rolandic regions.

While these studies used different techniques, they both suggest hypofunction in areas previously felt to be involved in maintenance of selective attention, planning, and freedom from distractibility. Clearly, further work that can be replicated is essential. What is thought provoking is the consistency of the evidence for hypofunctioning of subcortical and cortical regions in people who clinically present with excess activity. The data are consistent with theories that have postulated that stimulant medications improve attention and concentration by tightening inhibitory control. One would postulate that by tightening control, one increases function in the areas involved in maintenance of attention, leading to increased selective metabolism in those areas.

Finally, recent work by numerous researchers examined the effects of sugar on aggressive and inattentive behavior in normal children and children with ADHD. The common belief is that refined sugars exacerbate problem behavior in ADHD and learning-disabled children. However, no scientific study has documented consistent deficits following sucrose challenges. In one study, the children with ADHD were more aggressive than the control children and there was no change in aggression when the children with ADHD were given a challenge of sucrose after a carbohydrate meal (compared with aspartame or saccharin challenges). One measure of distractibility, the ability of the child to discriminate then respond to target or nontarget during a continuous performance task, showed a significant decrement 4 hours after the combination of a high-carbohydrate meal with a sucrose challenge.

Clearly, more work is necessary looking at morning nutrition and its role in problem behavior. Perhaps it is the carbohydrate load and not the sugar that is critical and what is being seen is the postprandial loginess of many people after high carbohydrate ingestion, particularly in a boring or minimally stimulating environment. This study has not laid to rest the issue of the effect of refined sugar on problem behavior, but it may have helped delineate the next step in the research spiral.

Further reading

Barkley R, DuPaul G, McMurray M (1990). Comprehensive evaluation of attention deficit disorder with and without hyperactivity as defined by research criteria. *J Consult Clinical Psych* 56:775–789

Lou H, Henriksen L, Bruhn P, Borner H, Nielsen JB (1989): Striatal dysfunction in attention deficit and hyperkinetic disorder. *Arch Neurol* 46:48–52

Wender E, Solanto M (1991): Effects of sugar on aggressive and inattentive behavior in children with attention deficit disorder with hyperactivity and normal children. *J Ped* 88:960–966

Zametkin A, Nordahl T, Gross M, King A, Semple W, Rumsey J, Hamburger S, Cohen R (1990): Cerebral glucose metabolism in adults with hyperactivity of childhood onset. *NEJM* 323:1361–1366

Basal Ganglia, Anatomy and Circuitry

Michael D. Crutcher and Garrett E. Alexander

Two different aspects of the anatomy and circuitry of the basal ganglia will be developed. First, the nuclei and neurotransmitters of the basal ganglia as well as the connections of these nuclei with the cerebral cortex and thalamus will be described. This description will emphasize the *serial* nature of basal ganglia connectivity. Second, a family of functionally distinct, *parallel,* segregated circuits that involve distinct subregions of each of these nuclei will be described.

Nuclei and basic circuitry

Traditionally, the *caudate nucleus, putamen,* and *globus pallidus* (GP) have been considered the primary components of the basal ganglia. However, because of similarities in embryological origin, cytology, and connectivity, several other nuclei are often considered part of the basal ganglia as well, including the *nucleus accumbens,* parts of the *olfactory tubercle,* the *substantia nigra* (SN), and the *subthalamic nucleus* (STN).

The nuclei of the basal ganglia can be grouped into three categories: input nuclei, output nuclei, and nuclei involved in intrinsic circuits among the basal ganglia nuclei. Most of the *inputs* to the basal ganglia are received by the caudate nucleus, putamen, nucleus accumbens, and olfactory tubercle. The caudate nucleus and putamen develop from the same telencephalic structure. As a result they are fused anteriorly and have identical cell types and histological appearance. Together they are referred to as the striatum or, more specifically, the neostriatum. Because of embryological and anatomical similarities, the nucleus accumbens and part of the olfactory tubercle have come to be considered as part of the striatum as well and are referred to as the ventral striatum. These four components of the striatum can be viewed as a single aggregate. Most of the *outputs* from the basal ganglia arise from the internal segment of GP (GPi) and the pars reticulata of the SN (SNr). There is evidence that GPi and SNr also have the same embryologic origin. They also have identical cell types and histological appearance and are often considered as a single entity. The remaining nuclei are involved in *intrinsic circuits* among the nuclei of the basal ganglia. They include the external segment of GP (GPe), the pars compacta of the SN (SNc), and the STN.

For many years the striatum was viewed as a relatively homogeneous structure. However, with the advent of new anatomical techniques a complex internal organization has been discovered. The striatum is composed of a mosaic of two compartments: cell *islands* (or striosomes) surrounded by a *matrix* compartment. Heterogeneities in markers for neurotransmitters, afferent inputs, and efferent projection neurons are often in register with the island-matrix compartmentalization. For example, striatal neurons that project to GPi and/or SNr are located within the matrix, while those that project to SNc are confined to the island/striosome compartment.

The anatomical connections of the basal ganglia with the cerebral cortex and thalamus, the neurotransmitters, and circuitry are shown in Figure 1. The most substantial inputs to the basal ganglia are the excitatory, glutamatergic projections to the striatum from the cerebral cortex. The striatum also receives input from the thalamus and an important dopaminergic projection from SNc. The two primary categories of cells in the striatum are the medium-sized, GABAergic, projection neurons, which have abundant dendritic spines, and the large, aspiny, cholinergic interneurons. Projection neurons are of two types: one type contains both

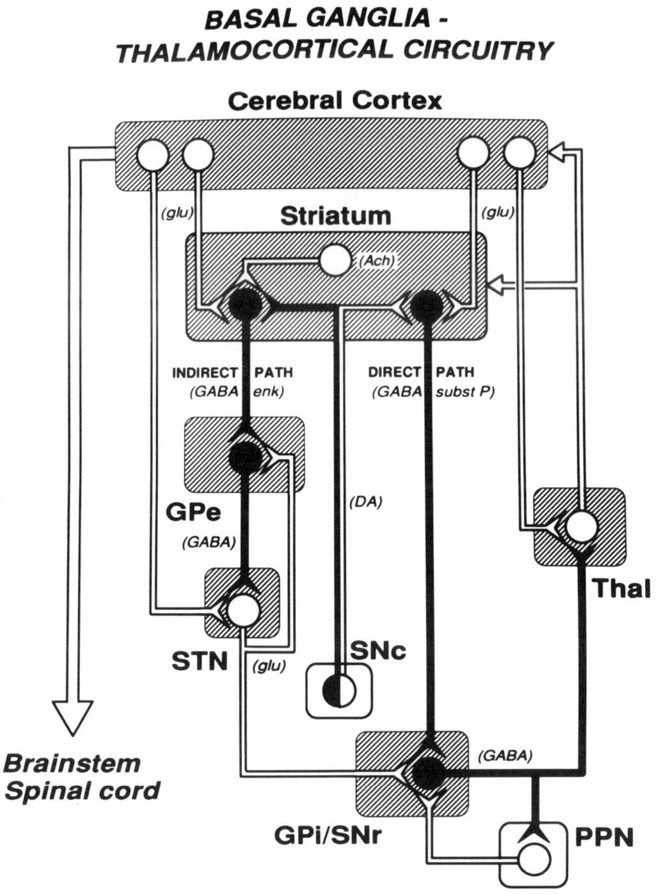

Figure 1. Schematic diagram of the connectivity and neurotransmitters of the basal ganglia–thalamocortical circuitry. Excitatory connections are open and inhibitory connections are filled. ACh, acetylcholine; DA, dopamine; enk, enkephalin; GABA, gamma aminobutyric acid; glu, glutamate; PPN, pedunculopontine nucleus; subst P, substance P; Thal, thalamus.

GABA and substance P and projects to the output nuclei of the basal ganglia (GPi and SNr); the other contains both GABA and enkephalin and projects to GPe. The cholinergic interneurons are thought to project preferentially to the GABA/enkephalin cells. Neurons in GPe have large, radially arranged dendritic fields and provide a purely GABA-ergic input to the STN. STN neurons in turn have glutamatergic projections back to GPe and to the output nuclei (GPi and SNr). The output neurons of both GPi and SNr are large neurons that send GABA-ergic projections to the thalamus as well as to the pedunculopontine nucleus (PPN) of the midbrain tegmentum. The PPN sends excitatory projections back to the basal ganglia.

By virtue of their high rates of spontaneous discharge, the basal ganglia output nuclei (GPi and SNr) exert a tonic, inhibitory effect on their target nuclei in the thalamus. This tonic inhibitory outflow from the basal ganglia appears to be differentially modulated by two opposing but parallel pathways. The *direct pathway* to GPi and SNr arises from the striatal neurons that contain both GABA and substance P. Excitation of these neurons, which are normally quiescent, by their cortical inputs tends to disinhibit the thalamus by inhibiting the inhibitory neurons in GPi and SNr. The *indirect pathway* to GPi and SNr from the striatum passes first to GPe, then from GPe to the STN, and finally to GPi and SNr. The high spontaneous discharge rate of most GPe neurons exerts a tonic inhibitory influence on the STN. Activation of the GABA/enkephalin striatal projection neurons tends to suppress the activity of GPe neurons and thereby disinhibit the STN, thus increasing the excitatory drive on GPi and SNr. This, in turn, results in increased inhibition of the thalamus. Consequently, the two striatal efferent systems (the direct and indirect pathways) appear to have opposing effects upon the GPi and SNr and thus upon the thalamic targets of basal ganglia outflow.

The role of dopamine within the basal ganglia appears to be complex, and many issues remain unresolved. There is evidence, however, that the dopaminergic projection from SNc may exert contrasting effects on the direct and indirect striatal GPi and SNr pathways. Dopamine appears to have a net excitatory effect on striatal neurons that send GABA/substance P projections to GPi and SNr via the direct pathway, and a net inhibitory effect on those that send GABA/enkephalin projections to GPe via the indirect pathway. Thus, in effect, the overall influence of dopamine within the striatum may be to reinforce any cortically initiated activation of a particular basal ganglia–thalamocortical circuit by both facilitating conduction through that circuit's direct pathway and suppressing conduction through the indirect pathway.

Functionally distinct, parallel circuits

The basal ganglia were long considered to be primarily or exclusively involved in the control of movement. However, recent anatomical, physiological, and clinical evidence has forced a reassessment of such a unitary role for the basal ganglia. For example, it is now apparent that the output of the basal ganglia is not funneled solely to cortical motor areas but rather is directed by way of the thalamus to most regions of the frontal neocortex. Also, it is now widely accepted that the basal ganglia contribute to a variety of behavioral functions in addition to motor control, including oculomotor, cognitive, and even "limbic" processes.

From the evidence that is currently available, it appears that there are multiple, parallel, segregated circuits through the basal ganglia. For each of these circuits, several functionally related cortical areas project to a distinct portion of the striatum. Each of these striatal subregions engages, in turn, restricted and largely nonoverlapping portions of GP, STN, SN, and the thalamus. Finally, the output of each of these circuits is directed via the thalamus to a specific receiving area in the frontal cortex (see Fig. 2). In addition to being anatomically segregated, each of these parallel circuits is functionally distinct. Several such circuits have been distinguished, including the "motor" circuit, the "oculomotor"

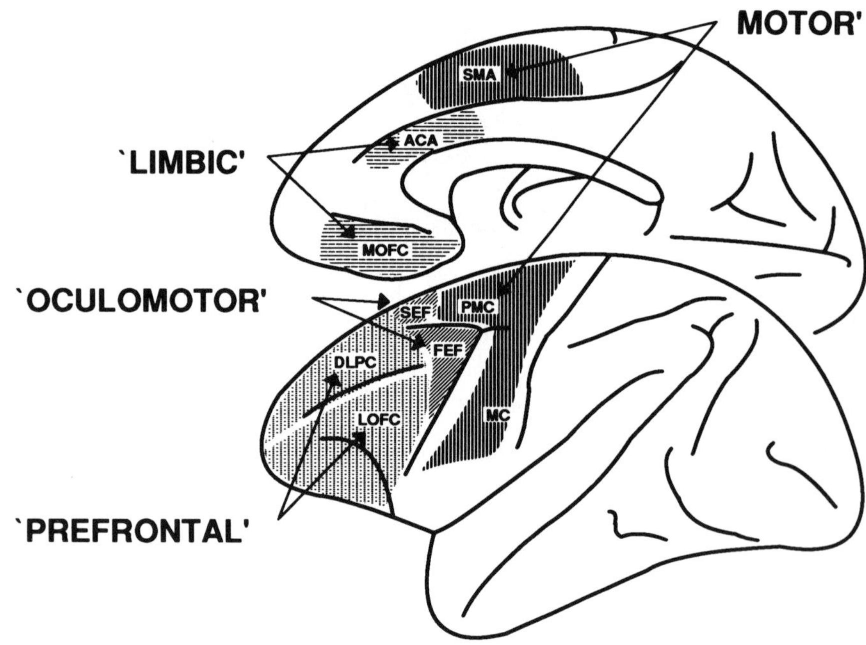

Figure 2. Frontal lobe targets of basal ganglia output. This figure illustrates schematically the cortical areas that receive the output of the separate basal ganglia–thalamocortical circuits. ACA, anterior cingulate area; DLPC, dorsolateral prefrontal cortex; FEF, frontal eye field; LOFC, lateral orbitofrontal cortex; MC, primary motor cortex; MOFC, medial orbitofrontal cortex; PMC, premotor cortex; SEF, supplementary eye field; SMA, supplementary motor area.

circuit, two "prefrontal" circuits (dorsolateral prefrontal and lateral orbitofrontal), and a "limbic" circuit. It is likely that with time additional circuits with additional behavioral functions will be described as well.

Motor circuit. In primates, the "motor" portion of the striatum includes most of the putamen. Several areas of the neocortex that are involved in the control of movement project to specific regions of the putamen. These include the primary motor cortex (MC), the premotor cortex (PMC), the supplementary motor area (SMA), the primary somatosensory cortex, and the somatosensory association cortex. The motor circuit is closed by its thalamic projections to the SMA and to parts of the PMC and MC. All of the connections of the motor circuit are topographically organized. For example, the "arm" areas of each of these motor and somatosensory cortices project to the same subarea of the motor putamen, while the "leg" areas project to a different subarea. Consequently, each component of the motor circuit is somatotopically organized (see Fig. 3). The motor circuit is involved in the preparation and execution of limb and body movements.

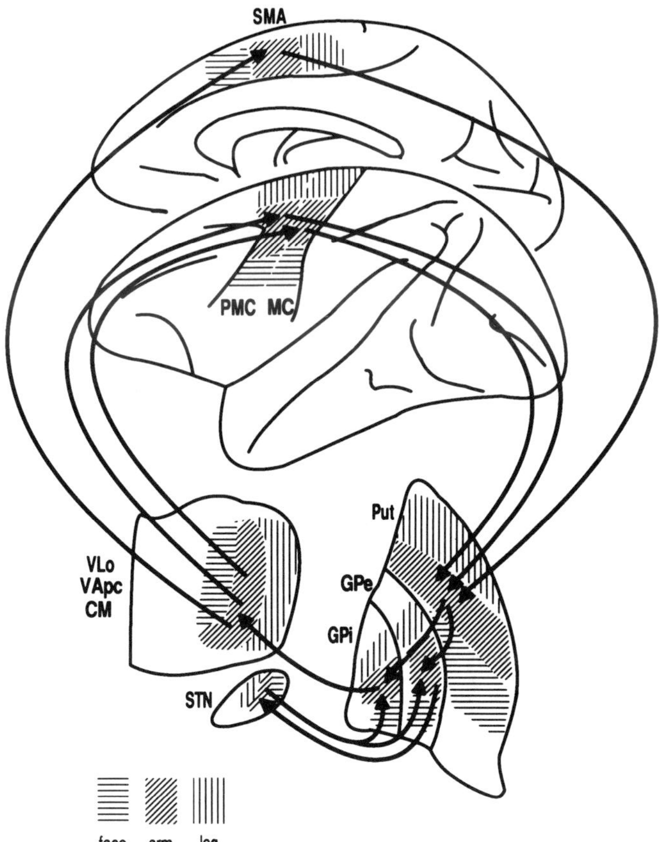

Figure 3. Somatotopic organization of the "motor" circuit. Somatotopy is maintained by virtue of topographically organized connections between each of the functionally distinct areas of these nuclei. The thalamic subnuclei engaged by the motor circuit are the oral portion of the ventral lateral nucleus (VLo), the parvocellular portion of the ventral anterior nucleus (VApc), and the centromedian nucleus (CM). Put, putamen.

Oculomotor circuit. The striatal portion of the basal ganglia–thalamocortical "oculomotor" circuit is in the body of the caudate nucleus, which receives projections from a number of interconnected cortical areas that are implicated in the control of eye movements, including the frontal eye fields (FEF), supplementary eye fields (SEF), dorsolateral prefrontal cortex, and posterior parietal cortex. This circuit is closed by thalamic projections back to the FEF and the SEF. This circuit also has a collateral projection from SNr to the superior colliculus, suggesting that this projection provides an important additional output pathway for the oculomotor circuit. Neurons have been found both in the caudate nucleus and SNr that discharge selectively in relation to fixation of gaze or to visually triggered or memory-contingent saccades.

Dorsolateral prefrontal circuit. The projection from the dorsolateral prefrontal cortex (DLPC), which includes cortical areas within and around the principal sulcus and on the dorsal prefrontal convexity, terminates within the dorsolateral head of the caudate nucleus and throughout a continuous rostrocaudal expanse that extends to the tail of the caudate. Portions of posterior parietal cortex that are interconnected with the DLPC have also been shown to project to the same region of the caudate nucleus. The thalamus sends return projections to the DLPC, thus closing the circuit. This circuit appears to be involved in cognitive and high-level processes such as spatial memory.

Lateral orbitofrontal circuit. The lateral orbitofrontal cortex (LOFC) projects to a ventromedial sector of the caudate nucleus that extends from the head to the tail of that structure. This part of the caudate also receives input from the auditory and visual association areas of the superior and inferior temporal gyri, respectively. The function of this circuit is poorly understood, but lesions of the LOFC or its projection area in the head of the caudate result in perseverative behaviors. This suggests that this circuit may be involved in switching behavioral sets.

Limbic circuit. The ventral striatum receives extensive projections from "limbic" structures, including the hippocampus, amygdala, and entorhinal, and perirhinal cortices and is therefore referred to as the limbic striatum. The ventral striatum also receives significant projections from the anterior cingulate area (ACA) and medial orbitofrontal cortex (MOFC), as well as from widespread sources within the temporal lobe. The limbic circuit is closed by thalamocortical projections to the ACA and MOFC. The function of this circuit is very poorly understood, but evidence suggests that it is involved in emotional and/or motivational processes.

Further reading

Alexander GE, Crutcher MD, DeLong MR (1990): Basal ganglia-thalamocortical circuits: Parallel substrates for motor, oculomotor, "prefrontal" and "limbic" functions. In: *Progress in Brain Research,* vol 85, Uylings HBM, Van Eden CG, DeBruin JPC, Corner MA, Feenstra MGP, eds. Amsterdam: Elsevier

Graybiel AM (1990): Neurotransmitters and neuromodulators in the basal ganglia. *TINS* 13:244–254

Parent A (1986): *Comparative Neurobiology of the Basal Ganglia.* New York: Wiley-Interscience

Smith AD, Bolam JP (1990): The neural network of the basal ganglia as revealed by the study of synaptic connections of identified neurons. *TINS* 13:259–265

Basal Ganglia, Models of Function: Normal and Disease

Thomas Wichmann and Mahlon R. DeLong

Functional anatomy of the basal ganglia

By virtue of their anatomical connections, the basal ganglia are viewed as major components of larger cortico-subcortical reentrant pathways that also include parts of the thalamus. In recent years, anatomical and physiological studies have provided evidence that the cortico-subcortical loops passing through the basal ganglia and the thalamus constitute a system of parallel circuits that are largely segregated, both structurally and functionally, permitting the basal ganglia to simultaneously participate in a number of different functions. According to their cortical areas of origin and termination in the frontal lobe, at least five of these circuits have been identified: the "motor" circuit is centered on the precentral motor fields, the "oculomotor" circuit on the frontal and supplementary eye fields, two prefrontal circuits on the dorsolateral prefrontal and lateral orbitofrontal cortex, and a "limbic" circuit on the anterior cingulate and medial orbitofrontal cortex.

Although these circuits originate and terminate in different cortical areas, and occupy different domains in the basal ganglia and the thalamus, they share certain anatomical features, which are outlined in a simplified scheme in Figure 1A. Each circuit comprises a number of separate cortical areas and different portions of the striatum, the external and internal segment of the globus pallidus (GPe, GPi), the substantia nigra pars reticulata (SNr), the subthalamic nucleus (STN),

and parts of the ventrolateral thalamus (VL). In each of these loops, the striatum serves as the "input" stage of the basal ganglia portion of the circuit (receiving inputs from cortical areas), whereas GPi, SNr, and ventral pallidum serve as "output" stages, sending inhibitory projections to the thalamus.

Two projection systems link the input and output structures of the basal ganglia: (1) a monosynaptic "direct" pathway between striatum and GPi and SNr, that is inhibitory and uses GABA and substance P as its neurotransmitters, and (2) an "indirect" pathway via GPe and the STN. In the indirect pathway, the projections between the striatum and GPe and between GPe and the STN are both inhibitory and GABA-ergic, whereas the STN–GPi pathway is glutamatergic. Activation of the direct pathway acts to inhibit the activity of the output nuclei and thereby to disinhibit thalamocortical projection neurons. In contrast, activation of the indirect pathway has a net excitatory effect on GPi/SNr activity and thereby acts to inhibit thalamocortical neurons. Dopamine, released from terminals of the nigrostriatal projection, appears to inhibit activity in the indirect pathway, whereas it facilitates that in the direct pathway.

Owing to its clinical importance and experimental accessibility, the "motor" circuit has received the most attention. This circuit arises from the precentral motor fields, and involves the putamen, "motor" portions of GPe, STN, and

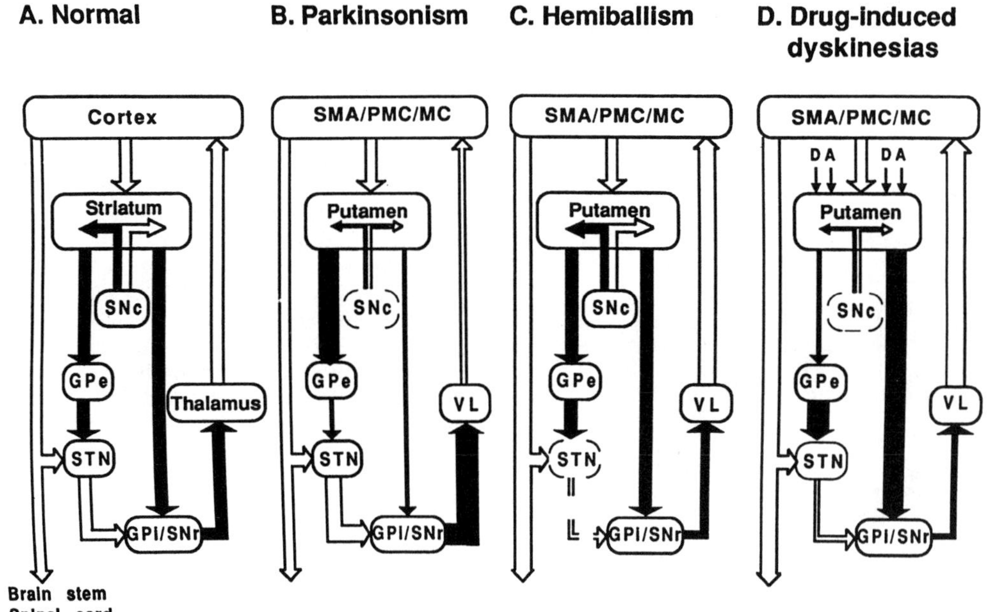

Figure 1. (A) Schematic diagram of the basal ganglia–thalamocortical circuitry. Inhibitory connections are shown as filled arrows, excitatory connections as open arrows. (B) Activity changes in the basal ganglia–thalamocortical circuitry in Parkinson's disease are shown. Note that lesions of SNc result in changes in the activity of basal ganglia structures, indicated by the thickness of the connections arising from these structures. Under these conditions, the basal ganglia output is increased. (C and D) Activity changes during dyskinesias due to a lesion of the STN (hemiballismus, C) and induced by the administration of a dopaminergic agonist in parkinsonism (D). Abbreviations: GPe, external segment of the globus pallidus; GPi, internal segment of the globus pallidus; SNr, substantia nigra, pars reticulata; SNc, substantia nigra, pars compacta; STN, subthalamic nucleus; SMA, supplementary motor area; PMC, premotor cortex; MC, primary motor cortex.

GPi/SNr, as well as parts of the ventrolateral "motor" thalamus. The circuit is somatotopically organized. Each somatotopic region contains groups of neurons that fire either in conjunction with the preparation for or with the execution of limb movements.

There is strong evidence that tonic as well as phasic output of this circuit is crucial in motor control. The tonic discharge of GPi/SNr neurons seems to be maintained within a narrow range that may determine the responsiveness of thalamocortical neurons. A decrease in tonic basal ganglia output may lead to disinhibition of these neurons, and thereby increase the overall amount of movements, whereas an increase of basal ganglia output may reduce the activity of thalamocortical neurons, and thus lead to a reduction of motor activity. Phasic GPi and SNr output may be involved in the execution of specific movements. Both direct and indirect pathways are probably phasically activated during movements. A phasic increase of inhibition of GPi via the monosynaptic direct pathway may lead to a brief disinhibition of thalamocortical neurons, which may induce increased activity in cortical areas, thus facilitating the movement. Phasic activation of the indirect pathway may have the effect of stabilizing the activity in other structures (e.g., the thalamus) or suppressing potential competing movements.

Pathophysiology of hypo- and hyperkinetic disorders

Most movement disorders of basal ganglia origin are believed to arise from malfunction of the "motor" circuit, resulting in alterations of GPi and SNr output, and disturbed motor performance. These disorders are characterized by either poverty of movement (hypokinetic disorders) or an excess of movement (hyperkinetic disorders). Hypokinetic disorders (e.g., Parkinson's disease) feature an impairment in movement initiation (akinesia) and reductions in amplitude and velocity of voluntary movements (bradykinesia). By contrast, hyperkinetic disorders (e.g., chorea or ballism) are characterized by involuntary movements (dyskinesias).

Hypokinetic disorders. The most common hypokinetic disorder is Parkinson's disease. In recent years, the understanding of the pathophysiology of this disorder has been greatly advanced by studies in primates treated with the dopaminergic neurotoxin MPTP. MPTP induces a parkinsonian syndrome that very closely resembles the clinical, pathological, and biochemical characteristics of the human disorder.

Studies of the neuronal activity in the basal ganglia in such parkinsonian animals have suggested the scheme of pathophysiological changes shown in Figure 1B. Loss of dopamine in the striatum seems to have opposite effects on the activity of the indirect and the direct pathway, eventually leading to increased excitation of GPi and SNr via the indirect pathway and to reduced inhibition of GPi and SNr along the direct pathway. In accordance with this scheme, it was found that GPe activity is reduced in MPTP-treated animals, whereas tonic and phasic discharge in the STN and GPi are increased. A causal link between these activity changes and parkinsonian motor signs is suggested by the beneficial effects of lesions of the STN in MPTP treated monkeys and of GPi in parkinsonian patients.

The increased tonic and phasic output from GPi and SNr may be responsible for bradykinesia as well as akinesia. In bradykinesia there is inability to perform large-amplitude movements with high velocity. Instead, movements are carried out discontinuously, with several small-amplitude segments. This may result from increased inhibition of thalamocortical neurons by excessive *tonic* output from GPi, reducing the overall responsiveness of cortical mechanisms. Increased tonic discharge in GPi may also prevent the faithful transmission of phasic reductions in GPi activity during movement execution to the cortex, thereby resulting in a reduced range of neuronal amplitude changes and scaling of movement.

Since *phasic* discharges in GPi may reflect feed-forward of motor commands from the cortex, abnormally large phasic output from GPi may be misinterpreted by the cortex and may in turn lead to reduced output to the agonist musculature. By the same mechanism increased gain in the feedback from proprioceptors may lead to slowing or premature arrest of ongoing movements.

Akinesia may result from increased tonic inhibition or decreased phasic disinhibition of thalamocortical neurons, rendering cortical projection areas less responsive to inputs normally involved in initiating movements. Some aspects of akinesia may also result from a disturbance of motor "set" functions, which are dependent upon the integrity of basal ganglia pathways. The degree of akinesia in humans as well as in MPTP-treated animals correlates well with the amount of dopamine loss in the caudate nucleus. Thus, besides malfunction of the "motor" circuit, the development of akinesia may also require dopamine depletion in the "nonmotor" circuits.

Hyperkinetic disorders. In contrast to hypokinetic disorders, hyperkinetic disorders appear to result from abnormally *low* GPi activity, due to reduced activation of GPi via the indirect pathway (Fig. 1C and 1D). Conceivably, thalamocortical neurons under such conditions become more responsive to cortical inputs or exhibit an increased tendency to discharge spontaneously, thus leading to involuntary movements.

The dyskinesias seen in ballism and Huntington's disease, as well as those induced by dopaminergic drugs, can all be explained by this mechanism. The term *hemiballism* characterizes involuntary limb movements most frequently due to lesions restricted to the contralateral STN. In metabolic studies of monkeys with experimental lesions of the STN, the activity both in GPi and in the ventrolateral thalamus is decreased, suggesting that GPi output may be reduced in this disorder. Furthermore, STN lesions lead to a significant reduction of tonic discharge in GPi and a decrease in the phasic responses of GPi neurons to limb displacement.

In early Huntington's disease, which is characterized by neostriatal degeneration leading to strong dyskinesias, there appears to be a selective loss of the striatal neurons that give rise to the indirect pathway. Subsequently, this may result in excessive inhibition of STN neurons, leading to abnormally low GPi activation. Similarly, metabolic studies in monkeys with dyskinesias due to treatment with dopaminergic drugs suggest that dyskinesias may be caused by excessive inhibition of the indirect pathway and by excessive stimulation of the direct pathway, both resulting in abnormally low GPi output.

Further reading

Albin RL, Young AB, Penney JB (1989): The functional anatomy of basal ganglia disorders. *Tr Neurosc* 12:366–375

Alexander GE, Crutcher MD (1990): Functional architecture of basal ganglia circuits: Neural substrates of parallel processing. *Tr Neurosc* 13:266–271

DeLong MR (1990): Primate models of movement disorders of basal ganglia origin. *Tr Neurosc* 13:281–285

Bell's Palsy

David Sternman and David M. Simpson

Facial paralysis has been recognized since antiquity, as shown in the sculpture, masks, and artwork of ancient society. *Bell's palsy* is the term reserved for acute peripheral facial nerve paralysis of unknown cause. It is named in honor of the distinguished Scottish physician, Sir Charles Bell (1774–1842), who first accurately described the condition. In 1829, he demonstrated that the seventh cranial nerve innervates the muscles of facial expression. The term "Bell's palsy" is often misused when loosely applied to facial paralysis of any type.

Anatomy

An approach to the diagnosis and treatment of facial paralysis requires a detailed understanding of the anatomy of the seventh cranial nerve (Fig. 1). This nerve consists of the facial motor nucleus and axons and the nervus intermedius, which contains sensory and parasympathetic components. The corresponding nuclei are located in the caudal pons. The portion of the motor nucleus that supplies the muscles of the lower part of the face receives corticobulbar fibers from the contralateral hemisphere. In contrast, the upper face receives fibers from both hemispheres. Therefore, a unilateral cortocospinal tract lesion spares the upper face. The facial nerve emerges from the pons and continues through the in-ternal auditory meatus. The motor fibers continue through the facial canal and exit the skull at the stylomastoid foramen. The stapedial nerve branches from the facial nerve within the canal. After emerging from the parotid gland the facial nerve arborizes to innervate the muscles of facial expression. The nervous intermedius enters the facial canal as well, where it expands to form the sensory geniculate ganglion and separates into two branches: (1) the greater petrosal nerve contains parasympathetic fibers destined for the lacrimal gland; (2) the chorda tympani has a sensory branch to the external auditory meatus and eventually joins the lingual nerve. It contains afferent fibers serving taste to the anterior two thirds of the tongue and parasympathetic fibers that innervate the submandibular and sublingual salivary glands.

Clinical features

Epidemiology. The incidence of Bell's palsy varies in the United States between 15 and 40 per 100,000 people per year. The sex distribution is equal except in the second decade of life, when females are affected twice as often as males. Bell's palsy is significantly less common before age 15 and after age 60. Pregnancy appears to increase the risk of Bell's palsy, particularly in the third trimester.

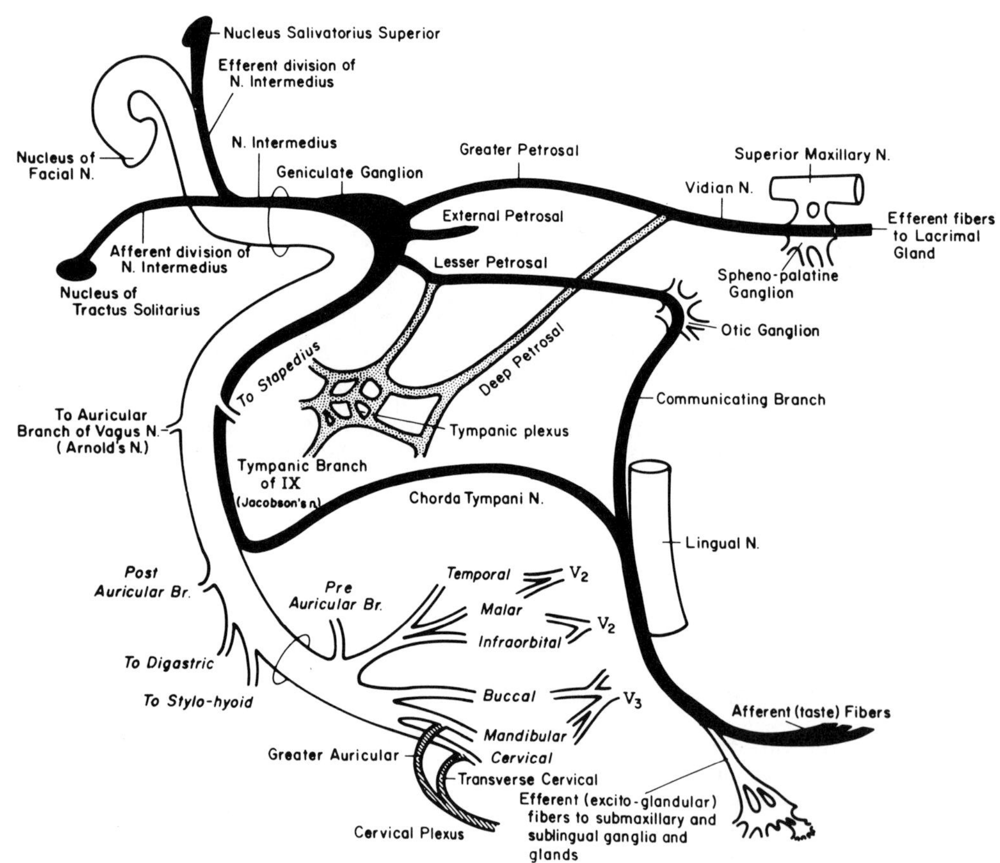

Figure 1. Schematic diagram depicting the anatomy of the facial nerve. (Reprinted from May M, *The Facial Nerve,* New York, Thieme Medical Publishers, Inc., 1986. Reprinted by permission.)

Signs and symptoms. Bell's palsy has often been considered a benign cranial mononeuropathy. However, there is data to suggest that Bell's palsy is often part of a cranial mononeuritis multiplex. Bilateral clinical involvement is uncommon. Electrophysiologically, subclinical pathology may often be demonstrated in the contralateral facial nerve. Vestibular dysfunction is present in 43% of patients; 29% of subjects with Bell's palsy have combined trigeminal and glossopharyngeal hypesthesia; and 11% have paresis of the superior laryngeal nerve.

Bell's palsy is commonly preceded by a viral prodrome and periauricular pain. Patients may complain of change in taste (57%), reduced tear flow (17%), and phonophobia (29%). Examination of taste, nasolacrimal reflex, and stapedial reflex demonstrates reduced function in 83%, 12%, and 71% of patients, respectively. This suggests that the most common site of involvement of the facial nerve is the tympanomastoid portion. Ten percent of patients have a positive family history of Bell's palsy.

Course and prognosis. The onset of facial paralysis is usually rapid. Complete paralysis occurs within a few hours and may become complete within a week. In one third of patients the paralysis is only partial. Of the patients with incomplete paralysis, 94% have total return of motor function, and the remaining 6% have only slight residual weakness. Overall, 71% of patients have a complete recovery, 13% have only slight residual palsy, and 16% of patients have fair to poor recovery. Complete return of facial strength occurs in the majority of patients from 3 weeks to 2 months after the onset of the palsy. If the remission is not complete by 4 months, then the chance for further recovery is poor. Virtually all patients with Bell's palsy have some degree of improvement by 6 months after the onset. Therefore, the diagnosis of Bell's palsy should be questioned in patients in whom improvement has not occurred by that time.

Certain sequelae may develop in patients with only partial recovery. These include synkinesis, spasm, contracture, tinnitus, and the "crocodile tears" phenomenon. The latter results from autonomic synkinesis, caused by faulty regeneration of parasympathetic fibers to the lacrimal and submandibular glands. Recurrent Bell's palsy occurs in 7–10% of cases.

The psychosocial aspects of facial paralysis should not be disregarded. Patients commonly experience feelings of self-consciousness, withdrawal, and depression in response to their appearance. Facial paralysis often presents an obstacle to social acceptance because of the emphasis that society places on physical attractiveness.

Differential diagnosis

Bell's palsy is a diagnosis of exclusion and is established only after known causes of acute facial paralysis have been ruled out. Over 80 different causes of facial palsy have been recognized (Table 1). In some series, up to 17% of individuals with acute facial palsy have been misdiagnosed as having Bell's palsy. In general, a thorough history and physical examination will serve to establish the correct diagnosis. Particular attention should be given to the onset and duration of symptoms, and to subtle involvement of other cranial nerves.

Certain systemic illnesses may present with acute facial palsy. Guillain-Barre syndrome is an acute demyelinating polyradiculoneuropathy that may cause generalized paraly-

Table 1. Differential Diagnosis of Facial Palsy

Guillain-Barre syndrome
Myasthenia gravis
Sarcoidosis
Diabetic cranial neuropathy
Infection
    Lyme disease
    Botulism
    Otitis media
    Herpes zoster (Ramsey-Hunt syndrome)
    Meningitis
    Human immunodeficiency virus
Neoplasm
    Schwannoma
    Cholesteatoma
    Glomus jugulare tumor
    Meningioma
    Carcinomatous meningitis
Cerebrovascular disease
    Milliard-Gubler syndrome
    Corticobulbar dysfunction

sis. It is a common cause of bilateral facial nerve palsy. Botulism classically begins as a cranial polyneuropathy and also spreads as a generalized paralytic disorder. These diagnoses can be confirmed by electrophysiologic studies and other laboratory tests. Bacterial, fungal, and mycobacterial meningitis may cause facial nerve palsy and should be suspected in the appropriate clinical setting.

Several diseases have a tendency to cause isolated facial nerve palsy. Lyme disease (borreliosis) must be considered, particularly in endemic regions, and can be diagnosed with antibody assays. Sarcoidosis is a multisystem granulomatous disorder that presents most frequently with pulmonary, skin, or eye lesions. Neurosarcoidosis occurs in up to 10% of patients. Facial palsy is a frequent presenting symptom and may occur alone or with other cranial neuropathies.

Diabetes mellitus is associated with acute unilateral facial nerve palsy. Isolated cranial neuropathy may sometimes be the presenting sign of diabetes. The prognosis for recovery is somewhat worse than that of Bell's palsy.

Herpes zoster of the geniculate ganglion, or Ramsey-Hunt syndrome, consists of acute facial palsy associated with a vesicular eruption in the external auditory canal and oropharyngeal mucosa. The eighth cranial nerve is often affected as well. Postherpetic neuralgia may develop, and the prognosis is poorer than that of Bell's palsy.

The facial nerve is subject to damage from acute or chronic otitis media as a result of the proximity of the middle ear to the facial canal. Facial nerve palsy also may occur in association with HIV infection, either in isolation or as part of a mononeuropathy multiplex.

Various tumors of the head and neck are liable to cause slowly progressive facial nerve palsy. Acute facial paralysis may occur when the tumor mass grows beyond a certain critical size, causing facial nerve ischemia. Neoplasms that originate close to the facial nerve include cholesteatoma, glomus jugulare tumor, meningioma, and schwannoma.

Finally, Bell's palsy must be distinguished from a supranuclear lesion of the facial nerve. In the latter condition the frontalis and orbicularis oculi muscles are less affected than those of the lower part of the face. There also may be preserved emotional movement in contrast to voluntary activity.

Diagnostic testing

Appropriate serologic and other laboratory studies should be performed to exclude the aforementioned disorders. Neuroimaging studies, particularly CT or MRI, may occasionally be necessary to exclude structural lesions of the facial nerve. This is particularly true in cases with an atypical, severe, or protracted course.

Electrophysiologic studies are helpful in determining both the pathophysiology of the facial nerve lesion as well as its prognosis. There are three essential components to these studies. The first is the nerve conduction study in which the facial nerve is stimulated at the stylomastoid foramen. The compound motor action potential (CMAP) is recorded at selected facial innervated muscles. The amplitude of the CMAP generally reflects the population of surviving axons, while the onset latency reflects the distal conduction of the fastest fibers. A significant amplitude reduction of the CMAP, when compared with the contralateral side, predicts a poorer prognosis for recovery.

The technique of direct facial nerve stimulation has limitations. The accessible nerve is distal to the more proximal site of pathology. Therefore, it may take up to a week for the distal study to reflect even severe proximal pathology. The blink reflex study avoids this limitation by employing a trigeminal facial reflex pathway that reflects the integrity of the afferent and efferent pathways to the brainstem. Finally, with needle electromyography, the status of muscle denervation and an estimate of axonal regeneration may be determined.

Pathogenesis

The cause of Bell's palsy is unknown, although a viral etiology has been proposed. In one study, complement-fixing antibodies to herpes simplex virus (HSV) were present in 100% of patients with Bell's palsy as compared with 85% of controls. However, none of the patients with Bell's palsy demonstrated a sufficient rise in convalescent-phase antibodies to clearly establish primary infection at the time of onset of Bell's palsy. HSV has been isolated from the epineurium of the facial nerve in a patient with acute Bell's palsy. Latent HSV has also been identified in the trigeminal ganglia of randomly selected cadavers. These observations suggest that Bell's palsy may be caused by reactivation of latent HSV within the facial nerve.

The spectrum of facial nerve pathology in Bell's palsy is unknown since few specimens have been available for study. Rare examinations of the facial nerve have revealed intraneural hemorrhage, edema, and cellular infiltrations. Edema within the rigid confines of the facial canal may lead to vascular insufficiency and hypoxia.

Treatment

Pharmacological. Several different medications have been employed in the treatment of Bell's palsy without clear benefit. These include cromolyn sodium, acyclovir, and vasoconstrictors. The most commonly used treatment is corticosteroid therapy. Several studies have suggested that prednisone reduces the incidence of severe denervation and other long-term sequelae. However, other investigators have reported that prednisone therapy does not alter the course of disease. A definitive resolution to this controversy would require a large controlled study, since Bell's palsy has a high rate of spontaneous remission.

Surgical. The role of surgical decompression of the facial nerve in the management of Bell's palsy remains controversial. This approach may be warranted in patients at high risk for poor spontaneous remission. However, it is not clear which criteria to apply in order to select these patients. Surgery is clearly of value for facial nerve exploration, biopsy, and decompression when the diagnosis of Bell's palsy is in doubt, especially if there is a suspicion of neoplasm. Surgery may also achieve facial reanimation. Several techniques have been developed in order to directly repair the facial nerve. In some patients, 75% of original voluntary facial movement can be restored, although spontaneous facial expression is minimal. One procedure employs the hypoglossal nerve to reanimate the distal facial nerve. Muscle tone is significantly improved and eye protection is almost always achieved with this procedure. In patients who cannot undergo nerve graft, regional muscle transfer using the masseter or temporalis muscle is partially effective in restoring facial strength and symmetry.

Ophthalmologic. Facial paralysis may lead to mild corneal abrasion or neurotrophic ulceration, perforation, and blindness. Prevention of corneal exposure and drying is accomplished by use of artificial tears, taping the eye closed at night, and protective glasses. Formal consultation by an ophthalmologist is indicated when the patient lacks Bell's phenomenon (the normal upward eye deviation upon lid closure), or has corneal anesthesia or dryness due to decreased lacrimation.

Psychological. An inquiry into the patient's reaction toward disfigurement may ease adjustment to this disability. While some patients need no psychological support, others may adapt poorly and require skilled psychiatric counseling.

Summary

Facial paralysis is a common disorder that may present a diagnostic challenge to general practitioners and specialists. It is hazardous to presume that a case of acute facial paralysis is (idiopathic) Bell's palsy without careful consideration of the differential diagnosis. Since the prognosis for recovery is generally favorable, treatment is primarily aimed at preventing secondary complications. The roles of specific pharmacological, surgical, or rehabilitative therapies in the management of Bell's palsy are controversial and await the performance of prospective controlled studies.

Further reading

Adour K et al (1978): The true nature of Bell's palsy: Analysis of 1,000 consecutive patients. *Laryngoscope* 88:787–801

Fisch U (1981): Surgery for Bell's palsy. *Arch Otolaryngol* 107: 1–11

May M (1986): *The facial nerve.* New York: Thieme

Peitersen E (1982): The natural history of Bell's palsy. *AM J Otol* 4:107–111

Blindsight, Residual Vision

Lawrence Weiskrantz

As there are as many as 10 parallel pathways from retina to brain targets, it is not surprising that animals in whom the major pathway—the geniculo-striate tract—has been interrupted can nevertheless be trained to make visual discriminations. In human patients striate cortex lesions produce field defects ("scotomata") that are blind phenomenologically. But using forced-choice "guessing" psychophysical techniques akin to the nonverbal methods of animal research, residual visual capacity can be demonstrated in at least some of these patients despite their apparent lack of awareness of the visual stimuli; this residual capacity has been called blindsight. Debates concerning interpretations based on diffusion of light onto intact fields, response bias shifts, or incomplete lesions can be put to rest on the basis of a number of control studies as well as demonstrations of blindsight in hemispherectomized subjects.

Since the original findings were reported some 15 years ago (see original *Encyclopedia* article), a number of recent advances are worth recording.

First, alternatives to forced-choice guessing techniques have been devised and exploited. The advantages are manifold. Many human subjects are uncomfortable with an instruction to "guess" about a stimulus that they avowably cannot actually experience, a discomfort shared by some experimenters who have to persuade the subjects to play this unusual game. Also, forced-choice psychophysical procedures are lengthy and tiring, often requiring thousands of trials. Alternative procedures avoid asking the subject to make any judgments about the blind field. There are two general stratagems: Inferences can be drawn about the capacity of the blind field by measuring the interaction of stimuli in the blind and the intact visual fields. The subject need only be instructed to respond to a stimulus in the intact field, but such a response might be facilitated or inhibited by a stimulus in the blind field. In one such study, the subjects were asked to make a saccade as rapidly as possible to a seen stimulus in the good field; a near simultaneous stimulus in the blind field, although not experienced by the subject, slowed the saccadic response, depending on whether the nasal or temporal hemiretina was stimulated. Similarly, a subject's key-press reaction time to the onset of a stimulus in the good field can be slowed or speeded up depending on the time interval between the stimuli to the blind and intact fields. The second stratagem is to use reflex responses to visual stimuli in the blind field. Changes in skin conductance is one such method that has been used successfully. More recently it has been shown, with appropriate computer averaging of responses detected by infrared scanning devices, that the pupillary response is exquisitely sensitive to contrast sensitivity, spatial structure, and spectral shifts in the intact visual field. Pupillary responses can also be obtained to visual stimuli in the scotoma, and they appear to correlate with results from forced-choice psychophysics in the same subjects. The psychophysical power of the method holds much promise, and has the added advantage that it could also be used with human infants and animals.

A second development stems from the implications of the electrophysiological evidence that different extrastriate visual areas in the primate are relatively specialized (e.g., V4 for color opponency and V5 [MT] for motion). Occipital lesions in man vary considerably in extent and disposition and almost invariably involve regions outside of striate cortex (VI). Whether or not V4, for example, is involved might be expected to determine the extent to which wavelength discrimination is possible in blindsight, as has been demonstrated in some patients. It has recently been shown, in fact, that in such patients a qualitatively normal spectral sensitivity curve is obtained by forced-choice guessing methods, with the peaks and dips in the photopic curve that are thought to reflect opponent processing; they also show a normal Purkinje shift with dark adaptation. The integrity of MT in another subject would seem to account for his excellent directional motion discrimination and the psychophysics of his response to transients. The modularity of extrastriate cortical regions may illuminate the varying profiles of blindsight subjects, but also be illuminated by them.

Finally, attention to fine anatomical detail and neuropathology has advanced the understanding of the basis of residual vision. For example, retrograde transneuronal degeneration of retinal ganglion cells following striate cortex lesions does not affect all classes of ganglion cells uniformly but is largely restricted to P-beta cells. These are the cells that project to the parvocellular layers of the LGN, and which are presumed to be essential for color opponency. Conversely, the electrophysiology of the P-alpha and P-gamma cells allows one to draw inferences about the types of psychophysical properties that might be found in blindsight, especially regarding the responses to transient stimuli. The importance of P-beta cells for color vision raises a paradox concerning the importance of their cortical targets after relay in the parvocellular LGN, because color opponent responses appear to be restricted to cortical neurons in the primate (although the universal negative has not yet been actively pursued for primate midbrain neurons). Because the LGN is usually considered to degenerate completely after striate cortex removal, the question arises as to how color discrimination might still be possible in some blindsight subjects. The paradox has been resolved by the finding that, with fine examination, not all LGN cells are found to degenerate: a small number remain and these project to extrastriate visual areas (e.g., V4). A final example of the attention to anatomical detail emerges from the finding that the inhibition of the saccadic response to stimuli in the intact hemifield by stimuli in the blind hemifield, referred to previously, depends on the stimuli in the blind field being projected onto the nasal retina, and does not occur when they fall on the temporal retina. The finding is accounted for by the heavier crossed than uncrossed projection of the optic nerve to the midbrain, the level at which the saccadic inhibition effect is presumed to be mediated. Various other psychophysical findings remain to be examined in terms of nasal versus temporal retinal stimulation.

A final note concerns recent discoveries that apparently similar "implicit" or "covert" residual capacities are found after a large variety of brain lesions in man, in such neuropsychological conditions as prosopaganosia, amnesia, aphasia, and dyslexia. Taken together they may allow more comprehensive understanding of the neurology of "awareness," a topic that also has stimulated interest by philosophers of mind as well as neuronal network theorists. Interest of experimental psychologists has also been directed to the study of a range of covert processes in normal subjects. It remains to be seen whether these are qualitatively similar seen in brain-damaged patients, but a common methodology aids the study in both domains.

Further reading

Cowey A, Stoerig P (1991): The neurobiology of blindsight. *Trends Neurosci* 14(4):140–45

Rafal R, Smith J, Krantz J, Cohen A, Brennan C (1990): Extrageniculate vision in hemianopic humans. *Science* 250:118–21
Stoerig P, Cowey A (1989): Wavelength sensitivity in blindsight. *Nature* 342:916–918
Weiskrantz L (1986): *Blindsight: A Case Study and Implications.* London, New York: Oxford University Press
Weiskrantz L (1990): Outlooks for blindsight: Explicit methodologies for implicit processes. *Proc R Soc Lond* 239:247–278

Brain Grafts, Genetic Engineering

Jasodhara Ray and Fred H. Gage

Neuronal transplantation provides a powerful novel technique for delivery of many pharmocologically active agents and enables one to address some fundamental questions regarding the basic principals of brain function and clinical applications in neurobiology. The mechanisms of action of implanted cells vary and include (1) delivery of trophic or tropic molecules, (2) secretion of neurotransmitters or hormones, (3) release of synaptically mediated neurotransmitters, and (4) functioning as a bridge for regenerating axons to grow from the proximal side of a lesion to the distal side. However, the types or ages of the donor cells that can be used for implantation depend on the long-term survival of these cells and extended production of functional agents *in vivo*. Fetal neuronal tissues have proven to be the most successful donor tissues for brain grafting and in many cases the extent of the success has been unequivocal and dramatic. The limiting factors for use of these cells are the number of cells available, the purity of the specific neuronal phenotype, and the need for immune suppression for graft survival and function. Finally, the ethical and political debate about the use of fetal tissues has hindered clinical applications of this experimental approach.

Alternative cell types, including Schwann cells and glial and peripheral neurons, have been used for grafting purposes. For example, autologous adrenal chromaffin cells as a source of dopamine have been used for treatment of Parkinson's disease. While some moderate success has been noted with fetal adrenal cells, adult chromaffin cells do not survive or function well in the absence of external trophic support. This general strategy of searching for easily accessible cell types that can be grafted intracerebrally to substitute for lost brain function requires that the other parts of the body can do without these cells and that the donor cells will survive in the graft for a reasonable amount of time and will make the missing molecule in sufficient quantities to replace the missing or defective functions.

Gene transfer into appropriate cell types and intracerebral grafting of these cells *in vivo* may meet the aforementioned requirements (Fig. 1). This approach requires the following basic steps: (1) selection of appropriate transgenes whose expression is correlated with the CNS disease, (2) preparation of donor cells from primary or established cell cultures, (3) selection or development of suitable vectors, (4) efficient gene transfer into donor cells, (5) *in vitro* selection and molecular characterization of the transgene to assess its successful production, (6) intracerebral grafting, and (7) demonstration that the transgene can have a desired phenotypic effect in the host animal. A combination of gene transfer and intracerebral grafting may prove to be a viable approach for treatment of certain CNS disorders such as Parkinson's disease.

Selection of appropriate transgenes

There are several criteria for the choice of appropriate transgenes to be used in various model systems. The complete cDNA sequence for genes of interest and the adequate marker to monitor the expression of genes must be available. Furthermore, gene products should result in a measurable function that can be monitored by established methodologies. For grafting into CNS two classes of transgenes are of particular interest. First, in a model system, the genes of growth factors can be used to study their role in the development or regeneration of the neurons or in delaying or preventing the neurodegenerative diseases. Second, the neurodegenerative loss of specific molecules can be replaced by genes that encode for neurotransmitters. In this model, the neurotransmitters' synthesis through multistep biosynthetic pathways may require the involvement of host enzymes (Fig. 2).

Preparation of appropriate donor cells

There are several criteria for choosing appropriate cells to be used for genetic modifications and intracerebral grafting. First, cells to be used for gene transfer should be readily accessible. Many established cell lines meet this criteria, including cells of neuronal (e.g., C6 glioma and NS20 Y neuroblastoma) and nonneuronal (e.g., fibroblasts-208 F and rat-1) origin or primary cells obtained from donor biopsies, including skin biopsies or from fetal or neonatal brains. A second criterion is that the cells should not only remain viable and express transgene *in vitro* but also *in vivo* after implantation into the brain.

Successful gene therapy of a CNS disorder requires the faithful formation of intercellular synaptic connections. Cells of neuronal origin may be useful for this purpose. However, the nonreplicating nature of the neuronal cells and their refractoriness toward the viral infection necessary for gene transfer have limited their use. Recent studies involving the immortalization of hippocampal neuronal cells suggest that the replicating neuronal cell culture system is now available for *in vitro* gene transfer and then *in vivo* implantation. Such neurons might be susceptible to efficient transduction by retroviral vectors or other viral vectors and, if they retain the neuronal characteristics, they may be able to establish synaptic connections with other cells after grafting into the brain. Alternatively, genes introduced into dividing neurons that are still at early stages of development can be used effectively. However, the establishment of synaptic connections *in vivo* by these cells is difficult to predict, as any developmental or functional commitments made by these cells before implantation and synaptic connection will affect this function.

The accessibility and the replicating nature of the nonneuronal cells provide a source of cells that may be used for grafting. However, their use precludes the development of specific neural connections to resident target cells of the host. Therefore, the phenotypic effects of donor cells or target cells *in vivo* would be through the diffusion of required gene products or metabolites through junctions ("metabolic cooperation") or through uptake by target cells of secreted donor cell gene products or metabolites. Alternatively, the donor cells can act as a toxin sink by expressing a new gene product and metabolizing or clearing a neurotoxin.

Immortalized cells or tumor cells have been used as donor cells for intracerebral grafting. Recently the focus has shifted to investigating the survival of autologous cells. Use of these cells has several advantages. They need not be CNS derived and their choice is based solely on their ability to be grown and manipulated *in vitro* and survive as a graft. In addition, many immunological problems related to graft rejection can

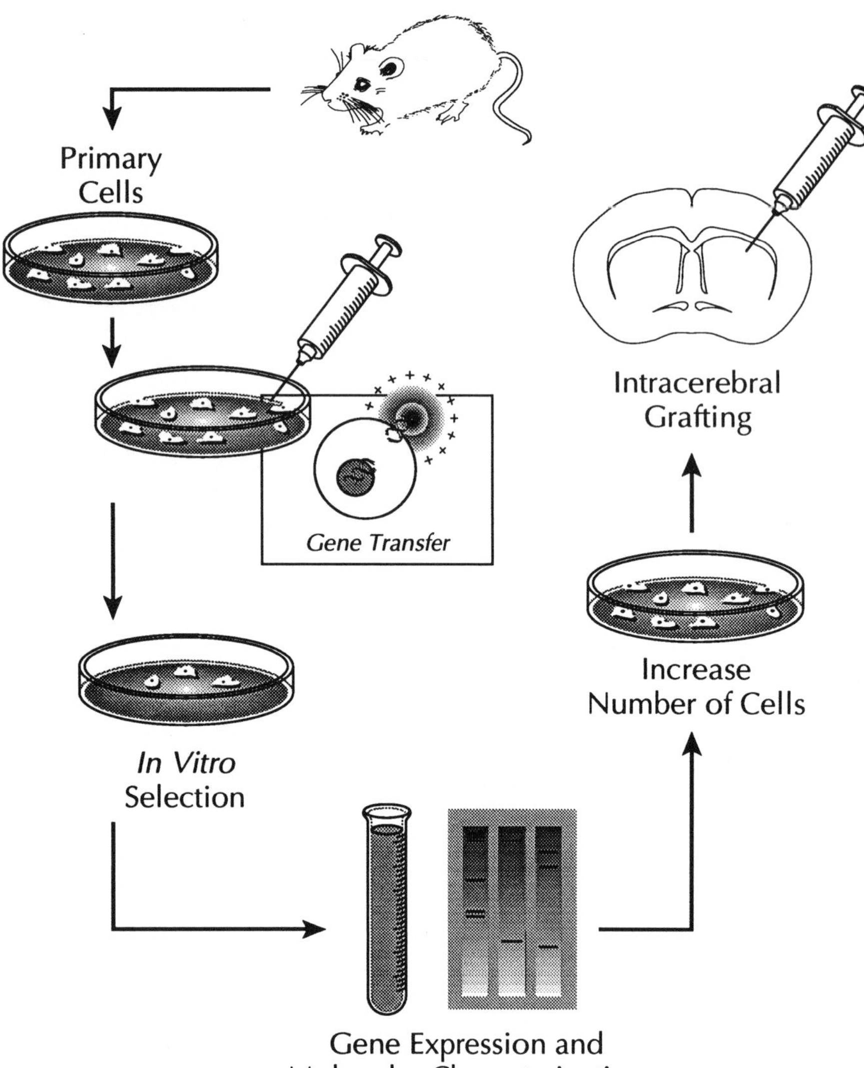

Figure 1. This schematic diagram illustrates the steps involved in the gene transfer and intracerebral grafting technique. Primary skin fibroblasts, obtained from skin biopsies, are genetically modified to express a gene of interest. After selection and assay to determine successful gene expression, the modified cells are grown in desired numbers and then implanted into the brain of the donor animal. (Reprinted with permission of Elsevier from FH Gage et al (1991): Genetically modified cells: Applications for intracerebral grafting. *Trends Neurosci* 14:328–333.)

be avoided. Although they are not the best suited cells for intracerebral grafting, skin fibroblasts have several advantageous features: (1) they are easy to obtain from skin biopsies, (2) they can be genetically modified easily, (3) they can survive as grafts in the CNS, and (4) they are secretory cells and thus constitutively secrete many transgene products.

Selection of suitable vectors and gene transfer methods

Eukaryotic cells can take up DNA spontaneously, with a portion of it getting integrated into chromosomal DNA. However, this is a very inefficient process. To overcome this problem, a wide variety of physical and chemical methods have been developed to facilitate gene transfer into eukaryotic cells *in vitro*. The methods include calcium phosphate precipitation, lipofection, electroporation, microinjection, and the use of retroviral vectors. Perhaps the most commonly used method is the gene transfer by retroviral infection. Neurotropic viruses as gene transfer vehicles have been developed for nonreplicating neuronal cells. However, technical difficulties of toxicity and latency are slowing the progress in their development.

Expression of transgenes *in vitro* and *in vivo*

The nature of the disease determines the approach of gene therapy that can be used for treatment of the CNS disorders. For example, the treatments may involve enhancing the production of specific gene products or supplying molecules that are beneficial for the compromised neurons. Two well-characterized models have provided an opportunity to observe functional effects after implanting genetically modified cells into the brain. In one model the genes of growth factors (e.g., nerve growth factor [NGF]) have been used to explore substances involved in the development and regeneration of neurons or their effect on preventing or delaying neuronal degeneration. In a second model the neurodegenerative loss of specific molecules has been replaced by genes encoding neurotransmitters or modulator-specific enzymes, for example the use of tyrosine hydroxylase (TH) in a model for Parkinson's disease. Clearly additional transgenes or combinations of transgenes will be investigated in the future.

Intracerebral grafting of immortalized genetically modified fibroblasts expressing NGF showed that the growth factor can provide trophic support to lesioned cholinergic neurons of the rat septum. Recently primary fibroblasts expressing

Neurotrophic Factor

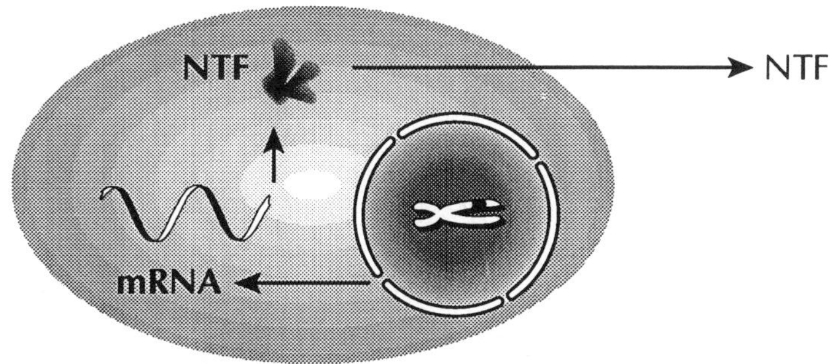

Neurotransmitter

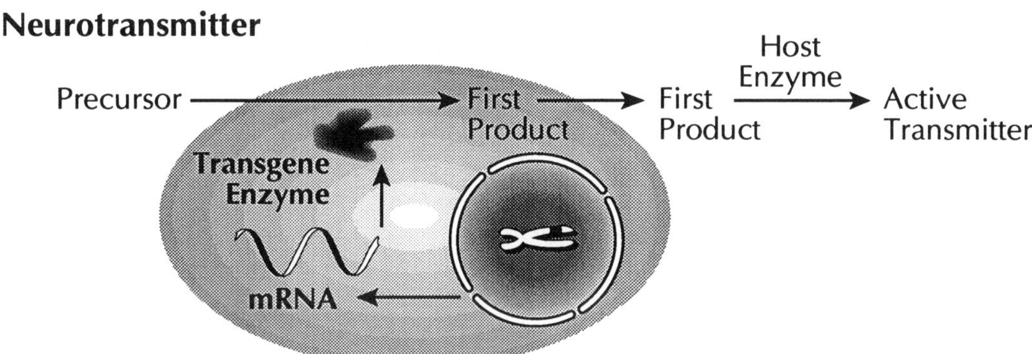

Figure 2. Schematic diagram showing the pathways for obtaining neurotrophic factors and neurotransmitters from genetically modified cells. For neurotrophic factors, mRNA is translated into a protein product that is released from the cells. However, for some neurotransmitters, the protein synthesized is an intermediate metabolite that, when released from the cells, interacts with host enzyme(s) to produce active neurotransmitters. (Reprinted with permission of Elsevier from FH Gage et al (1991): Genetically modified cells: Applications for intracerebral grafting. *Trends Neurosci* 14:328–333.)

NGF have been used successfully for implantation. Both immortalized and primary fibroblasts expressing TH have been used in a model for Parkinson's disease, and the catecholamine released from these cells reduced the apomorphine-induced rotations of dopamine-depleted rats.

Although some limited success has been achieved by the use of genetically modified cells, a number of issues still have to be resolved. It must be determined whether the state of the cells prior to grafting is important for the transgene expression and the survival of the cells in the graft. Further, under some conditions and with some particular cells it may be important to determine whether the suspended cells can be immobilized in a matrix, which optimizes the survival and decreases cell migration. In addition, the site of injection of cells (intraparenchymally or intraventricularly) may also be critical for the cell survival and gene expression. Finally, long-term cell survival and stable or, better yet, regulatable gene expression are also important issues to resolve.

Summary and conclusion

Grafting of genetically modified cells in the CNS has emerged as a viable method for gene therapy. The advantages of this method include the ability to customize the cells to produce factors of interest, focal delivery in the localized areas in the brain, and possibly, reduced immune rejection of the graft by the use of autologous cells. A number of studies have shown that genetically modified cells survive in the CNS, synthesize and release the desired products, and have functional effects. However, the survival of these cells in the brain depends on the cell types used, and achieving stable, long-term expression of the transgene remains a problem. Further improvement of the methodologies of gene transfer and grafting techniques, such as long-term survival of implanted cells and transgene expressions, may lead to novel strategies for treatment of neurodegenerative diseases.

Further reading

Gage FH (1990): Intracerebral grafting of genetically modified cells acting as biological pumps. *Trends Pharmocol Sci* 11:437–439

Gage FH, Fisher LJ (1991): Intracerebral grafting: A tool for the neurobiologist. Neuron 6:1–12

Gage FH, Kawaja MD, Fisher LJ (1991): Genetically modified cells: Applications for intracerebral grafting. *Trends Neurosci* 14:328–333

Miller AD (1990): Progress towards human gene therapy. *J Amer Soc Hematol* 76:271–278

Cachexia and Tumor Necrosis Factor

Ilene L. Bernstein

The macrophage-derived peptide, tumor necrosis factor, has been proposed as an important mediator of the cachexia associated with chronic inflammation and with cancer. This hormone, originally named "cachectin" because of its suspected role in cachexia of disease, was later found to be structurally identical to tumor necrosis factor (TNF), a hormone of interest because it was shown to be cytotoxic to certain tumor cell lines. This review will focus on TNF, but it should be noted that TNF is only one of a number of cytokines, including interleukin-1 (IL-1), which have been implicated in the response to inflammation, trauma, and malignant disease.

Cachexia is characterized by lowered food intake (anorexia) as well as a reduction in body weight through a loss of fat and protein mass. TNF was originally implicated in the cachexia syndrome by studies indicating that it interferes with lipoprotein lipase (LPL) activity in adipocytes.

Critical evaluation of the role of TNF in cancer cachexia has involved several related approaches. One approach has been to ask whether administration of exogenous TNF mimics critical features of the cachexia syndrome. It has also been important to ask whether elevated TNF levels are found to be associated with anorexia and cachexia in cancer patients and/or tumor-bearing animals. A final strategy has been to ask whether cachexia symptoms can be relieved by blocking TNF's effects by the administration of anti-TNF antibodies.

TNF administration and cachexia

A large number of studies have examined the effects of exogenous TNF on food intake and body weight, and results are generally consistent. In one typical study recombinant human TNF was administered twice daily to rats. High doses (100 µg/kg/day, intraperitoneally) were associated with clear and substantial reductions in daily food intake. Similar effects have been reported for mice. One puzzling aspect of these studies has been that effects of the hormone are not sustained. Rather, the typical observation is a rapid appearance of tolerance to the anorexic effects of TNF with food intake returning to normal in 2 to 4 days in spite of continued TNF injections. Tolerance development appears to be due to nonimmunological factors since an antibody response is not evident and since tolerance fades 2 to 3 weeks after termination of TNF treatments. The prompt development of resistance to the anorexic effects of exogenous TNF has raised questions about how TNF, acting alone, could be responsible for generating a sustained syndrome of cancer cachexia.

In contrast to the transient effects of repeated exogenous administration of TNF, long lasting reductions in food intake and body weight were observed in nude mice implanted with tumors that secrete recombinant human TNF. This occurred despite the fact that circulating levels of TNF were not higher in the tumor-bearing mice than in mice receiving exogenous TNF and displaying tolerance. These findings suggest that an underlying disease state interferes with the development of tolerance to TNF and/or that continuous infusions, provided by the tumor, were less likely to generate tolerance. Evidence, summarized presently, supports both these suggestions.

Cancer patients and tumor-bearing animals are more sensitive to catabolic actions of TNF than are healthy individuals. Also, the simultaneous release of other cytokines, such as interleukin-1, substantially increases the tissue response and toxicity of TNF. Evidence for synergy between TNF and other cytokines includes the observation that administration of IL-1 and TNF to healthy animals provokes hemodynamic collapse, decreased energy expenditure, and muscle protein degradation in doses where administration of either cytokine alone produced minimal effects.

Several studies have evaluated whether continuous infusion of TNF is less likely to produce tolerance to the anorexic and cachexic effects of the hormone than acute injections. When TNF was administered intraperitoneally to mice through implanted osmotic minipumps that produce a chronic, relatively constant infusion, tolerance still developed fairly rapidly. More recently, however, studies in which rats received TNF through continuous intravenous (IV) infusions yielded a different outcome. These rats received an amount of TNF identical to that of a group receiving bolus injections. Rats in the latter group displayed tolerance by the third day while those receiving continuous infusions continued to display anorexia throughout the 8 days of treatment and suffered considerably more weight loss and loss of body protein. These results support the hypothesis that tolerance is less likely to develop to chronic endogenous TNF production and therefore that TNF could mediate sustained reductions in food intake and body weight.

TNF production in disease states

A number of reports have appeared in the literature assessing TNF production in cancer patients and the findings tend to be contradictory. At least one study reported that a high proportion of cancer patients showed detectable levels of serum TNF, while other studies have failed to find TNF in serum samples from patients with advanced malignancies. In the one study where a significant proportion of serum samples from cancer patients contained TNF reactivity, weight loss data were available from about half the patients. There was no correlation between weight loss and serum TNF activity. The absence of evidence for an association between elevated TNF and cachexia symptoms raises questions about the role of TNF in this syndrome. Some have argued that tissue TNF production may be elevated without elevated levels in circulation. In this regard tissue macrophages and blood monocytes from patients and experimental animals with cancer have been found to produce TNF *in vitro* in much higher levels than controls. Negative and inconsistent findings when serum samples are examined may also reflect the low sensitivity of available assays for TNF. Interestingly, there are more consistent and positive reports of elevated circulating TNF levels in patients and animals with parasitic disease and with infections.

Anti-TNF antibodies and cancer cachexia

A critical test of the hypothesis that TNF is an important mediator of cancer cachexia is to evaluate whether neutralizing or blocking the effects of TNF with anti-TNF antibodies can reduce or eliminate cachexia symptoms. This is an attractive research approach because it also provides a potential therapeutic strategy. Mice with a transplantable MCA tumor were passively immunized every other day with a rabbit immunoglobulin fraction raised against murine cachectin/ TNF. Anti-TNF treatments produced a significant reduction in anorexia. Carcass protein and fat loss were also reduced by around 40%. Although, these findings are supportive of TNF as a mediator of tumor anorexia, it remains unclear whether the antibody's effects on food intake were direct or were secondary to its observed inhibition of tumor growth.

Mechanisms of TNF action

The long-term effects of TNF exposure on food intake as well as body weight and composition have only recently begun to emerge, in studies which circumvent tolerance development either through chronic infusions or escalating dosages. It has been noted, when comparing TNF-treated animals with pair fed controls, that weight loss due to TNF treatment is greater than can be accounted for by reductions in food intake alone. Also seen are reductions in body fat and protein, and evidence of negative nitrogen balance. Thus, TNF appears to have widespread effects on metabolism as well as on feeding behavior. Many of these effects may be mediated by direct action on peripheral tissues, while responses to central administration of TNF have also been reported. For example, microinfusion of TNF into the third ventricle, in rats, led to a suppression of food intake, presumably through a direct action on the central nervous system.

The suppression of voluntary food intake by TNF is of particular interest from a clinical perspective. Among possible peripheral mediators of this are TNF's effects on the gastrointestinal tract. TNF treatment is associated with a dose-related blocking of gastric emptying, which is evident within 30 minutes of administration. Distension of the stomach reaches its maximum size around 3.5 hours after injection and is still detectable 6 to 7 hours after an IV dose. It has been suggested that TNF suppresses food intake by inhibiting gastric emptying, a response which precedes an inflammatory response in intestinal tissue. It has further been demonstrated that TNF administration is associated with the development of strong aversions to the available diet. This finding is similar to reports of the effects of certain experimental tumors, and suggests that symptoms of malaise and nausea contribute both to the anorexia and to the learned aversions. Interestingly, brain stem lesions in the area postrema significantly attenuated the effects of TNF (and of tumors) on both food intake and on diet aversions. These results have been interpreted as supporting a parallel between the effects of TNF treatments and the effects of tumor growth on food intake, body weight, and diet preference.

Conclusions

Although the evidence cannot really be said to be conclusive, there is clearly considerable support for a role for TNF in cachexia. More research has been directed at cachexia symptoms in cancer, although the support may, in fact, be stronger for TNF's role in the cachexia of parasitic and inflammatory diseases. With the development of more sensitive and specific assays for TNF and more effective antibodies against TNF, we may expect a clearer picture of the role of endogenous TNF in cachexia syndromes. It is also likely that other cytokines, particularly IL-1, will prove to be important components in the body's response to inflammation and malignant disease.

Further reading

Beutler B, Cerami A (1987): Cachectin: More than a tumor necrosis factor. *New Engl J Med* 316:379–385

Darling G, Fraker DL, Jensen C, Gorschboth CM, Norton JA (1990): Cachectic effects of recombinant human tumor necrosis factor in rats. *Can Res* 50:4008–4013

Moldawer LL, Lowry SF, Cerami A (1988): Cachectin: Its impact on metabolism and nutritional status. *Ann Rev Nut* 8:585–609

Plata-Salaman CR (1989): Immunomodulators and feeding regulation: A humoral link between the immune and nervous systems. *Brain, Behav Immun* 3:193–213

Cerebellum, Messenger Molecules

Andrew M. Batchelor and John Garthwaite

A striking feature of the cerebellum compared with most brain areas is the simplicity of its cellular composition and anatomical organization (Fig. 1). In essence, within the cerebellar cortex, there are two major types of neurons (granule cells and Purkinje cells) and three classes of inhibitory interneuron; the structure has two main inputs (the climbing and mossy fibers) and only a single output (the Purkinje cell axons). This simplicity, coupled with the fact that the cells and the circuitry of the cerebellum are neatly organized into distinct layers, have long made this region an attractive one for experimentalist and theoretician alike. At one time, the cerebellum was considered to be a sort of neuronal machine that did some simple arithmetic with inputs and thereby produced an output. In some ways this is essentially correct, but we are now beginning to understand that communication between neurons is much more complex and displays greater levels of subtlety than was previously envisaged. At the heart of this new understanding is a more precise knowledge of the ways that neurons signal to each other, of the types of chemical that perform this function, and of the different ways that the neurons use the information.

Neurotransmitters

The primary messages are conveyed in a highly restricted space, the synaptic cleft. The release of the neurotransmitter is under strict control, and diffusional spread away from the synapse is prevented by efficient transport mechanisms into presynaptic terminals and glia. The way that the postsynaptic neuron responds is determined by the particular mix of receptor proteins present that the neurotransmitter can bind to and, hence, activate. In line with their status in the CNS as a whole, the transmitter employed in the excitatory synapses in the cerebellum is the amino acid glutamate, whereas its counterpart for inhibitory pathways is GABA (gamma aminobutyric acid).

Fast signaling through ionotropic receptors. Contrary to the earlier beliefs modeled on the neuromuscular junction, it is now clear that a single transmitter acts by means of an often bewildering array of receptors. The fastest effects are mediated by "ionotropic" groups of receptors that have an ion channel as part of their protein structure. Binding of the

Glutamate and GABA receptors in cerebellar synapses

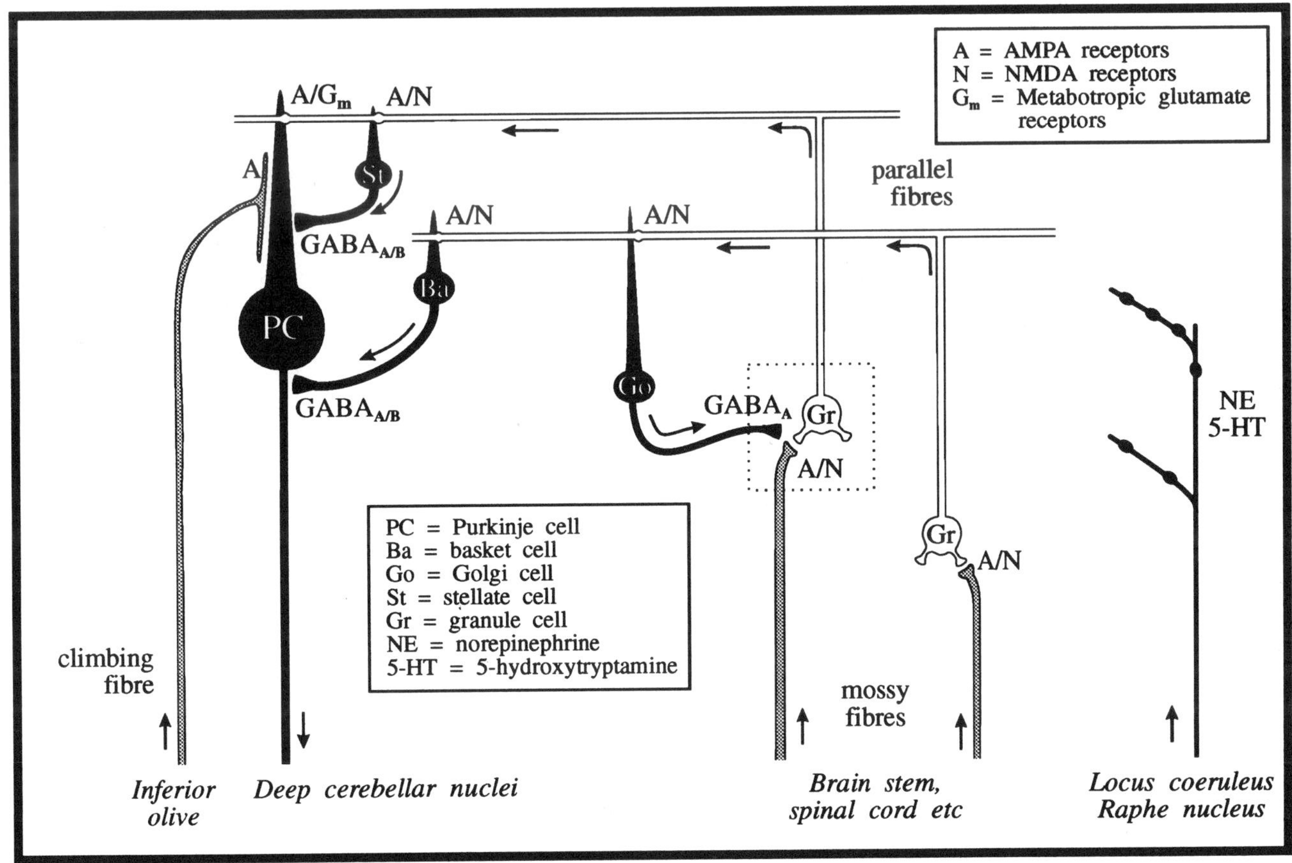

Figure 1. Schematic writing diagram of the cerebellar cortex indicating the subclasses of glutamate and GABA receptors that have been shown to be activated at the different synapses. Neurons whose function is inhibitory are shown in black; excitatory fibers and neurons are stippled or unfilled.

transmitter to a specific recognition site changes the shape of the protein so that the ion channel is opened. Depending on which ions are able to pass through the channel, the cell is made more, or less, excitable.

Glutamate acts on at least two broad types of ionotropic receptor, the so-called AMPA (α-amino-3-hydroxy-5-methyl-4-isoxszolepropionate) and NMDA (N-methyl-D-aspartate) receptors, the nomenclature being based on the names of selective agonists. On its release from mossy fibers, glutamate acts on both AMPA and NMDA receptors on granule cells; in contrast, at synapses formed by both climbing fibers and granule cell axons (parallel fibers) with Purkinje cells the NMDA variety is lacking. Activation of AMPA receptors induces a very fast inward current, rising to its peak in about a millisecond, and carried by sodium ions; because glutamate dissociates from these receptors very quickly and is then lost, the current is also very transient, lasting only a few milliseconds. AMPA receptors therefore bestow great speed on signals from mossy fibers to granule cells and from parallel and climbing fibers to Purkinje cells.

But despite the fact that the AMPA receptors at parallel fiber and climbing fiber synapses with Purkinje cells appear identical, the behavior of these neurons is markedly different depending on which pathway is activated. This is due to the anatomy of the connections. Purkinje cells possess a huge, elaborate dendritic arbor and the single climbing fiber supplying each neuron activates AMPA receptors at synapses dispersed over a large area of this tree. The ensuing dendritic depolarization causes the opening of special types of voltage-sensitive Ca^{2+} channels, causing secondary depolarizing potentials and a concomitant increase in the levels of Ca^{2+} within the dendritic cytoplasm. The climbing fiber input is therefore not only extremely powerful but its effects last for tens of milliseconds; moreover, the associated Ca^{2+} signal itself is likely to be an important activator of various intracellular mechanisms. In contrast, vast numbers of parallel fibers (up to 200,000) contact every Purkinje cell and the synapses are located on distal dendritic spines; an individual fiber therefore makes only a very small and highly localized contribution to the overall firing rate of the neuron.

NMDA receptors are prominently expressed at synapses between mossy fibers and granule cells and between parallel fibers and the inhibitory interneurons (basket, stellate, and Golgi cells). They differ from AMPA receptors in three main ways. First, the conductance of associated channels depends on the membrane voltage, such that at normal resting membrane potentials (-70 mV) only a small current will flow. This is due to a blockade of the channels by extracellular Mg^{2+}. With depolarization, the attraction of Mg^{2+} into the channel is reduced and the ionic conductance increases, reaching a maximum at about -40 mV. Second, the channels open rather sluggishly, 10 ms or so being needed for the peak current to develop; however, because glutamate dissociates only slowly from NMDA receptors (due to the high affinity with which it binds to them) the channels continue to open for hundreds of milliseconds. Third, NMDA receptor channels have a high effective permeability to Ca^{2+}. Because of the first of these properties, NMDA receptors do not participate much in low-frequency synaptic transmission but during high-frequency transmission, sustained depolarization, summation of NMDA currents, and (probably) fatigue of inhibition combine to bring the NMDA system into operation. The resultant electrical changes will likely be of importance but probably more so will be the associated inflow of Ca^{2+} and mobilization of second messenger systems.

Fast synaptic inhibition is mediated through ionotropic GABA receptors, or $GABA_A$ receptors. The channel here conducts Cl^-, the effects being to lower the resistance of the membrane (making excitatory currents less able to depolarize) and to hyperpolarize the membrane, for periods lasting tens of milliseconds. $GABA_A$-receptor–mediated inhibition operates at several different levels within the cerebellar circuit (Fig. 1). Especially powerful inhibition is exerted by basket cells on Purkinje cell firing. The synapses in question are numerous and are located close to the initial segment of the neurons where action potentials are generated. Stellate cells, in contrast, operate within the dendritic tree, where they contribute to the local integration of signals. Golgi cells produce both feed-forward and feedback inhibitory signals to granule cell dendrites. One interesting feature of Golgi cells is that many of them not only release GABA but also glycine. Yet classical inhibitory glycine receptors are not expressed by granule cells. Abundant NMDA receptors are, however, the significance being that glycine is a coagonist at NMDA receptors and is probably essential for those receptors to function. Golgi cells may thus have a dual role, both to suppress excitation and to regulate NMDA receptor activation. In order to further understand the workings of the Golgi cell, it will be essential to determine the precise way in which these two apparently functionally competing messengers are released and the kinetics of their actions on granule cells.

Slower signaling through metabotropic receptors. Glutamate and GABA also act on "metabotropic receptors," which are not coupled directly to ion channels but, instead, to G-proteins that, in turn, regulate the activity of intracellular enzymes generating second messenger molecules. A plethora of subsequent events are conceivable and, on electrophysiological grounds alone, the responses can last for seconds. For glutamate, the family of metabotropic receptors (mGluRs) is still growing; one is known to be coupled to the activation of phospholipase C, which generates two substances, inositol-1,4,5-trisphosphate and diacylglycerol. The former provokes the release of Ca^{2+} from intracellular stores and the latter stimulates the enzyme protein kinase C, which can phosphorylate many different proteins and so alter their properties. Ca^{2+} itself can also activate many enzymes and influence ion channels. This receptor, $mGluR_1$, is highly concentrated in Purkinje cells, where many of the other enzymes associated with this pathway are also enriched. It is probably responsible for a very slow (seconds) depolarizing wave observed following repeated activation of parallel fibers. Another metabotropic glutamate receptor, $mGluR_2$, is found in Golgi and granule cells and this one mediates a decrease in the levels of another intracellular messenger, cAMP.

GABA also acts at its own metabotropic receptor type, the $GABA_B$ receptor; these are particularly enriched in the molecular layer of the cerebellum, where the Purkinje dendrites and parallel fiber terminals are located. In Purkinje cells, $GABA_B$ receptor stimulation induces a slow hyperpolarization that is particularly prominent in young animals. There are also presynaptic $GABA_B$ receptors, whose activation leads to inhibition of transmitter release, on parallel fiber (and perhaps other) terminals.

Other neurotransmitters, including norepinephrine, 5-hydroxytryptamine, and dopamine, also play a part in the cerebellum. It is likely that they modulate fast transmission by interacting with second messenger cascades.

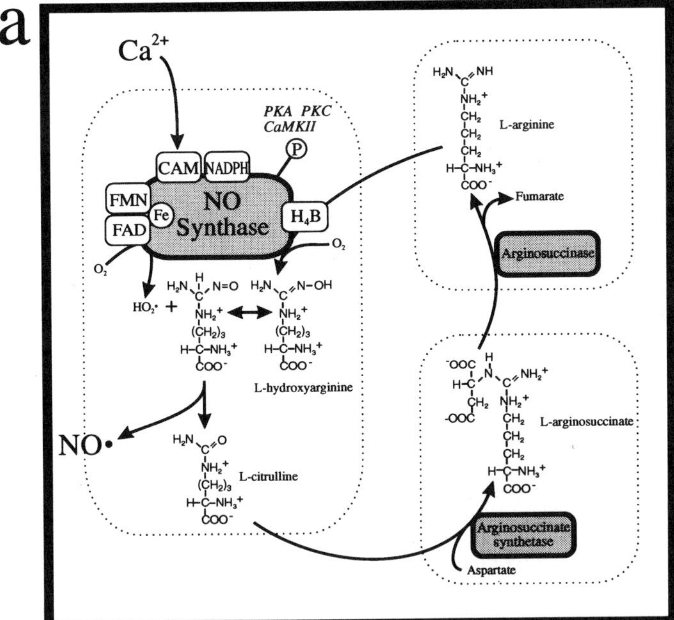

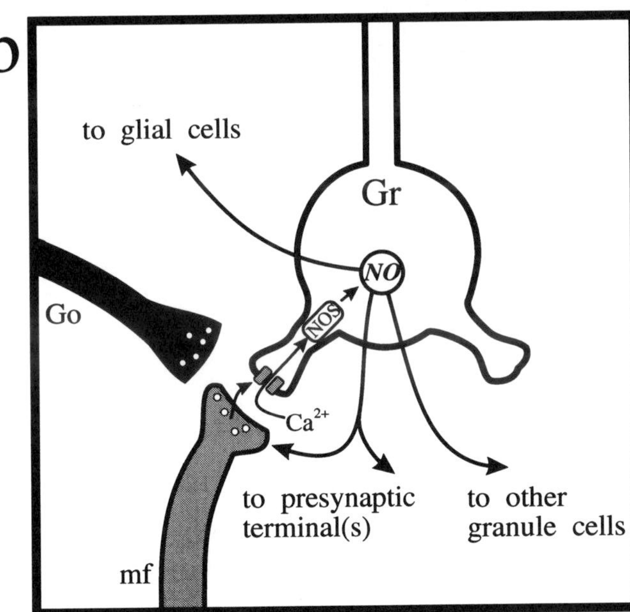

Figure 2. (a) Simplified reaction mechanism for the generation of NO by cerebellar NO synthase and the recycling of L-citrulline to the immediate precursor of NO, L-arginine. NO synthase requires molecular oxygen together with multiple cofactors, including calmodulin (CAM), NADPH, flavin mononucleotide (FMN) and flavin adenine dinucleotide (FAD), and tetrahydrobiopterin (H₄B). The enzyme contains iron (Fe) and can be phosphorylated (P) at different sites by protein kinase A (PKA), protein kinase C (PKC) and Ca²⁺- or calmodulin-dependent protein kinase II (CaMKII). Some evidence suggests that the three main components of the arginine cycle are located in different cells, represented by the dotted outlines. (b) Model for the generation of, and intercellular signaling by, NO at glutamatergic synapses between mossy fibers (mf) and granule cells (Gr); see boxed area in Fig. 1; Go = Golgi cell axon. The principal route for the Ca²⁺ influx needed to activate NO synthase (NOS) is provided by the NMDA receptor on granule cell dendrites. Following its formation, the NO diffuses freely to neighboring cellular elements.

Nitric oxide

Nitric oxide (NO) is a recently discovered messenger molecule that operates in a highly unconventional way. Though remarkably simple, NO is generated by a complex, multifunctional, Ca²⁺-dependent enzyme, known as NO synthase (NOS). NOS acts on one of the guanidino groups of the amino acid L-arginine to produce L-hydroxyarginine and then NO and L-citrulline (Fig. 2a). A major location of NOS is in cerebellar granule cells and the best defined stimulus for its activation is NMDA-receptor–induced Ca²⁺ influx. Elsewhere, NOS may be located presynaptically and be produced following action-potential–dependent Ca²⁺ influx. Possible sites of this latter mode of operation are in the climbing fibers and parallel fibers.

Unlike other second messengers, NO is a lipid soluble gas and so, like carbon dioxide and oxygen, it can diffuse very rapidly through cell membranes to act on neighboring cells. Indeed, this is the principal way it functions in the brain (and other organs). Unlike the point-to-point way in which neurotransmitters act, NO probably exerts its influence within a three-dimensional volume, perhaps in the 10 μm diameter range. Within this volume, the targets may be presynaptic nerve terminals, postsynaptic neurons, and even nonneuronal cells (Fig. 2b). Because of the high reactivity of NO, it may have several actions, but its established one is to stimulate the cyclic GMP synthesizing enzyme, soluble guanylate cyclase. Cyclic GMP can then have several effects including activation of ion channels, protein kinases, and enzymes regulating cyclic AMP levels.

Long-lasting alterations in synaptic strength: synaptic plasticity

The mainstay of the reductionist approach to learning and memory derives from the idea of Donald Hebb, that the strength of a synaptic transmission is modifiable and that changes to this property underlie the ability of the nervous system to adapt or learn with experience. The cerebellum is thought to partake in certain forms of motor learning, and one of the key modifiable synapses in the cerebellum is deemed to be the one between the parallel fibers and the Purkinje cell. Whether these synapses become enduringly modified or not depends on the other input, the climbing fibers. When parallel fibers alone are stimulated, the synapses become transiently (up to 30 minutes) strengthened, but if climbing fibers are stimulated at the same time, the efficacy of the parallel fiber synapses is selectively reduced and the effect lasts for hours or longer; this is termed long-term depression (LTD). LTD appears to be a result of a loss of sensitivity of AMPA receptors, but the underlying mechanism is poorly understood. NO and metabotropic glutamate receptors have both been implicated.

Further reading

Collingridge GL, Lester RAJ (1989): Excitatory amino acid receptors in the vertebrate central nervous system. *Pharmacol Rev* 40:143–210

Garthwaite J (1991): Glutamate, nitric oxide and cell-cell signaling in the nervous system. *Trends Neurosci* 14:60–67

Ito M (1984): *The Cerebellum and Neural Control*. New York: Raven Press

Ito M (1989): Long term depression. *Ann Rev Neurosci* 12:85–102

Conscious Experience, Neural Basis

Benjamin Libet

Conscious, subjective experience is accessible only to the individual having it. Consequently, the valid operational definition of it in scientific investigations is the introspective report by the subject, under conditions of reliability and credibility. That requires the availability of human subjects, awake and forthcoming, in whom simultaneous observations and manipulations of neural functions can be made. This combination of requirements imposes severe limitations on potential studies. Nevertheless, increasingly meaningful findings are being reported, taking advantage of newly available noninvasive techniques for studying brain function as well as those involving intracranial access and the experimental study of patients presenting lesions in the brain.

Effects of lesions in the brain

It was, of course, long known that lesions of the cerebral cortex that destroyed a primary sensory area resulted in severe sensory impairment. For example, with loss of striate cortex area 17 patients are blind; that is, they report being unaware of signals in the affected visual field. But further analyses are showing that these patients can detect signals without any reportable awareness, by pointing correctly to them in forced-choice tests; this is the "blindsight" phenomenon (see *Blindsight* by Weiskrantz, this volume). In a related manner, patients with lesions that abolish their reportable awareness of differences in color can correctly discriminate among different colors in forced-choice tests (Stoerig and Cowey, 1989); and with lesions that obliterate conscious identifiability of the shapes and locations of objects, patients can retain correct visuomotor guidance (of their hands) to these objects (Goodale et al., 1990). A similar discrepancy, between conscious and unconscious (or nonconscious) processes can be produced by lesions in the frontal-medial cortex (see, for example, G. Goldberg et al., 1981); patients exhibit the "alien-hand sign," in which the affected arm and hand perform spontaneous purposeful-looking movements that are not consciously willed or amenable to direct conscious control. A variety of other cerebral lesions can produce other interesting distortions of, or losses in, various kinds of awareness (as studied by A. R. Damasio and others; see also "Progopagnosia," this volume).

The foregoing kinds of findings in neuropsychology indicate cortical areas that are *necessary* for developing conscious experiences of sensory and volitional events. But they do not demonstrate that these areas are *sufficient* for this function; that is, they do not show where the conscious experience actually forms or arises. This distinction between necessity and sufficiency has at times been neglected, leading to potentially erroneous conclusions about the "seat of consciousness"; it presents a problem that is difficult to address experimentally. Also, they do not themselves provide evidence on the physiological activities and interactions of the neuronal groups involved in the conscious process. They do reinforce the point that behavioral performance cannot be taken to be valid primary evidence for conscious experience; the latter requires the introspective report of awareness by the subject.

Regional cerebral blood flow (rCBF) and metabolism

The ability to measure rCBF and relate this to psychological functions in awake human subjects was introduced over 20 years ago by Lassen and Ingvar. It has since been technically improved, as to localizability and quantitativeness, by the introduction of positron-emission tomography (PET scan) by L. Sokoloff and others, and by the coupling of this with the structural mapping by magnetic-resonance imaging (MRI). These techniques are essentially noninvasive, except for the administration of presumably harmless tagged agents, and have thus opened a highly productive avenue for studying brain activity in the awake and responsive human subject. However, while these methods tell us which areas or nuclei in the brain are changing in their rate of metabolism, they do not give information on the physiological dynamics involved; that is, the information is essentially topographical. The timing of the change is also still only possible to within some seconds, at best, in contrast to electrophysiological methods.

rCBF studies in which subjects are asked to perform conscious mental operations are of special interest to the issue of brain and conscious experience. In one such study, subjects were asked to imagine moving their fingers without actually moving them; a selective increase in blood flow appeared in the supplementary motor area (SMA; initially reported by Lassen et al., 1978). Such findings might bear on the question of whether and where a conscious mental function can initiate or influence neural processes. The time resolution of the rCBF method, however, as well as the absence of indicators of the precise relative timing of the onset of the mental imagining, vitiate any answers to that question. It has been shown experimentally that conscious intention to move, for example, actually *follows* onset of electrophysiological activity associated specifically with the volitional process (Libet et al., 1983).

Electrophysiological stimulation and recording

The numerous studies with extracranial (scalp) recordings of event-related potentials (ERP) over recent decades have not generally been sufficiently relevant to the issue of conscious experience. This is partly because the associated psychological tasks for the subjects were mostly those of detection of, recognition of, and responsiveness to signals, functions that could all potentially be performed unconsciously unless direct introspective reports of awareness are demanded. Scalp recordings also suffer from uncertainties of localization of the neuronal sources of the ERPs, although this is improving with recent technological advances. Whether the recording of magnetic rather than electrical fields will improve localizability significantly remains to be seen. Also, even when conscious reports are appropriately included, it is extraordinarily difficult to design experiments that may indicate causal interrelationships, as there is little possibility to manipulate one of the independent variables, the cerebral neural activity.

Intracranial electrodes can greatly improve both the localizability of electrical manifestations and the ability to manipulate neuronal functions in some quantitative fashions by electrical stimulation. On the other hand, the availability of such electrodes in loci suitable for study in awake, responsive patients is not common, and the kinds of experimental procedures permitted by the therapeutic and risk factors are greatly restricted. Fruitful experimental avenues have, however, been found. Stimulation can elicit conscious experienees, especially when applied in the cerebral sensory systems. It can also reversibly interfere with or "disrupt" normal cerebral functions. This latter ability can be used to investigate functions of cerebral areas where stimulation does not itself elicit any conscious responses; it has been employed fruitfully in recent years by Ojemann et al. to analyze the language functions of temporal (and other) cortical areas.

"Time-on" theory, for conscious and unconscious mental functions

Based on their earlier lines of evidence indicating a requirement for a substantial duration of neural activities in order to elicit a conscious experience, whether a sensation or intention (to move now), Libet et al. have devised and recently tested a theory to explain how the brain may distinguish between conscious and unconscious mental functions. The theory proposes the duration, or "time-on," of appropriate neural activities as one controlling factor in achieving this distinction: durations must exceed minimum values of up to about 500 ms to achieve conscious awareness of a mental process, but durations distinctly below those minimum values can mediate unconscious mental functions (without awareness). A specific experimental test of the proposal was carried out in relation to detection of a sensory signal with and without awareness (Libet et al., 1991). Electrical stimulus pulses were applied in the somatosensory (ventrobasal) thalamus of responsive human subjects. With intensity of pulses maintained at a constant near-liminal level, train durations of pulses were varied randomly in different trials. In each trial the stimulus was delivered during one of two lighted signals, in sequence, each turned on for 1 s. Subjects then indicated in which of the two lighted signals the stimulus was delivered, whether they felt any sensation or not; this forced-choice response indicates a degree of *detection* of the stimulus when the correct answers exceed the 50% level expected by chance alone. Subjects also then indicated their level of *awareness* of any sensation during the selected lighted period. The results clearly demonstrated (a) that a significant degree of detection occurred when stimulus train durations, as brief as 0.1 s, were insufficient to elicit any awareness of them; and (b) that an additional 300–400 ms of train duration, for the same repetitive inputs to the cerebral cortex, is required in order to convert from correct detection *without* awareness (an unconscious cognitive response) to correct detection *with* awareness (a conscious response).

The time-on theory and its experimental test provide an example of the kinds of directly relevant questions, hypotheses, and experimental answers that can be pursued fruitfully in this difficult area of research. This is true also of the other lines of study outlined previously. One may hope that increasing recognition of these feasible possibilities will lead to expanding investigations of this most challenging and fundamental issue: how the brain and conscious mind are interrelated.

Further reading

Goldberg G, Mayer NH, Toglia JU (1981): Medial frontal cortex infarction and the alien hand sign. *Arch Neurol* 38:683–686

Goodale MA, Milner AO, Jakobson LS, Carey DP (1990): Kinematic analysis of limb movements in neuropsychological research: subtle deficits and recovery of function. *Can J Psychol* 44:180–195

Lassen NA, Ingvar DH, Skinhoj E (1978): Brain function and blood flow. *Sci Amer* 239:62–71

Libet B (1973): Electrical stimulation of cortex in human subjects and conscious sensory aspects. In: *Handbook of Sensory Physiology*. Vol 2, *Somatosensory System,* Iggo A, ed. Berlin: Springer, pp 743–790

Libet B (1985): Unconscious cerebral initiative and the role of conscious will in voluntary action. *Behav Brain Sci* 8:529-566

Libet B (1989): Conscious subjective experience vs. unconscious mental functions: A theory of the cerebral processes involved. In: *Models of Brain Function,* Cotterill RMJ, ed. Cambridge: Cambridge University Press, pp 35–49

Stoerig P, Cowey A (1989): Wavelength sensitivity in blindsight. *Nature* 342:916–918

Weiskrantz L (1986): *Blindsight: A Case Study and Implications.* Oxford: Clarendon Press

Conus Venom Neuropeptides

Lourdes J. Cruz and Baldomero M. Olivera

Introduction

Conotoxins are small peptide toxins present in the venoms of predatory marine snails of the genus *Conus,* targeted with high affinity and specificity to key components of the nervous system. They are much smaller than most other polypeptide toxins, typically 10–30 amino acids in length. This is a size range in which chemical synthesis is feasible. In the hundreds of *Conus* venoms there is an unprecedented spectrum of different conotoxins. Not only are there classes of conotoxins that interfere with the function of different pharmacological targets, but even among toxins with the same general receptor specificity there is an unexpected diversity of molecular forms, highly divergent in sequence.

Recent advances in molecular neuroscience have revealed that every receptor and ion channel class has many different subtypes, which play different physiological roles. The conotoxins may be particularly useful as diagnostic reagents for distinguishing between subtypes of a general receptor class. These small peptides are often very narrowly targeted, so that they can discriminate between closely related receptor subtypes. Several conotoxins are already widely used by neuroscientists, in particular the ω-conotoxins, which inhibit the Ca^{2+} channels that control neurotransmitter release at synapses, and the conantokins, which inhibit NMDA (N-methyl-D-aspartate) receptors.

Biology of the cone snails

The genus *Conus* is a large group of tropical marine gastropods, comprising approximately 500 species. All *Conus* are predators that use venom to immobilize their prey. The snails sting the prey and inject their potent venom through a disposable harpoonlike tooth. A striking feature of *Conus* snail envenomation is the very rapid paralysis of prey after stinging has occurred. Some of the fish-hunting species literally immobilize their prey in 1 or 2 seconds.

Conus species can be highly specialized for particular prey, but the genus as a whole feeds on a wide variety of different marine animals. The spectrum of prey include vertebrates (fish), other molluscs (mainly gastropods), and a variety of worms (primarily polychaetes, but including hemichordates and echiuroids). Although most species are highly specialized, a few, particularly in more marginal environments for the genus, can be generalists that attack prey in several different phyla. The venom of a particular *Conus* species has clearly evolved to optimally paralyze its prey.

A number of the venoms are so potent that certain cone snails are dangerous to man. One species in particular, *Conus geographus* (see Fig. 1), has caused a number of human fatalities; over 30 have been recorded in the medical literature.

Toxicology of *Conus* venoms: The paralytic conotoxins

A number of different *Conus* venoms have been analyzed. The general picture that emerges is that each venom is a complex mixture of many different small peptides, most with three disulfide linkages (see Table 1).

Some peptides have a clear paralytic role in immobilizing the prey. The venoms of the fish-hunting *Conus* (a group

comprising approximately 50 species) are the best understood with regard to the detailed pharmacology of prey paralysis. A summary of some of the peptides in fish-hunting cone venoms for which both a biochemical and pharmacological analysis has been performed is shown in Table 1. Three of the peptides in Table 1 are potently paralytic to fish. The ω-conotoxins inhibit presynaptic voltage-sensitive Ca channels and prevent neurotransmitter release at the synapse. The α-conotoxins inhibit nicotinic acetylcholine receptors; these peptides are analogous in their biological activity to the α-neurotoxins of snakes such as α-bungarotoxin and cobratoxin. However, in contrast to those molecules, the α-conotoxins are very much smaller (13–15 amino acids versus 60–85 amino acids for the snake polypeptides). The μ-conotoxins inhibit muscle sodium channels directly; in contrast to the classical sodium channel toxins tetrodotoxin and saxitoxin, the μ-conotoxins discriminate with high selectivity between skeletal muscle and axonal subtypes of sodium channels.

A particularly useful characteristic of conotoxins is that a set of homologous toxins (such as the ω-conotoxins isolated from different species) exhibit considerable sequence divergence. Usually more than 30% of amino acids are different if any two ω-conotoxins from venoms of two different *Conus* species are compared; not a single sequence has been found in more than one venom. Consequently, the set of ω-cono-

Figure 1. Shell of *Conus geographus*. This fish-hunting species is believed to be responsible for most human fatalities due to *Conus* sting. Three major types of paralytic toxins are found in its venom.

Table 1. *Conus* Peptides with Known Molecular Targets

Class	Peptide	Sequence
α-Conotoxins: Acetylcholine receptor inhibitors	GI	E C C N P A C G R H Y S C *
	Consensus (7 peptides from 3 species)	- C C - P A C G - - - - C
ω-Conotoxins: Calcium channel blockers	GVIA	C K S *P* G S S C S *P* T S Y N C C R S C N *P* Y T K R C Y *
	MVIIA	C K G K G A K C S R L M Y D C C T G S C R S G K C *
	Consensus (12 peptides from 5 species)	C - - - G - - C - - - - - - C C - - - C - - - - - - C
μ-Conotoxins: Sodium channel blockers	GIIIA	R D C C T *P P* K K C K D R Q C K *P* Q R C C A *
	Variations (7 peptides from *C. geographus*)	P P R M K L
Conantokins: NMDA receptor antagonist	Conantokin-G or GV	G E γ γ L Q γ N Q γ L I R γ K S N *
	Consensus (3 peptides from 3 species)	G E γ γ γ R γ
Conopressins: Agonists of vasopressin receptor	Arg-Conopressin-S	C I I R N C P R G *
	Lys-Conopressin-G	C F I R N C P K G *

An asterisk indicates that the α-carboxyl group is amidated. Except for γ (γ-carboxyglutamate) and *P* (*trans*-4-hydroxyproline), the standard one-letter code was used for amino acid residues. Disulfide bridges are indicated by solid lines connecting cysteine residues.

toxins isolated from cone snails provides a powerful cohort of toxins for dissecting voltage-sensitive Ca channel subtypes. Several ω-conotoxins have become diagnostic tools for defining certain Ca channel subtypes.

In contrast to the general hypervariability in sequence found in conotoxins, one structural feature seems conserved: the position of the cysteine residues involved in disulfide bonding. Although there are many thousands of peptides found in *Conus* venoms, there are only a few disulfide-bonded arrangements used. Specific disulfide bonding would not normally occur in small peptides the size of the conotoxins. However, conotoxins are initially made as larger precursors that can fold into a fixed conformation. The Cys residues form disulfide linkages that "lock in" a specific conformation before the conotoxin precursor is processed to generate the small mature peptide.

Nonparalytic peptides in *Conus* venoms

Biochemical analyses of different *Conus* venoms invariably reveal that these are extremely complex; some venoms may have more than 100 different peptides present. A major fraction of such peptides do not directly paralyze prey when injected into test animals.

Many of these nonparalytic agents are peptides of the same general structure as the conotoxins, that is, small, relatively rigid peptides with a fixed conformation, highly enriched in cysteine residues that are involved in disulfide bonding. Although such peptides are not paralytic, it has been suggested that some of them may facilitate paralysis, by facilitating access of the paralytic conotoxins to their targets in the prey much more efficiently. An alternative biological explanation is that in certain cases at least, these peptides may be used defensively by the snail; for example, cases of cone-snail stinging of humans are clearly defensive.

The conantokins are among the most remarkable set of peptides that have been discovered so far that do not play a direct paralytic role. These are the only peptide ligands known to affect a subtype of glutamate receptors, the NMDA receptors. The conantokins are different from most other conotoxins in that they do not have multiple disulfide bonds; rather, they assume a rigid conformation because of a posttranslational modified amino acid γ-carboxyglutamate. In the presence of Ca^{2+}, the γ-carboxyglutamate residues apparently stabilize folding of these small peptides (17–27 AA) into stable α-helices. Thus, there may be additional strategies for assuming fixed conformations in small peptides apart from multiple disulfide bonds.

Prospects for the future

Conotoxins have several properties that make them of high interest in neuroscience. They provide a link between pharmacology and chemistry, on the one hand, and molecular genetics. Conotoxins are the products of genes; thus, in principle, it is possible to express and mutate conotoxins at the genetic level. On the other hand these peptides are synthetic chemicals with fixed conformations. The possibility of selecting conotoxin-like peptides with a desired pharmacological specificity *in vitro* by molecular genetic methods, and then analyzing the structure of the conotoxin-like peptide by chemical synthesis and 2D-NMR analysis, has exciting implications for the design of drugs and pesticides.

Further reading

Kohn AJ, Saunders PR, Wiener S (1960): Preliminary studies on the venom of marine snail *Conus. Annls NY Acad Sci* 90:706–725

Olivera BM, Gray WR, Zeikus R, McIntosh JM, Varga J, Rivier J, de Santos V, Cruz LJ (1985): Peptide neurotoxins from fish-hunting cone snails. *Science* 230:1338–1343

Olivera BM, Rivier J, Clark C, Ramilo CA, Corpuz GP, Abogadie FC, Mena EE, Woodward SR, Hillyard DR, Cruz LJ (1990): Diversity of *Conus* neuropeptides. *Science* 249:257–263

Olivera BM, Hillyard DR, Rivier J, Woodward S, Gray WR, Corpuz G, Cruz LJ (1990): Conotoxins: Targeted peptide ligands from snail venoms. In: *Marine Toxins: Origin, Structure, and Molecular Pharmacology.* Hall S, Strichart G, eds. Washington, DC: American Chemical Society

D

Depression, Childhood, and Treatment

Alan E. Kazdin

Depression as a disorder in children* is characterized by sad affect, loss of interest and pleasure in activities, diminished activity, feelings of worthlessness, and changes in appetite, weight, and sleep. The persistence of several symptoms for a week or two, or indeed much longer, and their impact on daily functioning serve to define the disorder. Within the last 20 years, research on adult depression has made major strides in understanding the nature of the disorder, alternative subtypes, family history, clinical course, treatment, and prevention. Within the last 10 to 15 years, research on depression in children and adolescents has advanced considerably as well.

Historical overview

The conceptualization of depression as a diagnostic entity in children has evolved over the last 50 years. The most dominant view proposed that children could not experience depression as a clinical syndrome. This view assumed that insufficiently developed personality structures prior to adolescence precluded the appearance of the disorder. In the 1970s, alternative views emerged. Salient among them was the view that children could experience depression but that core features (sad affect, loss of interest in activities) might not be evident. The notion of *masked depression* emerged to denote that depression might be manifest in other symptoms (e.g., hyperactivity, aggression, anxiety, bed-wetting, and school refusal) that overshadowed or replaced core features of depression. Eventually that view was abandoned largely because the masking of depression became difficult to test empirically. Moreover, evidence began to accumulate that depression could be diagnosed in children based on the presence and assessment of core features.

In the late 1970s and early 1980s, revised psychiatric diagnostic criteria elaborated the range of disorders that could be identified in infants, children, and adolescents. With the revision of psychiatric diagnosis, criteria for several disorders including depression were applied in unmodified form to children and adults. Several studies showed that the diagnostic criteria for depression could be reliably applied to children, adolescents, and adults.

In the last decade, advances in assessment have contributed greatly to research on childhood depression. Development of diagnostic interviews, questionnaires, inventories, and rating scales for depression have aided case identification and evaluation of the characteristics with which depres-

sion is associated. Several measures are available for children that focus on depression and associated features (e.g., hopelessness, helplessness, diminished self-esteem). Measures completed by children, parents, teachers, and peers are commonly used to assess childhood depression. Commonly used measures include the Children's Depression Inventory, the Children's Depression Scale, and the Center for Epidemiological Studies—Depression Scale. These measures require the rater to evaluate the presence or severity of various symptoms of depression and related characteristics (e.g., feelings of worthlessness, somatic complaints). In general, the development of psychological assessments has permitted evaluation of diverse characteristics of childhood depression such as cognitive functioning, peer relations, and family characteristics. In addition, several biological and psychophysiological measures (e.g., neuroendocrine function, sleep wave), adopted from research with adults, have helped to elaborate characteristics of depressed children.

Characteristics

Descriptive features. The findings that depression as a clinical syndrome can be diagnosed in children, adolescents, and adults does not mean that the manifestations of the disorder are necessarily identical. The syndrome encompasses diverse facets of affect, cognition, and behavior that would be expected to manifest themselves differently over the course of development. Perhaps surprisingly, the bulk of research has shown remarkable similarity in central and associated features of depression among children, adolescents, and adults. In relation to biological markers, several similarities have been identified between children and adults including hypercortisol secretion during depressive episodes and hyposecretion of growth hormone in response to insulin-induced hypoglycemia. In relation to cognitive functioning, depressed children and adults show negative thoughts about themselves, pessimistic views about the future (hopelessness), global attributions about their inability to influence their environment (helplessness), and low self-esteem. Many interpersonal behaviors including reduced social interaction and diminished nonverbal behavior and affective expression also have been evident in children and adults.

Some differences have been found between depression in children and adults. To begin with, selected symptoms are less likely to be present in children than in adults. For example, prior to adolescence, suicide is rare and hence is unlikely to be a part of the presenting symptoms among children seen for depression. In addition, other psychiatric disorders that may occur (be *comorbid*) with depression vary over the course of development. For example, in both children and adults, major depression and anxiety disorder often occur together. The specific forms of anxiety, however, vary among children and adults (e.g., school phobia, separation anxiety in children; agoraphobia, generalized anxiety disor-

---

*For present purposes the term "children" will be used generically to refer to youth ages 5–18. In cases in which the distinction between children and adolescents is relevant, separate terms will be used. Children less than 5 and indeed infants have been studied (e.g., Trad, 1987); much less is known about this age group in part because of the unique diagnostic and assessment issues raised at this age.

der in adults). Also, research has suggested that depression in children, especially boys, may be associated with antisocial and aggressive behavior. The reliability of such findings and their implications in relation to long-term course and treatment are matters of current research.

Prevalence and clinical course. The prevalence of depression in children has varied estimates in part due to different assessment procedures, criteria, and samples. In the United States, approximately 2–6% of children (ages 4–16) meet diagnostic criteria for major depression. Sex and age influence prevalence rates. Current research suggests that among prepubertal children, prevalence rates are similar for boys and girls. However, postpubertal differences suggest greater prevalence for girls than boys. Sex differences in rate and severity of depression appear to begin early in adolescence and to increase into adulthood.

Longitudinal studies have evaluated the course and prognosis of children with major depressive disorder. Several important findings have emerged. Among them is the finding that among children, recovery from episodes of depression is relatively high (>.90) within a period of 3–4 years. Children who are older at first onset appear to recover more rapidly than children with an earlier onset. Sex differences have also been reported, with greater persistence of depressive symptoms among boys than among girls. Current evidence suggests that depression in mild or severe forms for many youth reflect a chronic and recurring disorder. Although recovery rates are high within a given episode, early episodes of mild depression place youth at risk for more severe depression later. The level of risk depends on a host of factors including family loading for depression, family relations (e.g., attachment, bonding), academic and social competence of the child, and family stress that are now only beginning to be elaborated in research on childhood depression.

Etiology. Mood disorders consist of a set of disorders that vary qualitatively and quantitatively. No single cause has been identified to explain the different dysfunctions. Alternative models or conceptual views have been advanced to guide research on factors that contribute to depression. Rather than look for a single or primary cause of depression, the models draw attention to different facets of experience and levels of analysis. A variety of biological (e.g., genetic), psychological (e.g., cognitive processes, interpersonal behavior), and socioenvironmental (e.g., stress, social support) influences play a role in placing individuals at risk for depression. The models, developed from the study of adults, have only begun to be studied in relation to children. Unique influences of childhood, such as contact with parents and siblings who may be depressed, have yet to be elaborated.

Treatment

Treatments for childhood depression consist almost exclusively of variations of treatments developed for adult. Currently, the most commonly used treatments are psychopharmacological. Within this domain, imipramine has dominated research and practice and has been effective with major depressive disorder. Few clinical trials of alternative medications for childhood depression have been reported. Major questions remain to be addressed related to the responsiveness of children to placebos, the impact of emergent (side) effects including those that might influence development, and the responsiveness to medication as a function of age, sex, and type of mood disorder. The extensive literature on these issues in the treatment of adults provides a useful guide to stimulate research with children.

Very few studies of psychotherapy have been reported to date. The treatments are based primarily on cognitive and behavioral conceptions of depression, focus on increasing performance of activities, developing self-control, and reducing maladaptive cognitions. The results indicate that depressive symptoms can be reduced with treatment. Yet the paucity of studies and the lack of long-term follow-up preclude firm conclusions about the impact of treatment. Although various treatments successful with adults may be effective with children, there are features of children that suggest novel approaches. Unique influences on children (e.g., living with a parent who may have current or lifetime depression) and diverse situations in which children function (e.g., home and school) may warrant special attention.

Challenges for research

There are several challenges for evaluation of depression in children and adolescents. Initially, youth of different ages may vary greatly in their capacity to report on their emotions, cognitions, and behaviors. From the standpoint of assessment, this means that a given measure based on self-report may not be available for use for the full age range of interest (e.g., 4–18 years).

Second and related, assessment of clinical dysfunction among children usually relies heavily on child and parent report. Research reveals that child self-report and parent report often yield quite different information about the children's depression. Children do not invariably under- or overreport their depressive symptoms in relation to ratings completed by their parents. The discrepancy is reflected in very small or indeed sometimes zero correlations (*rs* range from .0 to .3) between children and parents in their ratings of the children's depression. Interestingly, both sources of information relate systematically to external criteria. For example, child ratings relate to hopelessness and suicidal ideation and attempt; parental ratings of the child's depression relate to the child's social behavior and school functioning. Current research is attempting to identify differences between ratings obtained from children, parents, teachers, and peers and how the information can be integrated to inform research and clinical practice.

Finally, identification of cases of depression and of children at risk for depression raises multiple challenges. Although depression is recognized among researchers and practitioners, parents and teachers who often are not responsible for case identification and referral may not detect signs of this disorder or view the signs they do see as worthy of intervention. Identification of children at risk for depression is more complex. Children at risk include those with family loading of affective disorder and those exposed to harsh conditions (e.g., abuse). As the range of risk factors is elaborated, the prospects of preventive interventions will increase.

Conclusion

Research in mood disorders represents an area where major advances have been achieved. Although the work has been conducted primarily with adults, theory and research developed in that arena have accelerated work with children. Many of the findings have conveyed similarities among central and associated features of depression across the developmental spectrum. This has been valuable in the case of depression where dysfunction for many individuals represents a lifelong disorder that can be identified early. Further work is needed to examine how development and dysfunction are intertwined and how depression might be treated early or averted altogether.

Further reading

Cantwell DP, Carlson GA, eds (1983): *Affective Disorders in Childhood and Adolescence: An Update.* New York: Spectrum

Hammen C (1991): *Depression Runs in Families: The Social Context of Risk and Resilience in Children of Depressed Mothers.* New York: Springer-Verlag

Kazdin AE (1990): Childhood depression. *J Child Psychol Psychiat* 31:121–160

Rutter MR, Izard CE, Read PG, eds (1986): *Depression in Young People: Developmental and Clinical Perspectives.* New York: Guilford

Trad PV (1987): *Infant and Childhood Depression: Developmental Factors.* New York: Wiley

Acknowledgment. Completion of this paper was facilitated by a Research Scientist Award (MH00353) from the National Institute of Mental Health.

Drosophila, Genes and Behavior

Jeffrey C. Hall

Introduction

Genetic studies of behavior in *Drosophila* were inaugurated many years ago, not long after the time when this organism's heredity began to be investigated in general. But fruit fly behavioral genetics did not really commence in earnest until the mid-to-late 1960s. Then, mutations disrupting the function of *D. melanogaster*'s nervous system began to be induced by chemical mutagens and isolated in systematic "screens" (whereas the early studies relied almost exclusively on behavioral characterizations of a small subset of mutants found due to the flies being visibly different from normal in morphology, color, etc.).

The screens resulted in mutants defective in the following categories of neural function: responses to visual stimuli, also to chemosensory ones, learning and memory, and biological rhythms. A further area of extensive genetic inquiry involves courtship and mating behavior, though extensive screenings for these kinds of mutants have not been carried out. Instead, visual, olfactory, learning, and rhythm genetics have each intersected with experiments in the area of reproductive genetics.

Almost all mutants resulting from the screens deemed worth saving for further analysis were normal in their external appearance. Some, however, proved to have neuroanatomical defects. These, in turn, seemed usually to be morphologically abnormal in the CNS, or possibly PNS (but not such that a sensory appendage or the like is merely missing), because of developmental defects, as opposed to late-onset degeneration of the neurons or ganglia in question. In any event, "functional" neurogenetics overlaps genetic and molecular studies that are aimed at unraveling the development of this organism's nervous system. In fact, developmental neuromolecular genetics involving *Drosophila* has become a large industry; it will not be discussed in this entry, except to mention some of the developmental-neural mutants that have been nicely applied as "tools" to ask physiologically or behaviorally based questions. In fact, certain such mutants were actually isolated on anatomical criteria per se (this approach began at least 10 years after the more standard one involving recovery of behavioral variants), in order that they might be applied in behavioral experiments. Moreover, the ready recoverability of brain-damaged mutants seems to have led to an argument about how behavioral genetic studies in *Drosophila* are beginning to say something about the nature of this organism's genome—a point that will be discussed further at the close of this entry.

Another general theme that may emerge concerns the "behavioral genes" that have proven to be surprisingly versatile. Thus, a given mutant isolated on one criterion has in several instances turned out to exhibit a sharply defined defect in what—at first glance—could be viewed as an unrelated aspect of the fly's neural functioning. This is different from the case of a mutant being generally "pleiotropic," whereby many kinds of behavioral defects would be exhibited because of general sluggishness or the like—defects that would not have to be revealed by some sort of concerted testing regimen. In fact, a not insignificant fraction of the behavioral mutants isolated in recent years were found, after mutagenesis, by just looking at putatively mutant adult flies. A fair proportion of such mutants turned out to overlap with those isolated (in other screens) by investigators looking for hypo- or hyperactive mutants. Some such variants have led to the identification of certain famous genes, which encode proteins that make up potassium and sodium channels. Yet, the pertinent *Shaker* (*Sh*) and *paralytic* (*para*) mutants do not really define behavioral variants, because the aberrant phenotypes are difficult to view in a context of "interesting" actions carried out by the wild type; but instead they simply refer to normal flies' well-being. Thus, these pathophysiological mutants will barely be discussed in this entry except for instances in which certain of them have been used as tools to ask behaviorally interesting questions. "Neurochemical genetics" is another area that has overlapped with behavioral genetics; yet almost all the recent work involving genes specifying neurotransmitter-metabolizing enzymes or neurotransmitter receptors have involved only biochemistry or molecular and developmental biology; so this area will not be discussed.

Visual mutants

Many dozens of mutants apparently defective in the function of *Drosophila*'s visual system have been isolated, by behav-

Figure 1. A courting pair of *Drosophila melanogaster*. These two adult flies are exhibiting reproductive behavior. At this particular stage of the courtship sequence, the male (which is on the left and is about 2 mm in length) extends one of his wings (or the other, at a given moment) and vibrates it to produce a "courtship song." The female (which is slightly larger than the male) eventually responds to these (and other) reproductive signals by being receptive to the male's mating attempts, within several seconds to a few minutes from the initiation of the sequence.

ioral tests, electroretingram (ERG) recordings, or noticing photoreceptor cell degenerations that occur postdevelopmentally. (Thus, this list would exclude the huge number of additional mutants recognized, recently or in the past, because of externally visible abnormalities whose etiologies involve eye formation during larval and pupal stages.) These visual-functional mutations define, however, only a few dozen genes, suggesting that the screens could be approaching "saturation"—whereby most of the genetic factors controlling photoreception, phototransduction, or signaling between these sensory cells and first-order interneurons have been identified by now, via the recovery of several independently isolated mutant alleles per locus. (Note that none of the other areas of neurofunctional inquiry, involving *Drosophila* genetics, have come anywhere near approaching such saturation; the reason is that almost all types of mutations except visual ones have been looked for on the fly's X chromosome only.)

Certain of the fly's visual mutations define functions that are widely known—or highly suspected—to be involved in the "pathway" of receiving and processing light stimuli. Thus, one did not really need certain physiologically defective mutants, which turned out to possess an altered or nonfunctional opsin gene (called *neither-inactivation-nor-afterpotential-E*, or *ninaE*), to know about that light-receptive substance. Nor did one need to move from the one case of a thoroughly blind *Drosophila* mutant to biochemical studies of a phospholipase-C (PLC) deficit in it, to cloning and sequencing of the pertinent *no-receptor-potential-A* (*norpA*) gene (confirming that it encodes a PLC), to suspect an importance of the relevant second-messenger system in phototransduction (at least for invertebrates). These accomplishments could be regarded as merely comforting, as are other more recent ones that have involved raw molecular studies of transducin- and arrestin-encoding genes in the fly. (This work has not yet led to anything experimental—e.g., from applying mutations—about the particulars of how these generally known entities function in *Drosophila* photoreceptors).

The special features of studying vision genetically are better exemplified by the following sorts of examples; these may have uniquely identified important components that act and interact as part of the transduction machinery, and which might not have been uncoverable otherwise (i.e., without screening for mutants). One case is the *transient-receptor-potential* (*trp*) gene, whose mutations cause photoreceptor potentials (elicited by relatively strong light) to decay back to baseline well before the light goes off. Isolating this gene, via genetics per se (as opposed to some kind of molecular "cross-homology"), disclosed a large, apparently membrane-bound protein that is hitherto unknown but would seem to play an important role in photoreceptor physiology. The other case involves a successful attempt to induce genetic variants that suppress photoreceptor degeneration caused by the *retinal-degeneration-B* (*rdgB*) mutations; the latter do so only when the late-developing or adult photoreceptors are exposed to light. Blinding (and ERG "null") *norpA* mutations suppress *rdgB*-induced degeneration as well (i.e., even in the light). Unsurprisingly, certain of the suppressor mutations (isolated by mutating an *rdgB* mutant) were orthodox *norpA* mutants; but one was special in that it was a mutation at the latter locus which, by itself, had apparently normal vision photoreceptor physiology; thus, its only salient property is suppression of the degeneration mutation. Moreover, that suppression was "allele specific," in that the *norpA^su* allele would only suppress the

particular *rdgB* mutation with respect to which the *su* variant was isolated. This phenomenon (well known in microbial genetics and molecular biology) suggests that the products of the two genes in question interact in a manner that is first glimpsed in an abnormal situation—wherein two wrongs make a right—but is likely to reflect an interaction of the normal *norpA*-encoded PLC and the protein produced by the wild-type allele of *rdgB*. The latter gene product is now known at the level of clone and sequence; and as for *trp*, is a novel and putatively membrane-bound substance, which is also likely to emerge as a genetically discovered component of the transduction mechanism. It will, however, be a challenge to figure out just what either of these two previously unknown kinds of proteins are doing, let alone the matter of proving the hypothesis that the NORPA and RDGB protein interact, and for what reasons. Indeed, neurogenetic cum molecular investigations are revealing a widening array of what some euphemistically call "pioneer proteins," and it will be interesting to see if the biochemical functions of these gene products can be systematically understood in the years to come.

Flies make more interesting responses to visual stimuli than just rushing toward light. Their visual system (PNS → CNS) is an intricate one of movement detectors. Mutants have helped dissect some of the components. Thus, the *optomotor-blind* (*omb*) mutant, isolated in regard to defective responses to rotating vertical stripes (i.e., subnormal yaw-torque), turned out to make poor responses with regard to "pitch" and "roll" stimuli as well. *omb* was also found to be an anatomical variant, devoid of all the "giant neurons" that usually arborize in a relatively central optic ganglion called the lobula plate. Another visual-system mutant is *lobula-plateless* (*lop*); it, however, was isolated based on anatomy per se, and retains the so-called horizontal subset of the giant fibers of this ganglion. The structure-function corollary came from finding that *lop* makes normal optomotor yaw responses, but very poor ones in terms of pitch and roll—owing, almost certainly, to an absense of the "vertical" giant neurons in this mutant.

Chemosensory mutants

There are a fair number of gustatory and olfactory mutants in *Drosophila*, though not as many as for visual variants. Some of the chemosensory mutants seem "specifically" defective, in the sense that they exhibit defective responses to one class of chemical. Certain of these have been shown to be subnormal in their physiological responses to the substances in question, via recordings made by applying electrodes to taste-receptive structures such as the distal legs or odor-receptive ones such as the antenna. Perhaps, then, some of these genes encode chemoreceptor molecules; if this turns out to be so, though, the molecular side of this story will lag behind that which is breaking open in vertebrates, via purely molecular experiments.

Other of the chemosensory mutants in *Drosophila* are more globally abnormal "taste-blind" or "smell-blind" mutants, which could possess defects in more "central processing" locations or have abnormal substructures on the pertinent portions of the sensory appendages just noted.

Little in-depth analysis (e.g., biochemical, molecular) of the mutants and genes identified in this area of fly neurogenetics has occurred. Yet some of the mutants have been useful tools for asking chemosensory questions in other areas. For example, two kinds of globally abnormal olfactory mutants have been shown, as males, to make very poor responses to female pheromones; yet these mutants courted females in a reasonably robust manner and were routinely successful in mating. In contrast, blind or optomotor-blind males do much worse (moment to moment), as they attempt to "track" moving females; though, given enough time, these visual defects are compatible with reasonable mating success. The only way to achieve behavioral sterility in this species by sensory deprivation was to thoroughly remove the flies' abilities to see, smell, and—from the male's side— produce certain courtship sounds.

From the purely genetic angle, an intriguing result has recently surfaced in the olfactory field: a certain *olfactory-trap-abnormal* (*ota1*) mutant, isolated after mutagenesis and using a device as alluded to in the variant's name, turned out also to exhibit visual defect (initially noticed at the level of the ERG). *ota1*, which is defective in responses to more than one kind of chemical, turned out to be mutated in the self-same gene as originally defined by the *rdgB* mutations; indeed, the latter are olfactorily defective as well as visually so (though neither *ota1* nor *rdgB* mutants exhibit any discernible degeneration of known chemosensory organs). This genetic connection between the control of, if you will, visual and olfactory processing has in a sense been backed up molecularly, because the products of the *rdgB* gene are not "eye specific" (they are still detectable in an anatomical mutant lacking eyes). An analogous example of genetically connecting (the control of) two different behaviors has come from studying a certain courtship mutant; it generates aberrant sounds, in comparison with those produced by a normal courting male as he follows a female and vibrates his wings at her: The *dissonance* (*diss*) mutation in question turned out to be altered in a previously known gene, *no-on-transient-A* (*nonA*), found in independent screens for visually defective mutants and named for its ERG abnormality. The hint of the connection came from noticing *diss*'s visual defects, though it has turned out that the other *nonA* mutants produce normal courtship songs.

Learning mutants

Drosophila exhibit a wide array of experience-dependent behaviors, which can be revealed by testing larvae or adults en masse and the latter individually. The best known learning paradigm involves exposing flies simultaneously to odors and electric shocks, which results in avoidance of the shock-associated odor; the memory of what to avoid lasts for hours. Using this system, learning and memory mutants were isolated by brute-force testing of descendents of mutagenized flies. Unfortunately, only three genes' worth of mutants were solidified in regard to genetic rigor and consistent behavioral abnormalities. However, two of them— *dunce* (*dnc*) and *rutabaga* (*rut*)—have been solved at the level of biochemistry; both involve cAMP metabolism. *dunce* was found to encode one form of *Drosophila*'s cAMP phosphodiesterase (PDE); the normal enzyme is distributed through all body regions but is dispensible for zygotic viability, almost certainly because there are at least two additional forms of the enzyme (i.e., encoded at other genetic loci). That the X-chromosomal region containing *dnc* was suspected to contain one of the loci was suggested by enzyme assays of partially aneuploid flies (e.g., three "doses" of the region leads to an appreciable increase in overall cAMP PDE activity); this is the same way that some of the salient neurochemically important genes in this species were zeroed in on.

dunce mutations are pleiotropic, as first shown by female infertility; indeed, two of the mutations at the loci were isolated on this criterion and later shown to exhibit poor learning. The etiology of that behavioral defect may have to do with *dnc*'s expression in the brain. This supposition stems from the one useful result of cloning this gene: the *dnc* sequence indicated an encoded PDE (an anticlimax); also, part of the cloned gene was used to engineer a "fusion protein," which lead to αDNC antibodies. These labeled the DNC protein throughout the CNS, but the immunohistochemical signals were by far most intense in the "mushroom bodies" of the larval or adult brain. That structure, a much studied one in insects, has long been suspected to be involved in learning. For *Drosophila*'s part, *dnc* mushroom body expression was interpretable because two different mushroom body mutants—isolated on that anatomical criterion—had previously been shown unable to learn.

rutabaga seems almost certainly to encode the catalytic subunit of an adenylate cyclase that is stimulated by calcium/ calmodulin. That stimulation is absent in the original (and by far most severely defective) *rut* mutant, and the overall levels of activity are lower than normal (though most markedly in the adult abdomen, where the levels in the mutant are still well above zero). Thus again, the cAMP-mediated second messenger system, which is of course known to be important in "cellular/biochemical" studies of learning in other organisms. The original experiments of this kind (notably in molluscs) involved tests of "simple" learning (e.g., sensitization). It is interesting in this regard that *dnc* and *rut*— isolated as associative learning variants—proved to be defective in simple learning as well.

Ultimately, stimulus-elicited rises in cAMP levels, related to experience-dependent responses, "get to" potassium channels in the relevant molluscan neurons. In this regard, another connection to *Drosophila* behavioral genetics concerns the fact that a *Sh* mutant, later proved molecularly to be a potassium channel variant, exhibits aberrant learning; this was shown both in the "shock-odor" system and by performing certain courtship tests as well. The latter involve altered reproductive behaviors that can be elicited by either the male or the female being exposed to certain previous courtship experiences; certain of the learning phenomena here are associative; others are sensitization- or habituation-like; all are abnormal in tests of males or females expressing *dnc* or *rut* mutations.

Biological rhythms

Over the 20 years of searching for rhythm mutants in *Drosophila*, by far the majority of the mutations that were isolated turned out to be at an X chromosomal locus called *period* (*per*). This is the only one that will be discussed here, because *per* has also been subjected to more analysis than any other "clock gene." The *per* mutants were recognized by shorter- or longer-than-normal circadian rhythm cycle durations, and by arrhythmicity, depending on the allele— respectively, per^s, per^L, and per^0. The circadian rhythm most commonly studied in experiments using *Drosophila* is that defined by rest-activity cycles underlying the adult's loco-motion.

The *per* mutations were shown to be pleiotropic in their effects on "time-based" phenomena. In one direction—from the circadian realm—*per* mutant males exhibit altered or absent cycles associated with the courtship song. As he sings, the rate that a wild-type male produces "pulses" (clicklike with the lengthening effects caused by per^L being the most dramatic. Other kinds of phenotypic defects (behavioral, physiological) have been reported for the various *per* mutants; some but not all have to do with temporal parameters; but many of these abnormalities have not been confirmed (let alone understood).

Yet there are some rough parallels between *per* pleiotropy and the fact that the mRNA and protein encoded at this genetic locus are expressed very widely. The gene products appear at most (perhaps all) stages of the life cycle and in a host of neural plus nonneural tissues. These patterns of temporal and spatial expression are so broad that, had the gene been discovered in this manner (and indeed many *Drosophila* genes are being identified from this molecular angle, as opposed to mutationally), one would not have guessed that *per* is a so-called clock gene, dedicated to the control of particular phenotypes (temporally related ones, as opposed to "everything").

The adult visual system is among the most prominent locations of *per* expression. This is especially puzzling, because the fly's circadian rhythms run quite nicely—and can be phase shifted by light—in animals that are anatomically blind (eyeless) or physiologically so (*norpA*). Thus, as in many organisms (though not mammals), the photoreceptors for *Drosophila*'s "circadian system" are extraocular. One further detail of *per*'s visual system expression is that the protein is found in nuclei of the photoreceptors of the retina and ocelli. This could eventually provide a clue as to the function of the PER protein. The nucleic acid encoding this material has been well sequenced, revealing yet another pioneer protein. But for now, the molecular studies on this clock gene have not yet been incisively informative as to the mechanism of PER's action, let alone how it is that biological factors like this one participate in time-keeping. (However, see the recent review by Takahashi, 1992.)

Conclusions

Most of the mutants mentioned here were isolated in behavioral screens that involved some sort of sophisticated testing of *Drosophila* actions and responses; the recovery of the other genetic variants discussed relied on reasonably skilled histological analysis. For the latter class of mutants, the investigators who were screening by sectioning came across a higher proportion of brain-damaged cases (per mutagenized chromosome) than they expected, implying that there may be more genes of this type than one might have guessed. In any event, most of the mutants mentioned in this entry would not have been noticed in searches for animals that are obviously abnormal, including dead (i.e., expressing "developmental-lethal" mutations). One could imagine, though, that the behaviorally and/or CNS-defective variants are expressing "mildly mutated" forms of the genes in question; this kind of argument goes on to say that a "null" mutation at such a locus would be lethal. Considering the many parts of the fly in which, and the many stages during which, gene products like those encoded by *dnc* and *per* are expressed, the notion that these could be vital factors is perhaps strengthened. Indeed, it has turned out that a few of the genetic loci that were originally identified by behavioral mutations are essential genes: later, lethally mutant alleles at these loci were deliberately isolated; or in other cases it turned out that such mutations had been recovered independently (by drosophilists interested in development as opposed to behavior).

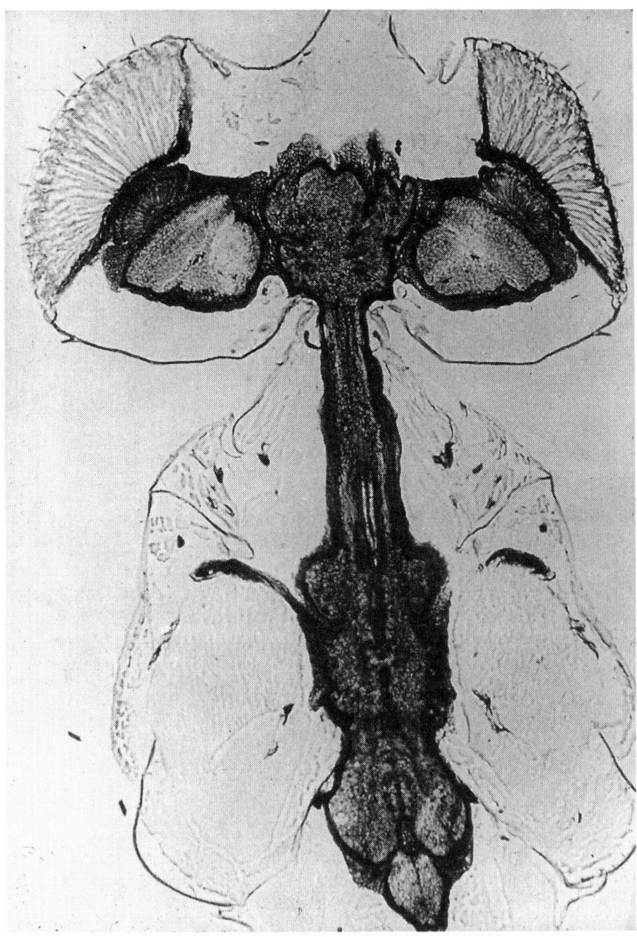

Figure 2. The central nervous system of the *Drosophila melanogaster* adult. This horizontal section (cut in a relatively ventral plane) was stained for activity of the neurotransmitter-metabolizing enzyme acetylcholinesterase, which is present in nearly all regions of the fly's nervous system: the ventral brain (as shown here at the top of the photo), flanked by the optic ganglia (which are in turn flanked by the unstained compound eyes); the connective thoracic/abdominal ganglia (which here are flanked by unstained muscles of the ventral thorax). To get a feel for scale, the width of the CNS in the head (measured from the lateral-most regions of the optic ganglia) is about 450 μm.

sounds) varies from ca. 25 to 35 per second; the variation is sinusoidal, with cycle durations of about one minute; per^S, per^L, and per^0 males sing normally in a given moment of time but have song cycles that are aberrant in a manner that parallels their circadian defects. The neural control of the song rhythm has been delved into in some genetic experiments; it was shown that temperature-sensitive action-potential mutations (e.g., $para^{ts}$) stop the "song clock" for the same duration that the mutant males are subjected to elevated temperatures. In contrast, localized turnoff of sodium-dependent action potentials in the pertinent region of the mammalian brain leaves the circadian clock running normally, though the animals' locomotor activity becomes arrhythmic for the duration of neurotoxin application.

Considering time spans considerably longer than a day, the *per* mutants are once more abnormal; these mutations alter the durations of *Drosophila*'s developmental stages,

However, a majority of the "behavioral genes" that have been delved into in sufficient genetic detail are nonvital loci. This has been demonstrated in a variety of ways: (1) achieving homozygosity for deletions of the gene in question (e.g., *dnc, per*) and being left with live flies that are "no worse" behaviorally than the most severe "point" mutants isolated previously; (2) deliberate isolation of many new alleles at a locus (as has been done for certain of the mutants found via the neuroanatomical approach), by rendering chemically mutagenized chromosomes heterozygous with one of the original "viable" mutations, with the subsequent demonstration that all new alleles are, when they are made homozygous, viable as well; (3) cloning of a given gene, followed by molecularly demonstrating that certain of the mutations in it are null (e.g., they generate no detectable mRNA or protein; or they are caused by a nonsense mutation), as has happened especially for factors defined by certain of the "visual-functional" mutations.

These results are slowly but inexorably expanding the "gene number" in *Drosophila*. That number has classically been fixed at 5000, a number still quoted by some developmental geneticists. By that, they mean that there are this many genes (this few genes, really) capable of mutating to lethality, whose etiology is almost always death occurring sometime between embryonic and pupal stages. But several of the behavioral and brain-damaged mutants—all of which had to survive to adulthood by virtue of the strategies for their isolation—are proving to be the "worst cases" in terms of what the corresponding genes are capable of mutating to. The functions specified by these loci could represent what the fly's genome contains, below the tip of the iceberg devoted only to its most "basic" developmental needs. Thus, there may be many thousands of genes that play "special" roles, as they participate in the construction and operation of the many subtle elements of the fly's nervous system. "Subtle" does not necessarily mean that these kinds of genes play demonstrably "minor" roles, whereby only a few neurons would express the encoded proteins. In fact, it was revealed in the foregoing sections that some of these genes are quite versatile in terms of phenotypes affected by the mutations, are expressed in many tissues in the nervous system and beyond, or both.

Finally, then, it is intriguing that behavioral genetic studies in *Drosophila* are doing a service beyond the limited aims of dissecting the fly's responses and actions. In addition, these investigations are beginning to suggest that most of this dipteran's genomic "coding capacity"—which is on paper rather more than 5000 genes' worth—might be actually used. Therefore, to get a functioning animal of this kind, there is a requirement for the action of and interaction among a number of factors whose complexity may rival that of the fly itself.

Further reading

Greenspan RJ (1990): The emergence of neurogenetics. *Sem Neurosci* 2:145–157

Hall JC (1985): Genetic analysis of behavior in insects. In: *Comprehensive Insect Physiology, Biochemistry, and Pharmacology*, 9, Kerkut GA, Gilbert LI, eds. Oxford, UK: Pergamon Press

Hall JC, Kyriacou CP (1990): Genetics of biological rhythms in *Drosophila. Adv Insect Physiol* 22:221–298

Heisenberg M, Wolf R (1984): *Vision in Drosophila: Genetics of Microbehavior.* Berlin: Springer-Verlag

Heisenberg M (1989): Genetic approach to learning and memory (mnemogenetics) in *Drosophila melanogaster.* In: *Fundamentals of Memory Formation: Neuronal Plasticity and Brain Function. Progress in Zoology,* vol 37, Rahmann G, ed. Stuttgart, FRG: Fischer Verlag

Smith DP, Stamnes MA, Zuker CS (1991): Signal transduction in the visual system of *Drosophila. Ann Rev Cell Biol* 7:161–190

Takahashi JS (1992): Circadian clock genes are ticking. *Science* 258: 238–240

Dystonia-Parkinsonism Syndrome, X-linked (XDP)

Ulrich Müller

The movement disorder *torsion dystonia* is characterized by involuntary, sustained muscle contractions affecting one or more sites of the body, frequently causing twisting and repetitive movements, or abnormal postures. According to clinical and genetic criteria, several "primary" forms of dystonia may be distinguished, including five autosomal dominant, one autosomal recessive, and one X-linked recessive forms. The latter is endemic to the Philippines, is associated with parkinsonism, and is referred to as the *X-linked dystonia-parkinsonism syndrome (XDP)*.

Phenotype of XDP

The mean age of onset of XDP is 35.0 ± 8.0 years. The disorder is initially focal and generalizes after a median duration of 5 years (range 1 to 11). The first symptoms may either occur in the lower extremities (36%), the axial musculature (29%), the upper extremities (23%), or the head (12%). The site of onset does not affect the course of the disease, which is severely disabling in the majority of cases. "Parkinsonian symptoms" including bradykinesia, tremor, rigidity, and loss of postural reflexes were described in approximately one third of a total of 42 cases studied. In those, parkinsonism was diagnosed as *definitive* (at least two parkinsonian symptoms including either bradykinesia or tremor) in 14%, *probable* (either tremor or bradykinesia alone) in 2%, and *possible* (rigidity and/or loss of postural reflexes) in 19%. More recent neurologic examinations of additional patients suggest that concurrent parkinsonism is even higher in XDP with one or more parkinsonian symptoms being present in at least 50% of the cases. There is no correlation between occurrence of parkinsonism and site of onset, the tempo of progression, or the general severity of the disorder. The older the age of onset, however, the likelier concurrent parkinsonism seems to be.

Neuropathology

Brains from diseased XDP patients have not been analyzed yet. Examinations of a few patients by computer tomography did not reveal gross abnormalities.

Prevalence

Prevalence of XDP is highest in the Philippine island of Panay, where it originated from a single mutation (founder effect). Conservative estimates suggest 1 in 4000 males to be affected in Capiz province, Panay. XDP has also been reported in Filipino immigrants to the United States and Canada. Presently, the disorder has only been diagnosed in people of Filipino extraction.

Treatment

Several medications have been tried in XDP patients including trihexyphenidyl, L-dopa, lorazepam, diazepam, and diphenhydramine. None of these, either alone or in combination, affected the course of the disease. Even in patients

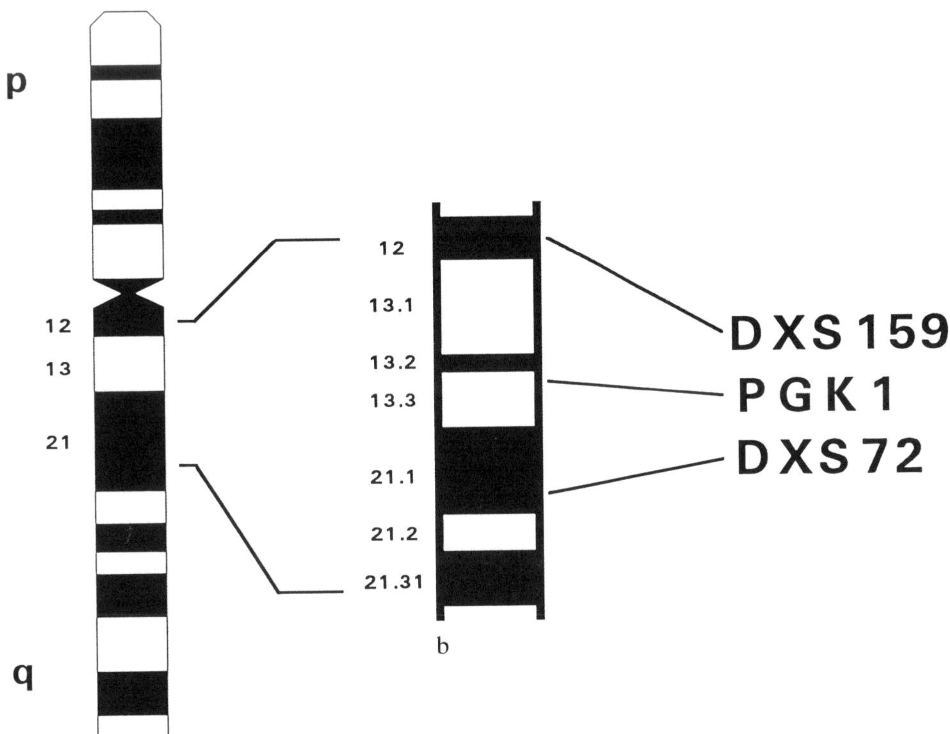

Figure 1. (a) Idiogram of the human X chromosome (Giemsa staining). *p* indicates the short and *q* the long arm. (b) XDP locus (*DYT3*) containing region is enlarged. *DYT3* lies between marker loci *DXS159* and *DXS72*.

with definitive concurrent parkinsonism (two treated), L-dopa did not result in noticeable improvement. Palliative and pragmatic treatments are also unsatisfactory.

Genetics

XDP is inherited as an X-linked recessive trait. Penetrance is high, approaching 100% by the end of the fifth decade. The disorder has primarily been described in males. In addition, three females with XDP, who apparently carried two copies of the defective gene, have been reported.

Linkage analysis assigned the XDP locus (*DYT3*) to a small region of the X chromosome long arm (Xq12–Xq21; see Fig. 1). Loci *DXS159*, *PGK1*, and *DXS72* of Fig. 1 display a high degree of "linkage disequilibrium" ("allelic association") with *DYT3*, indicating close proximity to the disease locus. The exact mapping of *DYT3* on the X chromosome is a prerequisite for the isolation of the gene. This is currently being attempted by isolating additional highly polymorphic DNA sequences from the region and testing their degree of linkage disequilibrium with respect to

DYT3 in many unrelated patients and in unaffected male controls. Once a region of complete allelic association is found, chromosome walks will be initiated in order to isolate transcribed DNA sequences. A candidate gene of XDP is expected to be expressed in the basal ganglia, malfunction of which results in parkinsonism and probably in dystonia as well. Final proof for it being the gene underlying XDP will come from the detection of a mutation in patients.

Further reading

Kupke KG, Graeber M, Müller U (1992): Dystonia-parkinsonism syndrome (XDP) locus: Flanking markers in Xq12–Xq21. *Am J Hum Genet* 50:808–815

Lee LV, Kupke KG, Caballar-Gonzaga F, Gebron-Ortiz M, Müller U (1991): The phenotype of the X-linked dystonia-parkinsonism syndrome. An assessment of 42 cases in the Philippines. *Medicine* 70:179–187

Müller U, Graeber M (1991): Dystonie-Parkinson-Syndrom. Molekulare Analyse einer X-chromosomal rezessiv vererbten neurodegenerativen Erkrankung. *Münch Med Wschr* 133(30): 456–460 (abstract in English)

Dystrophin and Duchenne Muscular Dystrophy

Hart G.W. Lidov and Louis M. Kunkel

Clinical features

Duchenne muscular dystrophy (DMD), also called pseudo-hypertrophic dystrophy or progressive muscular dystrophy, occurs in 1 in 3500 males and is the most common fatal X-linked human disorder. The disease was first described by Guillaume Duchenne de Boulogne in 1852 and the inheritance pattern recognized by William Gowers in 1879. A similar but less severe disease known as Becker's dystrophy (BMD) was described in 1955 and is now known to be an allelic variant.

Clinically DMD presents as hypotonia and skeletal muscular weakness in males between 2 and 5 years of age, usually as delay in learning to walk or run. Affected patients may rise from the floor by using the mechanically advantageous "Gowers maneuver," walk with a waddling gait due to hip girdle weakness, and develop pseudohypertrophy, usually of the calf muscles. Diagnostic studies include a serum creatine phosphokinase level, which is typically elevated more than 100 times normal, and muscle biopsy histology. Subsequently patients have progressive weakness of axial and proximal limb musculature and develop contractures of the elbows, knees, and hips. Most boys become wheelchair bound by age 10–12 years with thoracic musculoskeletal deformity as well as respiratory muscle compromise. Other associated features include gastrointestinal symptoms, presumably on the basis of smooth muscle involvement, and cardiac symptoms. Cardiomyopathy, at age 5–6 years, produces tachycardia and prominent R-waves (right axis deviation) on electrocardiogram. Progressive respiratory failure or, less commonly, cardiac failure result in death between late teens and the end of the third decade.

Case reports have suggested, and population studies have confirmed, that a mild nonprogressive cognitive impairment is a uniform feature of DMD; typically mean IQs are 82–85. The cognitive impairment is unrelated to the duration or severity of muscle disease and significant in comparison to patients with other chronic illness or other similarly severe neuromuscular diseases such as spinal muscular atrophy. The cognitive deficit appears to be a manifestation of the same genetic defect that produces the muscle disease and not an environmental epiphenomenon or a defect in a second nearby gene. No pathology has been established in the central nervous system.

Becker's muscular dystrophy is about one tenth as common as DMD but shares many of the same clinical features. Diagnosis is suggested by manifestations of weakness prior to the age of 30 years, and is supported by an elevated creatine phosphokinase (25–200 times normal), and myopathic features on electromyography and muscle biopsy histology. Typically BMD patients may be ambulatory into their second decade and live into their fifth or sixth decade. Cardiomyopathy is also a feature of BMD, and although studies of adequately large populations have not been done, some degree of cognitive impairment appears likely. Clinical severity and time course vary greatly, ranging from cases almost as severe as DMD to much milder forms; thus, the clinical distinction between DMD and BMD in some cases is rather arbitrary.

On muscle biopsy both DMD and BMD present features typical of a myopathy. As early as 1 year there is excessive variation in fiber size and a subtle increase in the thickness of connective tissue septae between myofibrils. With the passage of years large rounded hypertrophic fibers, fiber splitting, central nucleation, fiber necrosis, and basophilic regenerating fibers are prominent. There is marked increase in endomesial connective tissue, manifest as pseudohypertrophy, and cellular infiltrates composed of lymphocytes and macrophages as well as fibroblasts and myocyte nuclei. In an end-stage biopsy muscle fibers are sparse and atrophic and contractile elements are replaced by sheets of dense connective tissue and fat.

Molecular genetics

Despite exhaustive efforts over several decades, no attempts to carry the analysis of pathology directly to the biochemical or molecular level succeeded in identifying a consistent pathogenic abnormality. Beginning in the early 1980s several laboratories pursued a "top down" approach that began by identifying the genetic defect that resulted in DMD. This approach has been referred to as "reverse genetics" to indicate the identification of a genetic defect without reference to an identified gene product. Starting with genomic DNA from a DMD patient with a macroscopically visible chromosomal deletion, a subtractive hybridization technique was used to isolate DNA sequences present in normals but absent in this unique patient. Comparison with the DNA of many more affected individuals permitted the identification of a part of the gene responsible for DMD. A similar result was obtained by starting with DNA from a patient with DMD produced by a translocation involving the X chromosome. The gene when identified in its entirety displayed several remarkable features; it is the largest gene yet identified in the human genome, spanning 2.4 million base pairs. The mRNA that is transcribed from this gene is 14 kilobases and is transcribed from about 70 exons between which are unusually large expanses of intronic DNA.

Having identified the gene responsible for DMD it was possible to determine the abundance of the transcribed mRNA in various tissues. The highest levels are, not surprisingly, in skeletal muscle, cardiac muscle, and smooth muscle—the tissues that bear the brunt of the pathology in DMD. Smaller amounts are detectable in brain tissue, and negligible amounts are found in other tissues. Certain features of the gene product translated from the identified gene could be predicted based on nucleotide sequence alone (Fig. 1). This protein, for which the name "dystrophin" was coined, is composed of 3,685 amino acids and displays striking sequence similarities with the actin-binding proteins spectrin and α-actinin. Sequence homologies in the N-terminal region suggested that this region might function as an actin-binding domain. The central region of dystrophin is composed of 24 repeated units of 109 amino acids, a motif seen in spectrins, which suggests a rod-shaped cytoskeletal protein. Four proline-rich regions interrupt the rod domain and may function as hinges to impart flexibility. The C-terminal end is a unique sequence and highly conserved across species but the function remains to be determined. Ultrastructural studies suggest that the C-terminal end is membrane bound, and the recent description of glycoproteins

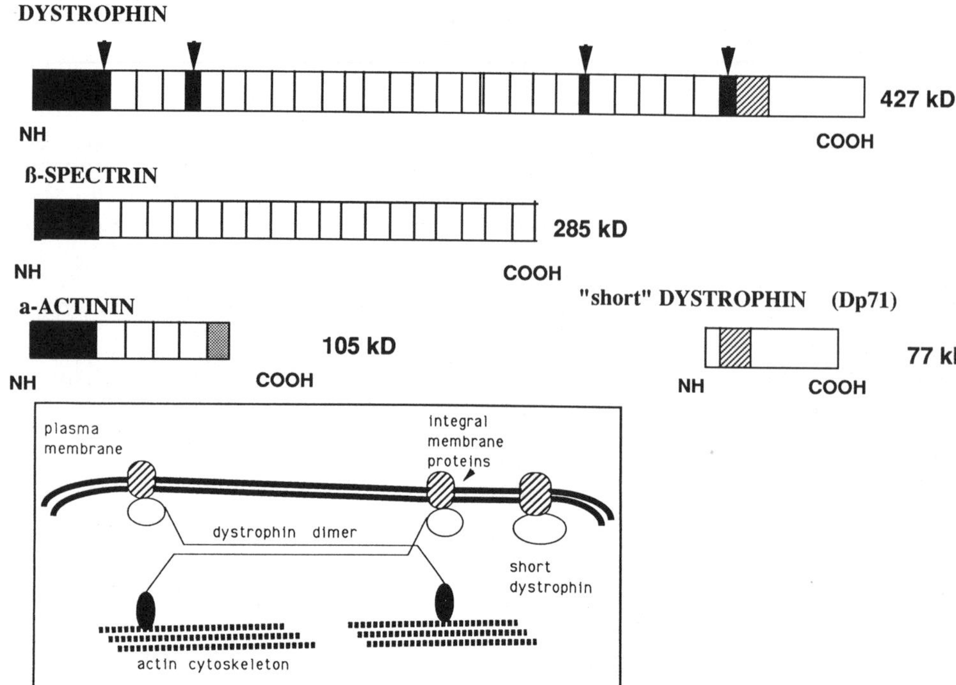

Figure 1. Schematic diagram of the domain organization of dystrophin and three related proteins. The proline-rich hinge regions of dystrophin are indicated by arrows. The actin-binding region is indicated by the black box, and the central repeated domains are indicated by the central open boxes. The C-terminal region unique to dystrophin and the cysteine-rich region are indicated by the terminal open box and the diagonal hatching. The insert shows the postulated relation between dystrophin, integral membrane proteins, and the actin cytoskeleton.

associated with dystrophin may indicate part of the membrane binding mechanism. The overall functional appearance seems to be consistent with a role as a cytoskeletal component that links plasma membrane or integral membrane proteins to the actin scaffold.

Using the predicted amino acid sequence of dystrophin, portions of the whole molecule were produced as bacterial fusion proteins or synthetic peptides, and antibodies were raised to these fragments. The antibodies have been used in Western analysis and immunocytochemistry to identify dystrophin gene products among the protein constituents of normal tissues. Immunoblotting has confirmed that dystrophin is a very large protein and is present at greatest abundance in skeletal, cardiac, and smooth muscle. Even in these tissues dystrophin is a rare constituent, estimated at 0.002% of total protein and 5% of membrane-associated protein. Smaller amounts (0.001–0.0001%) are present in brain tissue, and the trace amounts in other tissues probably reflect smooth muscle associated with blood vessels.

Immunocytochemistry at the light and electron microscopic level in skeletal muscle has shown dystrophin to be exclusively located beneath the plasma membrane of the muscle fiber, that is, the sarcolemma (Fig. 2). Especially dense accumulations are associated with regions of membrane specialization such as the neuromuscular junction and the myotendinous junction. At the neuromuscular junction dystrophin is preferentially located at troughs of junctional folds, separated from the acetylcholine receptor, and thus does not appear likely to be directly involved in anchoring the receptor molecule. No dystrophin is found associated with cytoplasmic components or the T-tubule system. The distribution in skeletal muscle, cardiac muscle, and smooth muscle is essentially similar in that the dystrophin molecule has a subsarcolemmal localization.

The occurrence of cognitive dysfunction in DMD, and the demonstration of dystrophin mRNA and protein in neurons and glia *in vitro,* raises the question of the localization in the nervous system. Although a critical localization in neurons is intuitively appealing, an alteration in blood vessels or glial cells might produce an indirect effect on neural function. Immunocytochemical studies show that dystrophin does occur in the neurons of the cerebral cortex, especially the pyramidal neurons of the hippocampus, and the Purkinje cells in the cerebellum (Fig. 3). Within each of these cell types dystrophin is distributed as punctate patches along the plasma membrane of the nerve cell body and along the dendrites. Cytoplasmic elements, axons, glia, and myelin are not im-

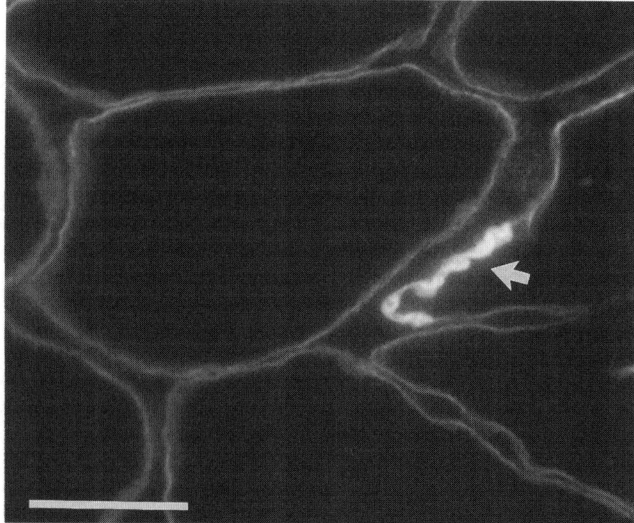

Figure 2. Dystrophin immunofluorescence in murine skeletal muscle. Muscle fibers show circumferential immunoreactivity beneath the sarcolemma. A neuromuscular junction is indicated by the white arrow. Scale bar = 30 microns.

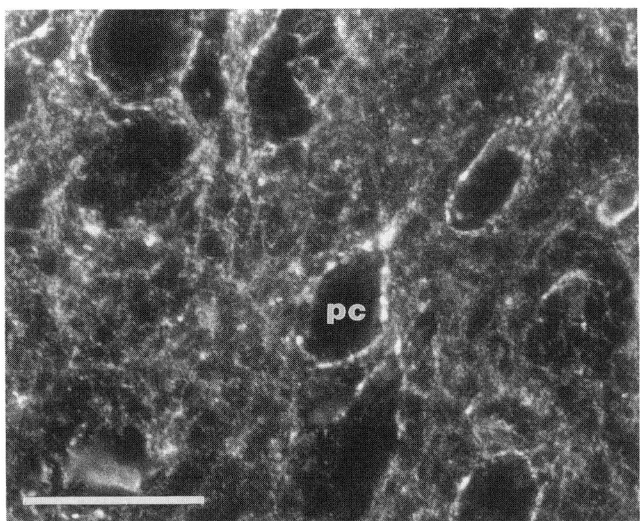

Figure 3. Dystrophin immunofluorescence in murine cerebral cortex. Perikarya of pyramidal neurons (pc) and proximal dendrites are outlined by punctuate immunoreactivity. Scale bar = 20 microns.

munoreactive. There are many types of neurons, such as the granule cells of the cerebellum, thalamic, hypothalamic, and striatal neurons, that appear not to contain dystrophin, or do so at a significantly lower level. The distribution of dystrophin at the light microscopic level would be consistent with a localization to postsynaptic apparatus at a restricted subset of synapses, and initial electron microscopic studies using immunogold-labeled cryosections suggest that this is the case. The localization to synapses, and to cerebral cortical neurons, is a likely substrate for the observed cognitive dysfunction.

Molecular–pathologic correlation

With the identification and sequencing of the dystrophin gene it has been possible to analyze in detail the genetic defects in patients with DMD and BMD. The extraordinarily large size of the gene alone is adequate to account for the high frequency of these diseases produced by new mutations. Approximately 65% of patients with either disease have deletions of parts of the gene, and although certain regions seem particularly vulnerable, deletions occur throughout the gene. Approximately 5% of patients have duplicated portions of gene, and the remaining 30% are believed to be point mutations or abnormalities in RNA processing signals. In general, deletions from particular portions of the dystrophin molecule do not result exclusively in one phenotype, either DMD or BMD. Furthermore, there has been no consistent evidence that cognitive dysfunction, or any other clinical feature, is preferentially associated with particular deletions.

Studies at the level of protein have demonstrated that, in general, patients who clinically have DMD have little or no detectable dystrophin, while patients who have BMD have either too little dystrophin or structurally altered dystrophin of abnormal size. In the majority of cases, mapping of the precise mutation shows that the deletion in DMD, wherever located and whatever the size, has altered the reading frame—the registration in which the triplet code in the RNA is read—and produces a "nonsense" product that is thought to be rapidly degraded. Conversely deletions in the same regions, even if relatively large, that do not alter the reading frame produce a structurally altered dystrophin molecule and are associated with the BMD phenotype. Presumably the presence of some dystrophin, even if in low abundance or of an abnormal size, is sufficient to produce a milder disease. A minority of patients are exceptions to this "reading frame hypothesis" either because of extraordinarily massive "in-frame" deletions or because RNA splicing mechanisms exist that circumvent the effect of some deletions that should produce a frame shift. Nevertheless, based on this molecular description of DMD and BMD it is possible to make a diagnosis of considerable prognostic significance to the patient in more than 95% of affected individuals.

Beyond empirical correlation the precise mechanism that results in the DMD/BMD phenotype remains speculative. The structural homology with cytoskeletal elements and the relatively uniform distribution beneath the sarcolemma suggest that dystrophin may stabilize the plasma membrane, especially during cycles of contraction and relaxation. Hypothetically, dystrophin-deficient muscle cells may suffer mechanical damage to plasma membrane, become leaky, permitting calcium entry, and resulting in cell death. Cumulative loss of muscle fibers and fibrosis follow as the regenerative capacity of muscle progenitor cells is overwhelmed and the connective tissue reacts to the chronic cell damage. Such a model would have several significant implications: (1) exercise may be deleterious to the DMD muscle, (2) the pathogenic process might be interrupted by replacing dystrophin or its gene, and (3) the progress of disease could be modified by preventing what may be reactive fibrosis.

Dystrophins: A family of proteins

Further studies have made it apparent that viewing dystrophin as a single molecular species is an oversimplification. Surveys of the mRNA transcripts from the dystrophin gene in several tissues demonstrate multiple alternatively spliced forms with alterations at both ends of the molecule. Of particular interest is a form incorporating an alternative first exon that has been identified in the brain. This exon is largely untranslated and codes for only about three new amino acids, so it cannot reflect a major structural alteration but suggests the basis for an independent regulatory process. This possibility has been further substantiated by the identification of a second promoter site associated with the alternative "brain" first exon. The distribution of the "muscle" and "brain" transcripts appears to be distinct—the former occurring in cultured glial cells and the latter in neurons. It is likely that there are further alternatively spliced forms, and that these also will have distinct distributions. In addition to these relatively similar alternatively spliced forms of dystrophin, there are significantly smaller dystrophin gene products (less than 100 kDa) which also have distinct tissue distributions. Some of these appear to be abundant in CNS, as well as in noncontractile visceral tissue. Whereas the full-size dystrophin transcripts may well be functionally and structurally similar, the small transcripts are C-terminal fragments of the larger molecule and might be predicted to have significantly different functions. Furthermore, deletions that would produce DMD or BMD may or may not, depending on location, affect the small dystrophin transcripts. Finally, at least one protein structurally similar to dystrophin, called *dystrophin related protein,* is encoded on chromosome 6. This protein also has a distinctive tissue and cellular distribution but is sufficiently homologous to dystrophin that it may be expected to have some functional similarity. Thus, although DMD and BMD are clearly caused by the absence of dystrophin, dystrophin appears to be only the index member of a family of more or less similar proteins. The functions and possible relation to disease states of most of the members of this family remain to be elucidated.

Further reading

Emery AEH (1988): *Duchenne Muscular Dystrophy,* 2nd ed. New York: Oxford University Press

Ervasti JM, Cambell KP (1991): Membrane organization of the dystrophin-glycoprotein complex. *Cell* 66:1121–1131

Hoffman EP, Kunkel LM (1989): Dystrophin abnormalities in Duchenne/Becker dystrophy. *Neuron* 2:1019–1029

Monaco AP, Kunkel LM (1988): Cloning of the Duchenne/Becker muscular dystrophy locus. In: *Advances in Human Genetics,* Harris H, Hirschorn K, eds. New York: Plenum, vol 17, pp 61–98

E

Extracellular Matrix, Laminin and Integrins

Michael J. Ignatius

Introduction

Nerve cell processes growing during embryogenesis or as a response to injury depend on a diverse set of interactions with cellular and acellular molecules in their environment to initiate, sustain, and direct their growth. Neurite outgrowth–promoting activities are found in three compartments: (1) on neuronal and nonneuronal cell surfaces (e.g., CAMs and cadherins); (2) in the soluble pool (e.g., neurotrophins, nerve growth factor); and (3) glycoproteins within the acellular, insoluble extracellular matrix (e.g., laminin and fibronectin and collagen). The failure of most of these elements to persist in the adult CNS may directly impact on the failure of this region of the nervous system to regenerate, while all are maintained to some degree in the peripheral nervous system, where repair of damaged axons is rapid. This chapter outlines our current knowledge of the extracellular matrix (ECM) and its functions within the nervous system, emphasizing laminin and its corresponding receptors within the integrin family of adhesion receptors as a model for these interactions.

The neuronal ECM

Within the developing nervous system the ECM was originally used to describe the acellular material visible as a dense fuzz or basal lamina in the electron microscope. This ECM divides or separates cell layers and axon bundles. Recent studies using immunohistochemistry to individual ECM components have broadened this designation to include any deposition of these insoluble glycoproteins, since it is often found distributed in patches of immunoreactivity speckled throughout areas of developing fiber tracts and cell columns. The most abundant ECM constituents in the developing nervous system are the glycoproteins collagen, laminin (LN), fibronectin, tenascin, thrombospondin, and vitronectin, as well as chondrotin and heparin sulphate proteoglycans. All of these molecules have apparent molecular weights in the range of 1000 kDa with macromolecular structures spanning several hundred nanometers. The emerging complexity of these components, with at least 12 collagens, multiple isoforms on LN, and splice variants of molecules like fibronectin, is beyond the scope of this work but underscores their diverse structure and potential for a variety of complex functions.

Laminin

The glycoprotein laminin (LN), a prominent component of the ECM, is thought to play a particularly important role during neuronal development because of its dramatic effects *in vitro*. Central and peripheral neuron attachment, migration, process outgrowth, survival, differentiation, and regeneration are all promoted on LN-coated substrates. Laminin has been shown to influence the differentiation of neuroblasts derived from chick neural tubes, and the transdifferentiation of frog retinal pigment epithelial cells into neurons. *In vivo,* LN immunoreactivity has been detected where axonal pathways are established in regions of both developing CNS and PNS. Serial EM reconstruction of developing fish optic nerve reveal the intimate association of retinal growth cones and the acellular, LN-enriched ECM, which in the adult appears to play a role in the regeneration of fish retinal axons.

Laminin is a cross-shaped, 1000 kDa glycoprotein composed of three chains: A1, B1, and B2. Isoforms of LN with their own distribution patterns have been identified, including merosin (A_M), an A-chain variant abundant in muscle and peripheral nerves, and S-laminin (B_S), a B1-chain variant, enriched in synaptic regions. All of these isoforms have potent effects on neuronal attachment and process outgrowth as well, although only motoneurons from the ciliary ganglion have been shown to bind S-LN. The regionalized distribution of these laminin isoforms raises the very exciting possibility that they serve unique functions, including the delineation of discrete axonal trajectories.

Integrins

Cell adhesion and recognition are fundamental to a variety of cellular functions in and out of the nervous system including tissue morphogenesis and repair, macrophage, leukocyte and immune system function, platelet thrombosis, and tumor metastasis. All of these interactions have been shown to be dependent in part on the integrin family of adhesion receptors, which bind extracellular matrix proteins and some cellular ligands, with at least one integrin expressed on every cell in the body. Integrins, named to describe their ability to *integr*ate cell-surface binding with cytoskeletal structures, are heterodimeric, transmembrane receptors of noncovalently associated α and β subunits. At least 15 integrins have been described: nine αs (1–8 and αvn) that can associate with the β1 subunit (e.g., α1/β1) and at least six other αs that pair with six other βs (2–7; e.g., αIIB/βIIIA). Ligand selectivities are established by the unique pairing of α and β subunits and require oligomeric integrity for function.

Integrins in the nervous system

Antibody perturbation studies *in vitro* have shown that β1-containing integrins are used by neurons for cell adhesion, migration, and process outgrowth on ECM proteins, such as laminin(s), fibronectin, and collagens I and IV. Injection of anti-integrin antibodies *in vivo* disrupts neural crest migration and can slow peripheral nerve regeneration. Viral-mediated antisense blockade of β1 expression can alter patterns of cerebellar cell migration along radial glia. Within individual neurons, integrins are evenly distributed, localized

Table 1. Integrins Expressed on Neurons

Name	MW (Nonreduced)	Amino Acid Number	mRNA Size	N-linked Sugars	Heterodimer Ligand Selectivities
$\beta 1$ (human)	110 kDa	778	4.2 kB	12	
$\alpha 1$ (rat)	200 kDa	1152	11.0 kB	24	LN/Col; Col
$\alpha 2$ (human)	150 kDa	1152	8.0 kB	10	LN/Col; Col
$\alpha 3$ (hamster)	150 kDa	1019	5.0 kB	11	LN/Col/FN
$\alpha 4$ (human)	140 kDa	999	5 or 6 kB	12	FN, VCAM
$\alpha 5$ (human)	150 kDa	1008	4.9 kB	14	FN
$\alpha 6$ (human)	150 kDa	1055	5.3 kB	10	LN
$\alpha 7$ (human)	150 kDa	?	? kB	?	LN
$\alpha 8$ (chicken)	160 kDa	1021	5.6 kB	17	?
αv (human)	140 kDa	1018	7.0 kB	13	FN, TB, VN

(LN) laminin; (FN) fibronectin; (Col) collagen; (VN) vitronectin; (TB) thrombospondin.

throughout cell bodies, axons, growth cones, and filopodia in punctate complexes termed focal contacts. Integrins and their associated ECM molecules are broadly expressed within the developing nervous system, in a regulated fashion, and individual α/β pairs show specific distribution patterns localized to subsets of cell types and discrete fiber bundles. Yet like ECM proteins, integrin expression is low in the adult brain. However, integrin expression is upregulated in human endothelial cells by treatment with tumor necrosis factor or retinoic acid and in a neuronal cell line PC12 with nerve growth factor.

A more detailed understanding of integrin function is derived from experiments where the attachment to ECM proteins by nerve cells expressing individual $\alpha/\beta 1$ heterodimer pairs are blocked by α-specific monoclonal antibodies. In this manner, a complex picture has emerged that reveals overlapping adhesive preferences for certain $\alpha/\beta 1$ heterodimer pairs (Table 1), as well as cell-type–specific variability in ligand selectivity for the same heterodimer pair. For example, five $\alpha/\beta 1$ receptors are used by cells to attach to substrates coated with LN, although to different regions and possibly different isoforms of LN with potentially unique response properties. Four of these are collagen receptors as well, and one, $\alpha 3/\beta 1$, in some cells is a receptor for fibronectin. Yet the capacity to bind multiple ligands appears now to be regulated by unknown cell-type–specific factors. $\alpha 1/\beta 1$ on two neuronal cell lines, one peripheral (PC12) and one central (B50), can attach to LN and at least two collagens, while rat superior cervical and dorsal root ganglia may use $\alpha 1/\beta 1$ for attachment only on collagen. More recently, $\alpha 1/\beta 1$ has been expressed in two nonneuronal cell types and shows the same variability in function, ruling out the formation of different receptors from alternate splicing. Instead, this result emphasizes the role of either posttranslational modification or accessory molecules in altering individual integrin function.

Integrin function can also be regulated by exogenous factors. For example, monoclonal antibodies to the receptor on platelets and lymphocytes, as well as neurons, can activate quiescent cell-surface receptors. Chick retinal ganglion cells, as they reach their targets, are no longer able to attach to ECM proteins, but application of Fab fragments to a domain on the $\beta 1$ subunit for 20 minutes rapidly promotes integrin-dependent binding. On macrophages, interferon-γ and phorbol esters activate the laminin receptor $\alpha 6/\beta 1$, which is coincident with phosphorylation and cytoskeletal association of the $\alpha 6$ subunit. The actual mechanisms of activation are poorly understood. So too are the mechanisms that inactivate receptors, although it is known that the process of differentiation can reversibly inactivate $\beta 1$ integrins on neurons or keratinocytes.

While it is apparent that extracellular structures can direct most aspects of ligand binding, cytoplasmic elements may also be critical for full function. Integrin α subunits and $\beta 1$ each contain short cytoplasmic domains of less than 40 amino acids, and for at least $\alpha 3$ and $\alpha 6$ these can be alternatively spliced, although with unknown effects on function. Deletion of the cytoplasmic domain of the $\beta 1$ subunit disrupts the colocalization of receptor and cytoskeletal elements, despite the presence of heterodimer pairs and retention of normal ligand binding. In contrast, the cytoplasmic domains of $\beta 1$ and $\beta 3$ can be interchanged without disturbing function, including heterodimer formation and targeting to focal adhesion plaques. Phosphorylation of the cytoplasmic domains of both integrin subunits have been reported. The $\beta 1$ subunit encodes a tyrosine phosphorylation site in its cytoskeletal domain, and while no such domain is seen in α subunits, phosphorylation of $\alpha 6$ in $\alpha 6/\beta 1$ heterodimers is associated with increased adhesion of macrophages to LN. Deletion of the cytoplasmic domain of another integrin, $\beta 2$, when paired with αLFA renders the receptor unresponsive to activation with phorbol ester. Yet truncation of platelet receptor αIIb$/\beta$IIIa increases ligand affinity, possibly by disrupting its interactions with cytoplasmic proteins that maintain a lower affinity state. Others have correlated loss of phosphorylation of integrins with differentiation-induced clustering of ligand/receptor complex, and purified phosphorylated receptor shows reduced affinity for ligand or talin.

In sum, integrin-dependent adhesion can be regulated by a variety of exogenous and endogenous factors and can therefore be viewed as a gating mechanism regulating the interactions between cells and the extracellular matrix. While fundamental to a variety of cellular processes outside the nervous system, they are likely to be of equal importance within it, acting in novel ways to aid in the assembly, and possibly the eventual repair of, the intricate connections in the adult brain.

Further reading

Bixby JL, Harris WA (1991): Molecular mechanisms of axon growth and guidance. *Ann Rev Cell Biol* 7:117–159

Hynes RO, Lander AD (1992): Contact and adhesive specificities in the associations, migrations and targeting of cells and axons. *Cell* 68:303–322

Reichardt LF, Tomaselli KJ (1991): Extracellular matrix molecules and their receptors: Functions in neural development. *Ann Rev Neurosci* 14:531–570

Springer TA (1990): Adhesion receptors of the immune system. *Nature* 346:425–434

F

Facial Expression and Emotion

Erika L. Rosenberg and Paul Ekman

Charles Darwin, in his book on the expression of emotions, wrote that facial expressions are innate, evolved behavior. In this century many social scientists have argued that facial expressions of emotion are language-like, socially learned, and culturally variable. In the last 25 years the first methodologically rigorous studies on the universality of facial expressions of emotion have supported Darwin's view. The major building blocks that support this position and new directions of research on the behavior and biology of emotional expression will be described.

Universality

Cross-cultural research on facial expressions of emotion has included Western, non-Western, literate, and preliterate cultures. Although the evidence on spontaneous emotional expressions is limited, several studies have shown that observers in different cultures apply the same emotion labels to photographs of faces depicting anger, disgust, happiness, fear, and sadness. The expression for surprise has been distinguished from all of the preceding emotions, but not from fear in one preliterate culture. The universal recognition of contempt is still a debated topic.

Comparisons between the expressions of great apes and humans have revealed similarities in facial morphology. Such interspecies commonalities are consistent with the universality findings, and with an evolutionary view of facial expressions of emotion.

Facial movement is under voluntary and involuntary control, allowing the opportunity for deliberate efforts to manage behavior to mask unintended emotional expressions. Experimental research has shown that cultural differences in the learned rules of expression management in social situations can produce the appearance of culture-specific facial expressions. Such cultural differences in display rules may account for the observations of both biologically based universal expressions and culturally specific emotional behavior.

Development

Facial expressions of emotion appear earlier in life than researchers had previously thought. By the age of two a child's repertoire of facial behavior includes the expressions for most of the basic emotions. Most of the debate in the developmental literature focuses on the age at which emotional expressions emerge, and whether all emotional expressions are present at birth or if there is a precise developmental sequence for the appearance of each affective expression. Infants demonstrate the ability to imitate expressions within a few hours of birth. Newborns show expressions that resemble adult disgust, smiles appear in the first few weeks of life, and anger appears by at least 4 months, although the components of some other negative emotion expressions are present by 1 month.

Three- to four-month-olds show differential responses to facial expressions in recognition tasks, supporting the findings on spontaneous expression. Empirical evidence suggests that between 4 and 8 months the ability to use expressions instrumentally emerges. During this period infants first learn to direct anger expressions at the source of frustration; by 7 months they will also direct them toward their mother to elicit help.

Measurement techniques

There are two different approaches for measuring facial expressions in muscular or anatomical terms. In one technique, human coders learn to recognize visually distinct facial actions that can singly or in combination account for all of facial movement. In such coding procedures, emotion interpretations are made on the basis of whether the actions or combinations of actions demonstrated are indicative of emotion on an empirical or theoretical basis. The advantage of this approach is that it is precise, and inferences about emotion are made after the facial behavior is coded. The disadvantage of this technique is that it is labor-intensive and insensitive to very slight changes in muscle tonus.

The other method is facial electromyography (EMG), in which surface electrodes placed over different regions of the face measure electrical discharge from contracting muscular tissue beneath the skin. The EMG signal lends itself to immediate recording, is not labor-intensive, and is sensitive to slight muscular movements that may not be visible even to the trained eye. One drawback is that EMG is highly obtrusive; the application of surface electrodes makes subjects aware of the facial measurement. Another disadvantage is that the recording selectivity of facial EMG is not muscularly specific, but rather regionally specific, and it is not yet certain whether EMG allows the differentiation of as many different emotions as can be done with measurement that relies upon observer scoring of visible muscular actions.

Voluntary and involuntary expression

Whether or not the face is an accurate source of emotional information depends on the situational context in which the expressions are elicited. When the individual is aware of being observed the face can provide misleading information to the untrained observer, as both deliberate as well as unintended expressions may be present. There is tentative evidence that some observers can distinguish between these types of information, but most cannot. Current research is developing methods to train people to differentiate deliberate from unintended emotional information on the face.

Most of the research comparing voluntary and involuntary emotional expressions focuses on the distinction between enjoyment smiles and the many types of nonenjoyment smiles (e.g., polite smiles, smiles put on to mask negative feelings, and compliance smiles). Enjoyment smiles can be distinguished from other smiles on the basis of timing and the presence or absence of the contraction of muscles around the eye (orbicularis oculi). As predicted by Duchenne de Boulogne, the neurologist who wrote in the last century, orbicularis oculi activity is present in enjoyment smiles but not in nonenjoyment smiles. Orbicularis oculi is a muscle that is difficult for most people to contract voluntarily; thus, it is a marker of enjoyment for most people when it occurs along with contraction of the zygomatic major (which raises the lip corners). Although smiling is the only mode of emotional expression for which the distinction between voluntary and involuntary contraction has been demonstrated, it is reasonable to expect that it will be possible to distinguish between voluntary and involuntary expressions of other emotions in a similar fashion.

Voluntary facial action as a generator of emotion responses

Facial expressions have been regarded as one of the peripheral response systems in emotion. In the past 8 years, however, researchers have found that voluntary contraction of facial muscles can elicit the physiological response of certain emotions, as well as the subjective experience of emotion for some people. This finding has been replicated five times. Differential patterns of autonomic activity distinguish between the expressions for several emotions. Not only do these patterns discriminate between positive and negative emotion expressions, but among negative emotions as well—anger, fear, sadness, and disgust each have a unique autonomic blueprint. Facially generated autonomic specificity has been replicated in non-Western cultures and in subjects from 20 to 70 years of age. There are new and not yet replicated findings of voluntary performance of universal emotion expressions also generating distinctive patterns of EEG activity. The mechanisms underlying the phenomenon of facially elicited physiological and experiential responses are the subject of debate, ranging from afferent feedback to direct links between the motor cortex and hypothalamic areas.

Expressions and physiology

The bulk of evidence on autonomic specificity of emotion comes from studies in which physiological reactions have been elicited by voluntary facial expression (as described previously). It is notable, however, that the autonomic patterns produced by voluntary facial expressions are enhanced when (a) the expressions most closely resemble the configuration of the universal expression for the associated emotion, and (b) when the subject reports that the emotion was experienced. Recent research has shown that these patterns of activity also occur when emotion is elicited by reminiscence. Whether these patterns hold up for emotion spontaneously generated in a social context remains to be seen. Three autonomic measures distinguish different types of emotions: heart rate, skin conductance, and finger temperature. On the basis of heart rate and skin conductance, distinctions have been made among the negative emotions of anger, disgust, and fear, and the positive emotion of happiness. Heart rate and finger temperature patterns also distinguish between four negative emotions. These findings challenge long-held but poorly substantiated claims in the literature that all emotions are characterized by an undifferentiated state of autonomic arousal.

There is also evidence of central nervous system differences among emotions, but to date it is more limited than the evidence for autonomic nervous system activity. The only replicated findings show inter- and intrahemispheric differentiation in EEG activity between emotions that may be distinguished on the basis of whether they are characterized by (a) positive versus negative valence or (b) approach versus withdrawal behavioral tendencies.

Brain damage and facial recognition

Research on the relationship between certain types of brain damage and the generation of facial expression has been equivocal. There have been problems with the evocation of emotion in some of these patient groups, as well as heterogeneity of lesion location. There is a substantial body of literature, however, on recognition of facial expressions of emotion. Patients with right temporal or parietal damage are poor at recognizing facial stimuli in general. The neuropsychological literature suggests that right temporoparietal regions are involved in recognition of emotional facial expressions, and that the ability to recognize emotional expressions may be separable from the ability to recognize faces in general.

Further reading

Davidson RJ, Ekman P, Aaron CD, Senulis JA, Friesen WV (1990): Approach-withdrawal and cerebral asymmetry: Emotional expression and brain physiology I. *J Pers Soc Psychol* 58:330–341

Ekman P, ed (1982): *Emotion in the Human Face,* 2nd ed. Cambridge: Cambridge University Press

Etcoff NL (1986): The neuropsychology of emotional expression. In: *Advances in Clinical Neuropsychology,* Goldstein G, Tarter RE, eds. New York: Plenum, pp 127–179

Levenson RW, Ekman P, Friesen WV (1990): Voluntary facial action generates emotion-specific autonomic nervous system activity. *Psychophysiology* 27:363–384

Fragile X Syndrome

W. Ted Brown and Edmund C. Jenkins

The fragile X syndrome is the most common inherited form of mental retardation. Its name derives from the appearance of a gap or break near the end of the X chromosome when cells from an affected individual are cultured in special media. This X-linked form of mental retardation was only recognized as a common and distinct entity in the late 1970s. At that time, Sutherland discovered that cell culture medium deficient in folic acid was necessary to visualize the marker chromosome. Subsequently, a number of undiagnosed families with two or more mentally retarded brothers were tested. In approximately 40%, the brothers were found to have the fragile X chromosome. Current estimates are that approximately 1 in 1000 males has the fragile X chromosome and 1 in 700 females is a carrier. Recent progress has led to the molecular isolation and initial characterization of a gene, fragile X mental retardation-1 (FMR-1), that underlies the syndrome. Molecular tests have been developed that allow for direct genomic Southern analysis as well as PCR diagnosis. Because fragile X is so common, all children with mental retardation of unexplained etiology should be evaluated for the syndrome. Screening of all pregnant mothers for their carrier status may also be appropriate.

Features of affected and unaffected transmitting males

Compared with individuals diagnosed with most other genetic and chromosomal syndromes, affected fragile X males are generally fairly normal in physical appearance. Some generally recognizable physical features that are variable in their presence include large or prominent ears, long narrow face, highly arched palate, hyperextensible joints, and large testicular volume (macroorchidism). Macroorchidism in childhood is difficult to distinguish, but most adult fragile X males have testicular volumes in the range of 30 to 120 ml compared with a normal mean of 17 ml.

Some affected boys with fragile X are only learning disabled. Frequently their measured IQ falls off as they get to be older. The majority of adult males have IQs in the range of 20 to 60 with a mean of about 35. Fragile X is associated with infantile autism and is probably the single most common specific biomedical etiology. About 10% of autistic males have the fragile X syndrome and approximately 15% of fragile X males are considered autistic. Many fragile X males demonstrate autisticlike behavioral features even though they do not fulfill strict diagnostic criteria for autism. Also, they commonly exhibit hyperactivity, speech delay, repetitive speech patterns, a relative lack of expressive language, and significant deficits in social interaction with individuals outside their family. Common neurologic features in fragile X males include a static central nervous system encephalopathy without focal lateralizing signs and impairment of fine motor coordination. About 20% of fragile X males have a history of seizures, which are usually transient or are well controlled with anticonvulsants. Neuropathological information indicates that brain structure is grossly normal. Microscopic analysis suggests immaturity of dendritic processes and decreased synaptic contacts in cortical areas. Volumetric MRI studies suggest a decreased ratio of posterior to anterior cerebellar vermis and increased volume of fourth and posterior ventricles. Similar MRI findings have been reported in autistic individuals. There is currently no specific

pharmacological therapy, although stimulant-type drugs that decrease hyperactivity have been used often with good results. Early intervention and special educational therapies are recommended to minimize developmental delays.

The fragile X syndrome shows an unusual X-linked inheritance pattern in several respects. Normal grandfathers of affected grandchildren have frequently been identified who are completely nonexpressing carriers. They are cytogenetically negative for the marker chromosome. They are normal physically and mentally unaffected. Since they are nonpenetrant for the gene mutation and yet transmit the mutation, they are referred to as normal transmitters (NT). All their daughters (and none of their sons) receive their X chromosome and are also nonexpressing. However, among the sons of the daughters of NTs, about 40% are affected, while 10% are also NTs.

Carrier females

Overall, approximately 56% of female carriers are cytogenetically positive and/or mentally impaired, while 44% are negative and unimpaired. Approximately 90% of mentally impaired females are positive on cytogenetic testing, while 25% of mentally unimpaired carriers exhibit the fragile site. About 30% of carriers are mentally impaired; one third are considered mentally retarded, while two thirds have a mild learning disability. Affected females tend to have lower scores in arithmetic, digit span, block design, and object assembly, while they may have increased verbal performance scores and do very well academically. There may exist subtle defects on emotional development based on mild neurocognitive functioning deficits. Shy and socially withdrawn behavior are common among the affected sisters of fragile X males. Other variable deficits include socially inappropriate comments, inappropriate affect, poor modulation of verbal tone, tangential speech and odd communication patterns. Mild schizotypal features and remitting recurrent depressive episodes have been described among some female carriers.

Molecular insights into the fragile X mutation

Recent progress has been made in understanding the molecular nature of the underlying mutation and in developing molecular diagnostic tests. The site of the mutation was localized to a small region of the X chromosome that contains a gene identified as FMR-1. As illustrated in Figure 1, this gene is variable in length and contains a (CGG)n repeat sequence that undergoes a tremendous length amplification in affected males. The normal FMR-1 gene contains approximately 30 CGG repeats but ranges from approximately 5 to 50. Carrier females and NT males have an enlargement of this region to a range of approximately 50 to 200 repeats, which is referred to as a premutation. Premutation alleles are meiotically unstable. Offspring of carrier females but not NTs tend to have amplifications of this region to values ranging from 200 to 2000, referred to as the full mutation. The probability of this amplification to the full mutation occurring is correlated with the size of the premutation. Generally, all individuals with the full mutation exhibit the chromosomal fragile site at Xq27.3 and frequencies tend to correlate with the average size of the amplification.

The Fragile X Region

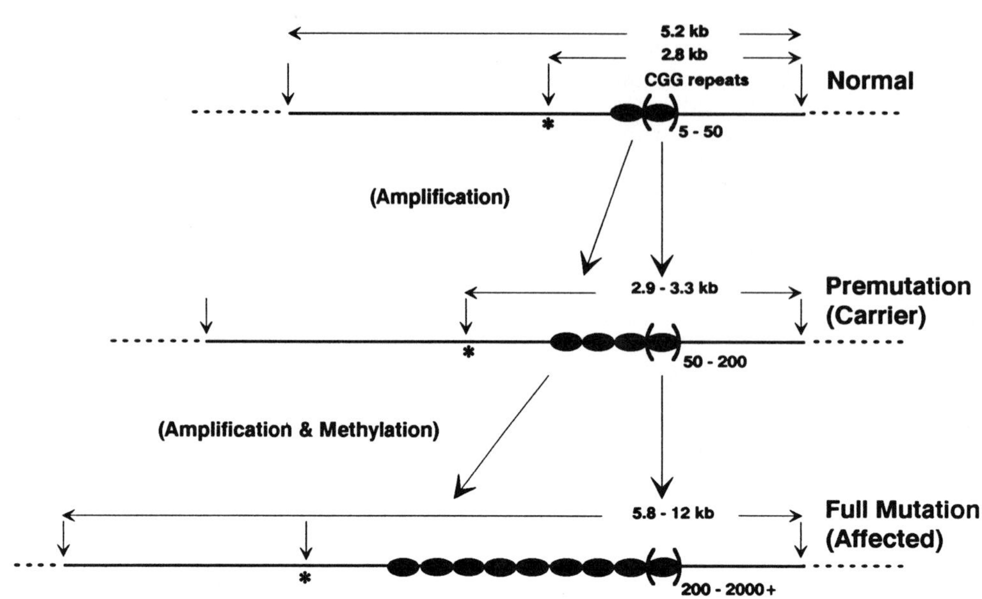

Figure 1. Part of the fragile X region that includes the gene FMR-1. The gene contains a CGG repeat that is amplified to a premutation in carriers. It undergoes further amplification and methylation to result in the full mutation in affected individuals. Using the probe StB12.3 and following a double digest with the restriction enzymes EcoRI and EagI, restriction fragments are generated by cleavage at the small vertical arrows. DNA methylation prevents cleavage (by EagI) at the position marked * (a CpG island). Only the fragments containing the CGG repeat region are detected. In normal individuals, a 5.2 kb fragment is detected in methylated chromosomes and a 2.8 kb fragment is detected in unmethylated chromosomes. The amplification of the CGG repeat leads to an increase in the size of these fragments. In affected individuals, not only is the CGG repeat region amplified, but the DNA is also methylated. This methylation prevents cleavage at the methylation sensitive (*) restriction site.

Approximately 250 bps upstream to the CGG repeat region is a CpG island that shows specific methylation in affected individuals. Such CpG islands are often found a short distance above gene coding regions and are usually associated with gene promoter regions. Full mutations are associated with methylation of the upstream CpG island, which correlates with loss of expression of the FMR-1 gene mRNA. Full mutations are mitotically unstable. They often include a heterogeneous set of bands and smears and may include a small proportion of premutation size alleles, a situation referred to as mosaicism.

The isolated FMR-1 cDNA has been used to identify a 4.8 kb mRNA in human brain. This size would correspond to a protein of about 1600 amino acids. The CGG repeat within the coding region of the gene would suggest a protein containing 30 continuous arginine residues. Such a protein region would be extremely basic and may represent a DNA binding region. However, it is uncertain that this region is coding. The FMR-1 sequence appears to be highly conserved through evolution including organisms as diverse as *C. elegans* and yeast (Verkerk et al., 1991).

Molecular diagnosis of fragile X

Molecular diagnostic testing including prenatal diagnosis can be conducted by either of two methods. The first method is direct genomic Southern blot analysis, which uses a probe that flanks the CGG repeat region. One such probe, StB12.3 (Rousseau et al., 1991), uses a double digestion with two restriction enzymes. As illustrated in Figure 1, digestion with the first enzyme (EcoRI) produces a 5.2 kb band on a Southern blot. The second methylation sensitive enzyme (EagI) cuts unmethylated DNA at the CpG island, producing a 2.8 kb band, but leaves methylated DNA uncut. Thus, a normal female produces both a 2.8 kb band, reflecting her active unmethylated X chromosome, and a 5.2 kb band, reflecting her inactive, methylated X chromosome. Premutation alleles generally produce bands in the range of 2.9 to 3.2 kb for the active X and 5.3 to 5.7 for the inactive X. Normal carrier females generally have two doublets on analysis. Affected males generally have bands or smears in the 5.8 to 9 kb range. The second method employs the use of the polymerase chain reaction (PCR). PCR analysis of fragile X mutations is rapid and uses small amounts of starting DNA for analysis. One PCR method, such as we have developed (Pergolizzi et al., 1992), allows for full mutations to be visualized.

It appears that all affected individuals with the full mutation have a carrier mother. Such carrier mothers typically have the premutation but may have the full mutation and be unimpaired. They also appear to nearly always have a parent who is a carrier. Although the transition from premutation to full mutation appears to be quite common, the transition from normal allele size to premutation size appears to be quite rare. This indicates that the new mutation frequency is quite low. The development of methods for the direct and PCR molecular diagnosis should allow for rapid and cost-effective screening programs to be established. This will help to diagnose individuals and families with the syndrome. It will assist with carrier detection and prenatal diagnosis. Further understanding of the function of the FMR-1 gene will undoubtedly lead to insights into the genetics of mental deficiency and may eventually result in new therapies.

Further reading

Hagerman RJ, Silverman AC (1991): *Fragile X Syndrome: Diagnosis, Treatment, and Research*. Balto, Md: The Johns Hopkins University Press

Pergolizzi RG, Erster SH, Goonewardena P, Brown WT (1992): Detection of full fragile X mutation. *Lancet* 339:271–272

Rousseau F, Heitz D, Biancalana V, Blumenfeld S, Kretz C, Boué J, Tommerup N, Van Der Hagen C, DeLozier-Blanchet C, Croquette M-F, Gilgenkrantz S, Jalbert P, Voelckel M-A, Oberlé I, Mandel J-L (1991): Direct diagnosis by DNA analysis of the fragile X syndrome of mental retardation. *N Eng J Med* 325:1673–1681

Verkerk AJMH, Pieretti M, Sutcliffe JS, Fu YH, Kuhl DPA, Pizzuti A, Reiner O, Richards S, Victoria MF, Zhang F, Eussen BE, van Ommen GJB, Blonden LAJ, Riggins GJ, Chastain JL, Kunst CB, Galjaard H, Caskey CT, Nelson DL, Oostra BA, Warren ST (1991): Identification of a gene (FMR-1) containing a CGG repeat coincident with a breakpoint cluster region exhibiting length variation in fragile X syndrome. *Cell* 65:905–914

G

Gelsolin

Bertrand Y. Tuan and David J. Kwiatkowski

Gelsolin

Gelsolin is an 81 kDa actin-filament–modulating protein that serves an important role in the regulation of the actin filament architecture during motile processes. Gelsolin was first isolated from rabbit macrophages and is found in all mammalian cells. It is unique among mammalian proteins in that in addition to the cytoplasmic form, a separate plasma gelsolin isoform is synthesized specifically for secretion by a variety of cell types, particularly muscle. Overexpression of gelsolin in mammalian *fibroblasts* has been shown to increase translocational motility.

Gelsolin has been found in cultured sympathetic neurons, at high levels in oligodendroglia of developing rabbit brain, and at lower levels in the adult central nervous system. It is thought to serve an important role in the motility of neuronal growth cones.

The role of gelsolin and actin in cellular motility

Protrusive cell movement involves extension and retraction of pseudopodia, which are extensions of a hyaline cellular structure that excludes cellular organelles. This region of the cell, the cell cortex, displays a gel to sol transition during these motile activities. Actin filaments are the primary constituent of the cell cortex, and ultrastructural analysis has shown they occur as a network of short (an average length of 0.5 to 1 μm) branched filaments. The importance of the actin filament architecture in cell structure and protrusive activity has been documented by studies using cytochalasin B, which disrupts the assembly of actin monomers into filaments. When cytochalasin B is applied to cells, pseudopods and ruffles cease to be formed and organelles are no longer excluded from the cell cortex. Lamellae along the periphery of the neuronal growth cone exhibit both protrusive and ruffling motility. Cytochalasin B eliminates stainable actin microfilaments from the leading lamellae and reversibly blocks growth cone motility.

The organization of the actin filament architecture is determined and dynamically modulated by numerous proteins that interact with actin. These proteins are of several functional types, including those that cross-link actin filaments and others that regulate actin filament length and assembly.

Actin is a globular monomeric protein that assembles into double helical filaments, in a process characterized by a relatively slow nucleation of two to three monomers into a nascent protofilament followed by rapid addition of more monomers during filament elongation. Fragments of myosin can be bound to an actin filament and reveal that the filament is polarized, with a pointed and a barbed end. Monomers attach preferentially to the barbed end. Gelsolin interacts with actin in three ways: it severs preformed actin filaments, caps the rapidly growing (barbed) end of actin filaments, and assembles actin monomers into nuclei to accelerate the formation of actin filaments. The mechanism by which gelsolin severs actin filaments does not involve proteolysis or exogenous energy and appears to occur in two stages: an initial stage of binding to the side of an actin filament, followed by kinking of the filament and displacement of the actin–actin contacts within the filament by gelsolin–actin binding.

Micromolar Ca^{2+} levels activate the severing and other functions of gelsolin, while the phosphoinositides PIP and PIP_2 inhibit all of its interactions with actin. Figure 1 illustrates our current model of how agonist stimulation of cell surface receptors causes the assembly and disassembly of actin filaments through Ca^{2+} and polyphosphoinositide regulation of gelsolin.

Cell movement is believed to occur through repetitive cycles of this process controlled in a temporospatial manner within small regions of the cytoplasm. Other proteins interacting with actin, especially myosin 1 and actin-monomer–binding proteins, may provide important additive and cooperative effects.

This model was constructed based upon a variety of *in vitro* evidence of gelsolin's role in actin architecture regulation. Recently this model has been further validated in experiments in which mouse fibroblasts overexpressing gelsolin displayed increased translocational motility.

Structure and function analysis of gelsolin

Gelsolin binds two actins with high affinity, and binding regions and activation sites within the protein have been defined by structure-function analysis.

Analysis of the amino acid sequence of gelsolin (determined from its cDNA) indicates it is composed of six homologous repeats (domains 1 through 6) of an ancestral sequence that is likely to represent a primordial actin-binding sequence motif (Fig. 2). These six subunits of gelsolin have subsequently diverged functionally to have different activities. Homologous severing proteins comprising one half the size of gelsolin are found in lower organisms; severin in the slime mold *Dictyostelium discoideum,* is one example of such a half-gelsolin. Villin, found in intestinal microvilli, shares the six-fold motif with gelsolin but has an additional C-terminal domain that gives it the ability to cross-link actin filaments into bundles in the presence of low Ca^{2+} concentrations.

C-terminal deletional mutagenesis of gelsolin led to the identification of a short fragment consisting of domain 1 plus an additional 11 amino acids (total length 135) that had no nucleating activity, but Ca^{2+} and PIP_2 regulated severing activity comparable to that of full-length gelsolin. Domain 1 by itself has no severing activity, and with other studies, these results indicated the importance of the first 11 residues of domain 2 (KHVVPNEVVVQ) in the PIP_2 binding of gelsolin and its initial interaction with actin filaments during severing.

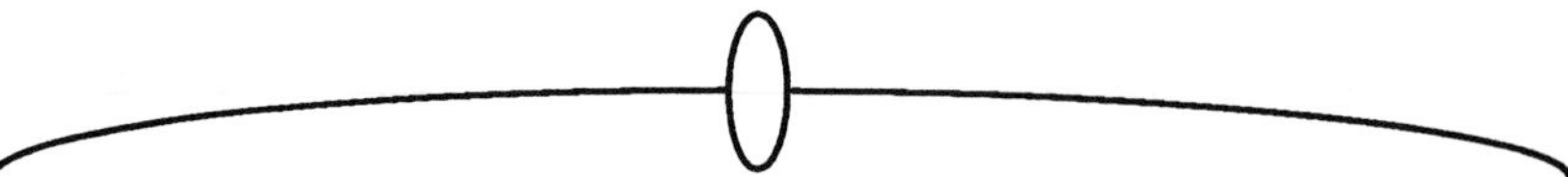

1. Binding of agonist to receptor hydrolyzes PIP2 to DAG and IP3

2. IP3 releases Ca++ stored in intracellular vesicles.

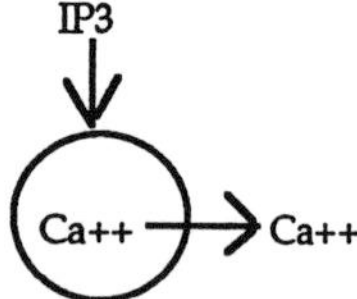

3. Released calcium (and decreased PIP2) activates gelsolin to:
   a. sever and then cap the rapidly growing end of actin filaments:

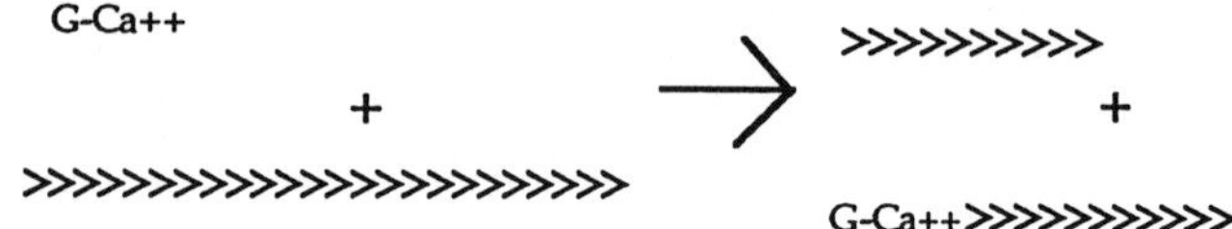

   b. nucleate actin monomers:

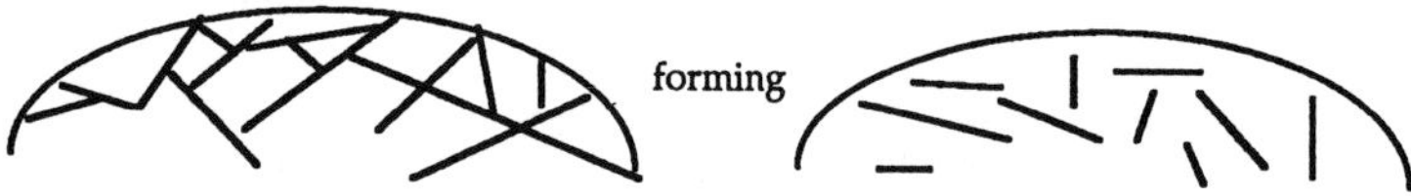

4. Resulting in:

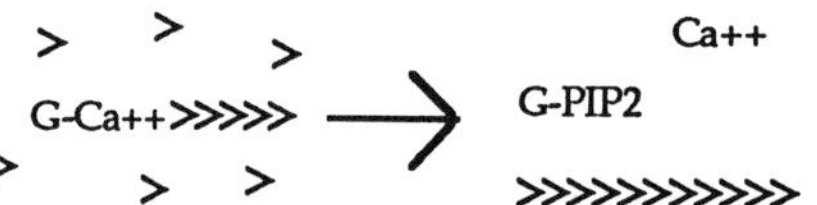

forming

Later:
5. Ca++ falls to resting levels

6. PI and PIP become phosphorylated

7. a. Gelsolin stops severing actin filaments and falls off the barbed end allowing rapid elongation at the newly uncapped barbed end

   b. Gelsolin no longer caps nucleated monomers freeing up a pool of uncapped barbed ends facilitating elongation of filaments

   c. Uncapped actin filaments and nuclei rapidly elongate at the barbed end extending filaments and powering protrusion.

forms

Figure 1. Signal transduction leads to remodeling of the actin filament cytoskeleton via gelsolin activation. PI: phosphatidylinositol; PIP, PIP2: phosphatidylinositol phosphate and 4,5-bisphosphate; IP3: inositol 1,4,5-triphosphate; DAG: 1,2-diacylglycerol; >: actin monomer; G: gelsolin.

Plasma gelsolin

Plasma gelsolin is an isoform of gelsolin secreted by many cells, particularly muscle cells. Both forms of gelsolin are encoded by a single gene through the use of alternative transcription initiation sites. The only structural difference between plasma and cytoplasmic gelsolin is the presence of 25 additional amino acid residues at the N-terminus of mature plasma gelsolin.

The precise function of plasma gelsolin is unknown, but it may play a role in scavenging actin filaments released into the bloodstream during processes of tissue injury. Plasma gelsolin levels are decreased in a rabbit model of oleic-acid–induced lung injury, in patients with malaria who are actively hemolyzing erythrocytes with release of actin into the circulation, and in patients with the adult respiratory distress syndrome.

Gelsolin is composed of six repeated domains with a 25 a.a. plasma extension peptide on the secreted protein. Proteolytic cleavage sites tend to occur between each structural domain.

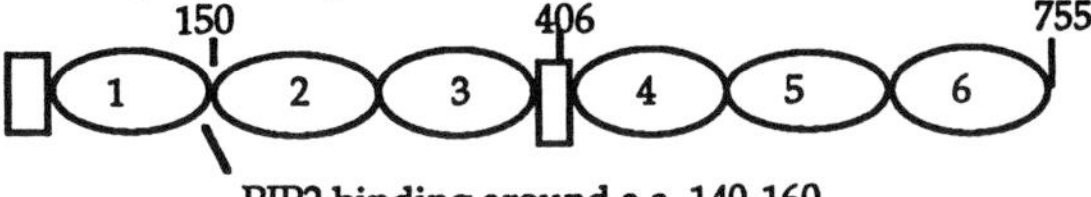

PIP2 binding around a.a. 140-160

A 160 residue gelsolin derivative has Ca++ regulated severing and barbed-end binding activity

Loss of another 11 a.a. leads to loss of severing activity

Dictyostelium severin has 45% amino acid identity with domains 1-3 of gelsolin

Figure 2. Structure and function of gelsolin and related proteins. Ovals represent the homologous repeated domain of gelsolin, which is also found in other actin-binding and -severing proteins.

Profilin, with weak homology to domain 1 of gelsolin, binds to actin monomers

Gelsolin: Genetics and Finnish amyloidosis

The gelsolin gene has been mapped to the long arm of chromosome 9, band q32–34, and a RFLP within the gene provided the first evidence for genetic linkage of torsion dystonia (DYT1) to a DNA marker. However, subsequent studies have excluded gelsolin as the cause of dystonia.

Over the past two years, evidence has accumulated that the amyloid protein responsible for Finnish-type familial amyloidosis (FAF) is a variant fragment of the gelsolin molecule. FAF, also known as familial amyloid polyneuropathy type IV, is an autosomal dominant disorder characterized by corneal lattice dystrophy, cranial neuropathy, and distal sensimotor neuropathy. Purification of the amyloid protein from these patients revealed two components that were identical to residues 173–225 and 173–243 of cytoplasmic gelsolin with the exception of the substitution of asparagine for aspartic acid at residue 187. This difference was shown to correspond to a point mutation in the gelsolin gene (G^{654} to A^{654}) in affected individuals, and segregation analysis has confirmed that the mutation occurs only in those affected with the disease. Whether the amyloid deposition in FAF reflects defective proteolytic degradation of patients' mutated gelsolin, as occurs in other amyloidopathies, or whether this accumulation reflects an alteration in the functional activity of gelsolin and its interactions with actin are questions that remain to be answered. This type of amyloidopathy is relatively common in Finland (incidence about 1 in 10,000) but quite rare outside of Finland, with only a few cases reported in the United States.

Neuropathologic examination in one patient with FAF who died of dementia demonstrated, in addition to typical neuronal plaques and neurofibrillary tangles, cortical and classical Lewy bodies more commonly found in association with Parkinson's disease. The Lewy bodies from this patient, as well as others with Parkinson's, are immunoreactive with antiserum to the FAF amyloid subunit of gelsolin.

Further reading

Cunningham C, Stossel T, Kwiatkowski D (1991): Enhanced motility in NIH3T3 fibroblasts that overexpress gelsolin. *Science* 251:1233–1236

Hartwig J, Kwiatkowski D (1991): Actin-binding proteins. *Curr Opin Cell Biol* 3:87–97

Maury C (1991): Gelsolin-related amyloidosis. Identification of the amyloid protein in Finnish hereditary amyloidosis as a fragment of variant gelsolin. *J Clin Invest* 87:1195–1199

Stossel T (1989): From signal to pseudopod: How cells control cytoplasmic actin assembly. *J Biol Chem* 264:18261–18264

Way M, Weeds A (1990): Actin-binding proteins. Cytoskeletal ups and downs. *Nature* 344:292–294

Gilles de la Tourette Syndrome and Self-injurious Behavior

Mary M. Robertson

Self-injurious behavior (SIB) may be categorized by both the type of patient and clinical context in which it occurs, and is recognized as occurring in many disorders, including Gilles de la Tourette Syndrome (GTS). In the American Psychiatric Association's revised diagnostic and statistical manual (DSM-III-R), SIB is acknowledged as an impulse disorder, and is self-inflicted, nonaccidental injurious behavior, variously referred to as self-mutilation, self-injury, self-destructive behavior, or deliberate self-harm. GTS is a disorder characterized by the presence of both multiple motor tics and one or more vocal tics (phonic tics, vocalizations or verbalizations), and in some cases, there may be obsessive-compulsive (OC) thoughts and behaviors, attention deficit disorder and hyperactivity, echophenomena (copying behaviors), coprolalia and copropraxia (inappropriate uttering of obscenities and making of obscene gestures), as well as a variety of other associated behaviors.

SIB in the context of GTS has been acknowledged ever since the syndrome was first described, and subsequently a number of case reports describing SIB in GTS have been documented. In addition, several investigations involving nearly 900 clinic and TSA (Tourette Syndrome Association) populations have evaluated SIB in the context of GTS, with SIB being found in some 33–53% of GTS individuals. The most common types of SIB encountered in case reports and the large studies were patients punching, hitting and/or striking themselves, pinching themselves, head banging, biting lips, cheeks, tongues and extremities, body-to-hard-object banging, repeated and forceful pushing of a sharpened pencil into the ear canal, pressing vigorously on the eyeball, scab pulling, repetitively sticking pins under the skin, poking sharp objects into body, picking at sores, scratching parts of the body, putting hands through windows, touching hot objects, poking the umbilicus with the forefinger, violent head shaking, filing of teeth, attempting to dislocate joints, repeated digging of the forefinger into the hollow of the cheek, tooth extraction and severe eye injuries.

It has been shown that SIB (hitting of self) or hitting of others was the second most frequent complex movement in some 35% of such movements found in over a hundred patients in one study. It has also been shown that GTS adult patients have much higher scores than those of normal control populations on standardized psychiatric rating scales; in addition, SIB is significantly related to hostility and obsessionality, the severity of GTS motor symptoms and a past psychiatric history. Head banging has been shown to be associated with attending a special school, being aggressive, having severe GTS, and being psychiatrically disturbed. The same study also reported four GTS patients who had severe SIB resulting in serious eye injuries. They were all refractory to treatment; one died as a result of severe head shaking, while a second required psychosurgery (limbic leucotomy). In this context it is important to note that eye SIB is invariably associated with psychosis such as schizophrenia, but none of the GTS patients were, however, psychotic.

In general, SIB is related to the severity of GTS symptoms and psychopathology, but both epidemiological and family or pedigree studies have indicated that SIB occurs in milder forms of GTS, and in fact, in a pedigree study SIB and OC features distinguished cases from noncases. This suggests that SIB may be found in GTS of all severities.

In psychiatric hospital populations between 1 and 4% have SIB. On the other hand, SIB among mentally retarded or learning-disabled people occurs in between 7 and 9% of such individuals, increasing to 14–19% of institutionalized retarded or disabled children, and occurring in as many as 40% of institutionalized psychotic children. The incidence of SIB in the GTS cohorts is therefore much more than in psychiatric populations and somewhat similar in frequency to that in the learning disabled, being found in some 17 to 53% of GTS individuals.

In which other conditions is SIB found, and what are the links, if any, between those disorders and GTS? SIB resulting in mutilative lesions has been described in a variety of conditions such as the Lesch-Nyhan syndrome (LNS), the learning disabled (individuals of low intelligence), neuroacanthocytosis, psychotic disorders such as schizophrenia, and depressive illness, but in one large study no evidence of an association between SIB in GTS and any of these disorders was found.

SIB and GTS may share biochemical abnormalities. The dopaminergic system has been most implicated in GTS and has been reported as being abnormal in SIB in both animal experiments and human studies. Neuropeptides have also been implicated as being abnormal in SIB. Of interest in this context is that patients with SIB have been successfully treated with naloxone and naltrexone, suggesting that they play a role in the regulation of the endorphin/enkephalin systems and can alter SIB. It is of interest that recently both overactivity and underactivity of the endogenous opioid system have been postulated in GTS, as has dynorphin, and there have been reports of a significant reduction in GTS tic symptomatology with treatment by naloxone. Abnormalities of serotonin have also been demonstrated in GTS and possibly SIB. The basis for consideration of a role for the serotonin system in SIB is the similarity of some of its features (compulsivity) to obsessive-compulsive disorder (OCD). Some early studies have reported severe eye SIB occurring in patients with OCD. In addition, compulsive lip biting in a person of normal intelligence has been reported. Others have, moreover, suggested a compulsive mutilatory syndrome that possibly involved the hypothalamus/serotonergic pathways.

In conclusion, SIB is encountered in various clinical syndromes, including GTS. Taking the review of the literature into account, it would appear that this behavior in GTS may well be underreported to date. Results suggest SIB can occur in mild cases, but in general, the clinical correlates of SIB are the severity of GTS symptoms and psychopathology. In particular, GTS patients with SIB and self-mutilators in other studies score highly on obsessionality and hostility measures. This is particularly interesting in that links have been demonstrated in GTS patients between obsessionality and hostility and copro- and echophenomena, core features of GTS. Thus, SIB may be part of the syndrome in some patients that would fit in with the suggestion that GTS may well be a heterogeneous condition.

The types of SIB in GTS patients are, in general, not typical of that encountered in patients with LNS, neuroanthocytosis, schizophrenia, depression, or personality disorders, but are nonspecific and somewhat similar to that found in learning-disabled populations. Of importance is that most

GTS patients are of average intelligence. There are, however, difficulties in comparing the results of SIB in GTS with other studies investigating SIB, as there are many varying definitions of such behaviors. In the majority of cases of deliberate SIB the more usual means is by ingestion of toxic substances or drugs; such cases were not included in most studies.

Some GTS patients incur serious injuries, perhaps exemplified in some studies by severe sequelae such as blindness, cavum septum pellucidum cavities, and death. As regards a biochemical substrate for GTS patients with SIB, the most likely areas implicated appear to be the dopaminergic, serotonergic, and endogenous opioid systems.

Another question to be addressed is that of management, but no studies have to date investigated the treatment of SIB in GTS. However, as SIB in one cohort was related to both severity of GTS symptoms and psychopathology, one possibility is that with treatment of the manifestations of GTS or the psychopathology, the SIB may itself disappear or reduce as a consequence. Based on biochemical abnormalities that have been suggested in both GTS and SIB, and the author's clinical experience, it is suggested that the following agents be used alone or in combination: dopamine antagonists (sulpiride, haloperidol, pimozide), some serotonin reuptake inhibitors (chlomipramine, fluoxetine, fluvoxamine), clonidine, naloxone and naltrexone, lithium, and finally beta-blockers. Psychosurgery must be borne in mind as a lifesaving treatment in the most severe of cases. Stereotactic bilateral limbic leucotomy with electrocoagulative probe lesions made in both lower medial quadrants of frontal lobes and separate lesions made in the anterior cingulum have been described as successful in the treatment of severe SIB in GTS.

Clearly, further studies investigating SIB in GTS need to be conducted to validate and replicate previous results and address the question of management.

Further reading

Robertson MM (1989): The Gilles de al Tourette syndrome: The current status. *Br J Psychiatry* 154:147–169

Robertson MM, Gourdie A (1990): Familial Tourette's syndrome in a large British pedigree. Associated psychopathology, severity, and potential for linkage analysis. *Br J Psychiatry* 156:515–521

Robertson MM, Yakeley J (1992): Obsessive compulsive disorder, self injurious behavior and the Gilles de la Tourette syndrome. In: *Handbook of Tourette's Syndrome and Related Tic and Behavioral Disorders*, Kurlan R, ed. New York: Marcel Dekker

Robertson MM, Trimble MR, Lees AJ (1989): Self-injurious behaviour and the Gilles de la Tourette syndrome: A clinical study and review of the literature. *Psychol Med* 19:611–625

Glial Growth Factors

Hideo Kimura

Neural tissue consists of two major classes of cells, nerve and glia. Glia surround neurons and are thought to support them structurally and metabolically. The first studies of growth factors for glial cells examined the activity of identified fibroblast mitogens on primary cultures of rodent glia. More recent work includes the effects of growth factors on glial proliferation and diversification in the central nervous system (CNS) and on Schwann cell proliferation in the peripheral nervous system (PNS). These three areas of investigation will be addressed separately.

Glial cell mitogens in the CNS

The glial cells of the CNS are classified into two major groups, oligodendrocytes and astrocytes. Oligodendrocytes wrap around axons to form a myelin sheath. Astrocytes play an important role in guiding the development of the nervous system and in controlling the chemical and ionic environment of the neuron. Recent developments in the culture of pure populations of astrocytes and oligodendrocytes by J. de Vellis and his colleagues and in the use of chemically defined media have led to the identification of growth factors for these two classes of cells.

Epidermal growth factor (EGF), one of the first growth factors to be identified, was isolated from the submaxillary glands of mice as a mitogen for epidermal cells. EGF is a 6000 dalton protein proteolytically cleaved from a longer precursor protein. EGF stimulates DNA synthesis in astrocytes, and binding sites for EGF have also been found in oligodendrocytes. It also stimulates the differentiation of oligodendrocytes as defined by large increases in the accumulation of the myelin basic protein.

Basic and acidic fibroblast growth factors (bFGF and aFGF) were isolated from pituitary and brain and are proteins of 16,500 and 15,500 daltons, respectively. They have distinct isoelectric points but are structurally related, sharing about 55% amino acid sequence identities. Originally, both FGFs were identified as mitogens for fibroblasts. Later, FGF was found to be both a mitogen and a differentiation factor for a variety of mesoderm- and neuroectoderm-derived cells. bFGF has been studied extensively and is a potent mitogen for oligodendrocytes and astrocytes. In ad-

dition, it modulates the induction of astrocyte-specific differentiation markers such as a glial-specific intermediate filament protein, glutamine synthetase and S100 protein. bFGF also stimulates the migration of astrocytes and increases the rate of extension of cellular processes typical of mature astrocytes.

Glial cell diversification in the CNS

The optic nerve consists of the axons of retinal ganglion neurons that project from the eye to the brain. Since this preparation contains numerous glial cells but no nerve cell bodies, it has been used as a source of glial cells to study their development. In this preparation astrocytes are classified as type 1 and type 2 on the basis of differences in morphology, antigenic phenotype, and response to growth factors. Oligodendrocytes and type-2 astrocytes develop from a common, bipotential progenitor cell called O-2A, whereas type-1 astrocytes develop from a distinct precursor cell. The growth factors outlined in Table 1 determine the cell fate and the timing of differentiation of glial progenitor cells derived from the optic nerve.

PDGF. Platelet-derived growth factor (PDGF) was discovered as a mitogen for smooth muscle cells and fibroblasts; subsequent studies have shown that PDGF is synthesized by a large number of different cell types. PDGF is a 30,000 dalton dimer composed of disulfide-bonded A and B polypeptide chains with similar amino acid sequences. All possible dimeric forms of PDGF have been identified in different cells. Several experiments suggest that type-1 astrocytes secrete PDGF as a mitogen for O-2A cells. First, purified PDGF mimics the effects of type-1 astrocyte conditioned medium. Second, the conditioned medium of type-1 astrocytes contains bona fide PDGF and type-1 astrocytes express mRNA encoding PDGF A chain. Third, antibodies against PDGF inhibit the ability of conditioned medium of type-1 astrocytes to stimulate O-2A progenitor cell proliferation.

CNTF. Ciliary neurotrophic factor (CNTF) was first identified from heart cell conditioned medium as a survival factor

Table 1. Properties of Glial Growth Factors Whose Corresponding Gene or mRNA Have Been Cloned

Name	Structure	Source	Biological Activity
EGF	6000 dalton	Submaxillary glands	Mitogenic for astrocytes Differentiates oligodendrocytes
bFGF	16,000 dalton	Pituitary and brain	Mitogenic for astrocytes, oligodendrocytes, glial progenitor cells Mitogenic for Schwann cells in the presence of cAMP
PDGF	30,000 dalton homo- and heterodimer	Platelet	Mitogenic for glial progenitor cells Mitogenic for Schwann cells in the presence of cAMP
CNTF	20,000–25,000 dalton	Eye and sciatic nerve	Differentiates glial progenitor to astrocytes
TGF-beta	25,000 dalton homodimer	Platelet	Mitogenic for Schwann cells
SDGF	31,000–35,000 dalton	Schwannoma cells	Mitogenic for astrocytes, weakly for Schwann cells

for chick ciliary ganglion neurons. It was purified from chick eye and rat sciatic nerve and is a 20,000 to 25,000 dalton acidic protein. In the optic nerve, oligodendrocyte development occurs autonomously, while type-2 astrocyte development requires inducing signals. Several lines of evidence suggest that the type-2 astrocyte-inducing protein from the optic nerve is either CNTF or a closely related protein. CNTF and the inducing protein are similar in size and have a similar tissue distribution, and purified CNTF induces O-2A progenitor cells to express an astrocyte-specific marker, glial fibrillary acidic protein (GFAP). This biological activity of CNTF on O-2A progenitor cells is indistinguishable from that of extracts from optic nerve.

FGF. Recently it was found that bFGF maintains a high rate of mitosis in O-2A progenitor cells and that the continued presence of bFGF reversibly blocks their differentiation. O-2A progenitor cells express the PDGF receptor, but the level of PDGF receptor decreases as the cells differentiate. Because bFGF maintains a high level of expression of PDGF receptor on O-2A progenitor cells, it is likely that bFGF activates a signaling pathway for PDGF and keeps O-2A progenitor cells proliferating to prevent their premature differentiation.

Schwann cell proliferation in the PNS

Schwann cells are glial cells supporting axons in the peripheral nervous system. During development they proliferate in response to contact with axons, ensheath them, and eventually synthesize a myelin sheath that surrounds axons. Schwann cells also have the capacity to proliferate during the wound repair processes that accompany nerve regeneration. Because of the potential role of Schwann cells in nerve regeneration, there has been a great deal of interest in the proteins that regulate their division. In culture, the division of rat Schwann cells is very slow even in the presence of serum, but their DNA synthesis is stimulated by a limited array of soluble growth factors. The development of the methods to obtain highly purified cultures of normal rat Schwann cells by J. Brockes and M. Raff has led to the discovery of several mitogens.

Crude extracts of bovine pituitary and brain contain a potent mitogenic activity for Schwann cells. This mitogen has been partially purified from bovine pituitary and the activity designated glial growth factor (GGF). GGF was reported to be a 31,000 dalton basic polypeptide, but has yet to be sequenced. More recently, several known growth factors were shown to elicit Schwann cell division. Transforming growth factor (TGF-beta), a 25,000 dalton homodimer originally isolated from platelets, potentiates the proliferation of Schwann cells. Schwann cells express TGF-beta mRNA, suggesting that TGF-beta may also act as an autocrine growth factor for Schwann cells. PDGF and FGF are only weakly mitogenic by themselves but synergize with agents that increase intracellular cyclic AMP such as cholera toxin, forskolin, and dibutyryl cyclic AMP to promote the rapid proliferation of Schwann cells.

Schwann cells secrete one or more autocrine mitogens into their culture medium and human schwannoma extracts contain a mitogen for Schwann cells. A cell line recently established from a spontaneous tumor from the sheath of the rat sciatic nerve has been used as a source of a protein that is a potent mitogen for astrocytes and a weak mitogen for Schwann cells. The purified mitogen, designated schwannoma derived growth factor (SDGF), is a glycoprotein with a molecular weight of between 31,000 and 35,000 daltons. SDGF is homologous to human amphiregulin and belongs to the epidermal growth factor family.

The size, biological activities, and basic nature of SDGF are similar to the pituitary-derived glial mitogen GGF. In spite of these similarities, the mitogenic activity of SDGF on Schwann cells is weaker and of slower onset than that of GGF. Although SDGF is not GGF per se, SDGF may be a part of the GGF activity. For example, GGF activity from pituitary could contain two or more components of similar molecular weights that copurify. It is therefore possible that SDGF is a component of GGF and that a synergistic interaction with other proteins is required to potentiate the rapid growth of Schwann cells.

Summary

The study of glial growth factors was initiated several years ago by examining the effect of identified growth factors for fibroblasts on pure cultures of glial cells. Most of the mitogens for fibroblasts also potentiate the growth of glial cells, suggesting that both cell types have the same or related signal transduction pathways for these growth factors. The availability of a clonal Schwann cell line has allowed the purification and cloning of one new glial mitogen (SDGF). Experiments from several laboratories have shown that pairwise combinations of growth factors elicit mitogenic responses in glial cells that are severalfold higher than the additive effects of the individual proteins. These data and those emerging from the work on optic nerve glial lineages suggest that combinatorial effects of several growth factors may direct the division and differentiation of glial cells in the nervous system.

Further reading

Kimura H, Fischer WH, Schubert D (1990): Structure, expression and function of a schwannoma-derived growth factor. *Nature* 348:257–260

Raff MC (1989): Glial cell diversification in the rat optic nerve. *Science* 243:1450–1455

Ratner N, Bunge RP, Glaser L (1986): Schwann cell proliferation in vitro. *Ann NY Acad Sci* 486:170–181

Saneto RP, de Vellis J (1987): The use of primary oligodendrocyte and astrocyte cultures to study glial growth factors. In: *Neuronal Factors,* Perez Polo JR, ed. Boca Raton: CRC Press, pp 175–195

Glucose-Sensitive Potassium Channels

Sylvie Zini, Valérie Crépel, and Yehezkel Ben-Ari

Potassium conductances are present in all excitable cells, since they contribute to the resting membrane potential and control repolarization. Potassium channels represent the most diverse group of ion channels so far identified; they include voltage-dependent and calcium- or other ion-activated K^+ channels, receptor-coupled K^+ channels, or stretch-activated K^+ channels. Some of these channels are coupled to G-proteins that mediate the formation of second messengers. There has been a considerable interest recently in an additional class of potassium channels that are inhibited by intracellular ATP (K-ATP). These glucose-sensitive potassium channels have been shown to play an important role in the regulation of insulin secretion. In addition to the pancreatic β-cells, channels with similar properties are present in skeletal and smooth muscles and in cardiac cells. Because of their sensitivity to intracellular ATP, K-ATP channels constitute a unique mechanism to modulate excitability according to the metabolic state of the cell. These important properties have led to the development of pharmacological compounds that could become very useful, notably in cardiovascular therapeutics.

We shall briefly review the principal properties of K-ATP channels and their modulation by pharmacological agents.

Physiology of ATP-sensitive potassium channels

In 1983, Noma reported the existence, in isolated cardiac cells, of a subtype of K^+ channels that are inhibited by ATP. ATP-sensitive potassium channels have been now observed in various tissues of different species including β-cells of Langerhans islets, smooth or skeletal muscles, and in the brain. The unitary conductance of K-ATP channels varies between 40 and 80 pS. At physiological intracellular ATP concentration (in the mM range), K-ATP channels are usually closed, and they open when the cell is submitted to a metabolic stress: a decrease of intracellular ATP concentration induced by exposing cells to hypoglycemia or hypoxia or by maintaining inside-out patches in ATP-free solution, produce an increase of a K^+ conductance that is completely inhibited by intracellular application of ATP.

How ATP inhibits the potassium conductance of such channels is not completely elucidated. The blockade of the channel by ATP does not require the presence of Mg^{2+} or the hydrolysis of the nucleotide to ADP, and nonhydrolyzable analogs of ATP are as effective as ATP, suggesting that G-proteins do not participate in the mechanism of action. ADP and AMP, as well as purine or pyrimidine nucleotides, also block K-ATP channels, but their efficacy is considerably weaker than ATP. Furthermore, ATP has a dual effect on K-ATP channels: the nucleotide is essential to block the activity of the channels but its hydrolysis (and the presence of Mg^{2+}) is also required to maintain channels' activity after their opening and to avoid a run-down phenomenon.

Pharmacology of ATP-sensitive potassium channels

ATP-sensitive potassium channel blockers. The pharmacological characterization of K^+ channels has been largely based on the use of various blockers, including non-selective ones such as tetraethylammonium or 4-aminopyridine, and more selective ones such as hypoglycaemic sulfonylureas. These drugs, used for a long time as treatment of type II

diabetes mellitus, are known to decrease K^+ permeability. Patch-clamp experiments performed on cardiac cells and pancreatic β-cells have shown that tolbutamide (a sulfonylurea of first generation) and glibenclamide (a sulfonylurea of second generation) effectively block K-ATP channels. Glibenclamide is the most potent of all sulfonylureas with a half-maximal inhibition constant (IC_{50}) in the nanomolar range. The therapeutic effect of sulfonylureas is largely mediated by a blockade of K-ATP channels of pancreatic β-cells, thus inducing a depolarization, activation of voltage dependent Ca^{2+} channels, and secretion of insulin.

Electrophysiological studies have shown that sulfonylureas also block K-ATP channels of other tissues, including vascular smooth muscles and cardiac cells. Furthermore, binding studies have demonstrated the existence of specific high- and low-affinity binding sites (their equilibrium dissociation constants [Kds] are about 0.1 and 100 nM, respectively) for sulfonylureas in pancreatic β-cells, in various smooth muscles, and in the brain. Binding sites to glibenclamide have been partially purified and three membrane proteins of 140, 65, and 43 kDa have been identified by photoaffinity labeling with iodinated glibenclamide. Excellent correlations have been made between the affinity of sulfonylureas for binding sites and their potency to block K-ATP channels or to stimulate insulin secretion. Nevertheless, since the receptor has not yet been cloned, it is not possible to assert whether the sulfonylurea binding sites correspond to the K-ATP channel or whether they are closely associated to the K-ATP channel activity.

ATP-sensitive potassium channel openers. More recently, selective K^+ channel openers (PCOs) have been developed and found to produce membrane hyperpolarization and relaxation of various smooth muscles. These compounds are of particular interest for their clinical application to treatment of cardiovascular diseases, bronchial asthma, or irritable bladder syndrome. The type of K^+ channels activated by PCOs corresponds often to a class of K^+ channels sensitive to ATP.

Typical K-ATP channel openers are represented by several chemically unrelated compounds such as cromakalim (a benzopyran derivative), pinacidil (a guanidine derivative), or RP 49356 (a thioformamide derivative). These compounds produce a relaxation of bronchial, vascular, and intestinal smooth muscles, an effect that can be competitively antagonized by sulfonylureas. These compounds have been developed for their hypotensive or bronchodilator properties *in vivo*. Interestingly, at effective concentrations, these agents do not activate K-ATP channels of pancreatic β-cells and do not inhibit insulin release, a phenomenon that may suggest different subtypes of these channels. In contrast, the benzothiadiazine derivative diazoxide, a well-established hypotensive drug, exerts its activity, at least in part, by opening K-ATP channels in vascular smooth muscle, while its frequent hyperglycemic side effects could be due to the opening of K-ATP channels of pancreatic β-cells.

Functional role of ATP-sensitive potassium channels

Pancreatic β-cells. Pancreatic β-cells are electrically excitable cells that release insulin when depolarized. Earlier observations have shown that an increase in extracellular glucose produces an increase in input resistance of these cells due to a decrease in their membrane permeability to K^+. In inside-out patches of pancreatic β-cells, intracellular ATP produces a direct inhibition of K^+ channel opening. In contrast to most other tissues, some K-ATP channels are opened in pancreatic β-cells under normal blood glucose concentration. Hyperglycemia enhances the cell metabolism and thus increases the intracellular ATP concentration. Such an increase in intracellular ATP will induce closure of K-ATP channels, depolarization of the cell membrane, and activation of voltage-dependent Ca^{2+} channels, allowing Ca^{2+} entrance and secretion of insulin by exocytosis.

In view of this mechanism, it has been proposed that hypoglycemic sulphonylureas promote insulin release via the binding to specific receptors that lead to the blockade of K-ATP channels, thus mimicking the effect of glucose. Diazoxide, in contrast, induces an opening of the channels that hyperpolarizes the cell membrane, and thus it inhibits the secretion of insulin. Moreover, it has recently been shown that hormones like galanin or somatostatin also open K-ATP channels via a pertussis toxin-sensitive G-protein, which may represent an additional mechanism of modulation of these channels. K-ATP channels are present in human nondiabetic pancreatic β-cells. Figure 1 is a schematic diagram of the mode of operation of K-ATP channels.

Heart and smooth muscles. In physiological conditions, no activity of K-ATP channels is detectable in patch-clamp experiments from cardiac cells. In this tissue, the high intracellular ATP concentration blocks the channels efficiently, and thus K-ATP channels do not contribute to the resting membrane potential, in contrast to pancreatic β-cells. The activation of K-ATP channels in cardiac cells is generated by a decrease in intracellular ATP level subsequent to anoxia or ischemia. In these conditions, the opening of K-ATP chan-

nels is responsible for the shortening of the plateau of the ventricular action potential and of the reduction in heart muscle contractions. These events are interpreted as a protective mechanism aimed at preventing an excessive rise in intracellular calcium and consequent irreversible cell damage. On the other hand, the opening of K^+ channels increases the threshold of excitation and thus slows the activity of the pacemaker. PCOs like cromakalim, pinacidil, or RP 49356 are able to shorten the action potential duration of cardiac cells, an effect that can be antagonized by sulfonylureas. Moreover, parenteral or oral administration of PCOs produces dramatic hypotension due to vasodilatation caused by the relaxation of vascular smooth muscles following their increase in K^+ permeability. This relaxant effect is competitively antagonized by sulfonylureas, but in contrast to the pancreatic β-cells where glibenclamide is active in the nanomolar range, submicromolar (or higher) concentrations of this drug are necessary to block the action of PCOs and to induce relaxing effects. These observations suggest the existence of different K-ATP channels in pancreatic β-cells and vascular tissues. Furthermore, PCOs differ in their biological activity: the same dose of diazoxide can induce a hypotensive and a hyperglycemic effect, whereas cromakalim elicits only a hypotensive activity without hyperglycemic effect.

The characterization of the various classes of PCOs is essential for the development of new antihypertensive agents devoid of hyperglycemic side effects. In fact, cromakalim, pinacidil, and other K-ATP channel openers are already implicated in important preclinical investigations related to their application in cardiovascular diseases. Cromakalim or pinacidil also relax bronchial smooth muscles and inhibit the spontaneous contractile activity of the urinary bladder, presumably by opening K-ATP channels of these smooth muscle cells. These interesting properties suggest a potential use of these drugs for the treatment of asthma and irritable bladder syndrome.

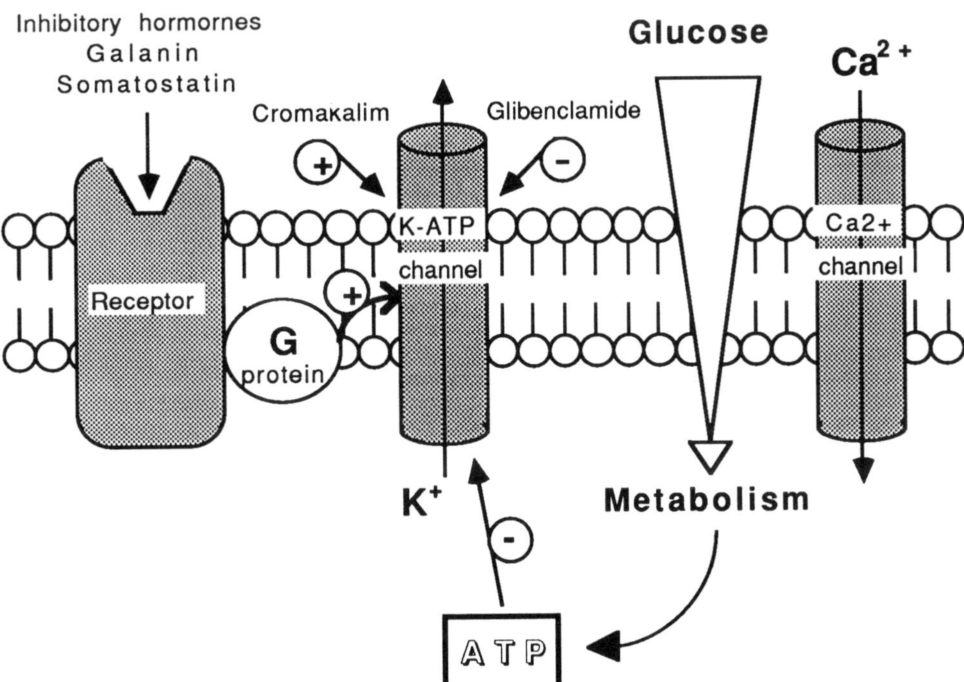

Figure 1. Schematic model of the possible regulation of ATP-sensitive potassium channels in pancreatic β-cells islets.

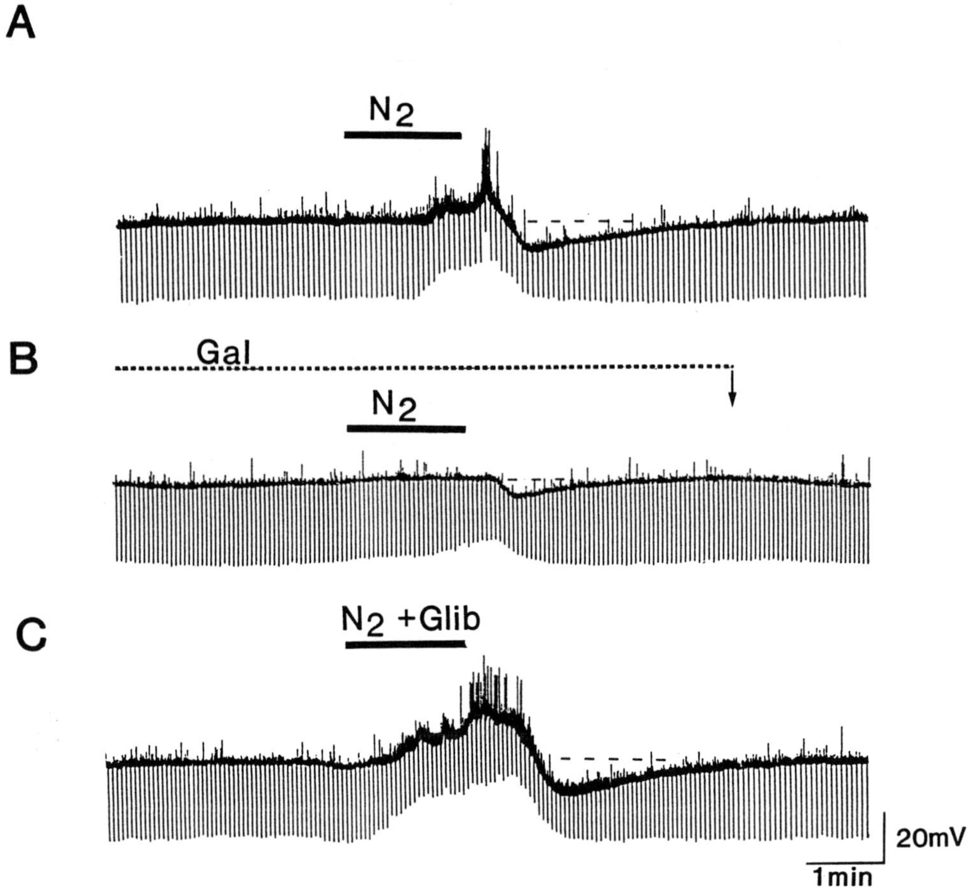

Figure 2. Modulation of the anoxia-induced depolarization by galanin and glibenclamide in rat hippocampus. Intracellular recordings of a CA3 hippocampal neuron in a slice maintained in artificial cerebrospinal fluid (ACSF). The anoxic depolarization produced by a 90 s perfusion with ACSF saturated in N_2 (trace A) was reduced by 1 μM galanin (Gal, trace B) and augmented by 10 μM glibenclamide (Glib, trace C). Both agents had no effect on resting membrane potential or input resistance in normoxic conditions. The anoxic hyperpolarization that precedes the depolarization was not conspicuous in this recording; the postanoxic hyperpolarization is due to the activation by reoxygenation of a Na^+-K^+ electrogenic pump.

Central nervous system. Specific binding sites for sulfonylureas are widely distributed in the brain, with the highest concentration in the substantia nigra, the globus pallidus, and the motor cortex. ATP-sensitive potassium channels' activity is more difficult to observe in the brain in comparison to the periphery, and the coexistence of receptors for sulfonylureas and K-ATP channels has not been demonstrated in most brain structures. Nevertheless there is some evidence to suggest the existence of such channels in the central nervous system. Patch-clamp experiments have shown ATP sensitivity of K^+ channels in cultured neurons from neonatal brain. In the ventromedial hypothalamus, which is implicated in the regulation of food intake, neurons are depolarized by glucose, presumably via a K-ATP channel mechanism. Moreover, it has been reported that an increase in extracellular glucose concentration or application of sulfonylureas facilitates the release of GABA from nerve terminals in the substantia nigra, an effect that is inhibited by PCOs. ATP-sensitive potassium channels have also been studied in the hippocampus, a region that is particularly vulnerable to anoxia and ischemia. A high concentration of binding sites have been found in the CA3 region, where they are located presynaptically on mossy fiber terminals. Electrophysiological data indicate that agents acting on K-ATP channels may modulate effects of anoxia: in intracellular recordings in hippocampal slices, a brief anoxic episode produces a biphasic response including an initial hyperpolarization, followed by a depolarization. The initial hyperpolarization is due to the opening of K^+ channels, but not of the

K-ATP channels type. In contrast, the depolarization is abolished by relatively high concentrations of diazoxide or by the hormone galanin, while glibenclamide produces the opposite effect, suggesting an implication of K-ATP channels (Fig. 2). It is known that anoxia induces an important release of glutamate from nerve terminals. It has been suggested that the fall in intracellular ATP, consecutive to the anoxia, produces an activation of K-ATP channels located on presynaptic terminals, which will reduce excitotoxic glutamate release and thus protect the cell.

Although the evidence for K-ATP channels in the brain is circumstantial and requires more pharmacological data, several lines of evidence do suggest that K-ATP channels play an important role in the central nervous system. These interesting observations suggest that PCOs might be considered in the future as useful tools for the treatment of cerebral ischemia.

Further reading

Ashcroft FM (1988): Adenosine 5'-triphosphate-sensitive potassium channels. *Ann Rev Neurosci* 11:97–11

Ben-Ari Y (1990): Galanin and glibenclamide modulate the anoxic release of glutamate in rat CA3 hippocampal neurons. *Eur J Neurosci* 2:62–68

Cook NS, (ed.) (1990): *Potassium Channels, Structure, Classification, Function and Therapeutic Potential.* Chichester: Ellis Horwood Series in Pharmaceutical Technology

Noma A (1983): ATP-regulated K^+ channels in cardiac cells. *Nature* 305:147–148

G-Proteins and Neuronal Signal Transduction

Mark R. Brann

The primary function of neurotransmitter receptors is to detect (by binding) the presence of transmitter released from neuronal terminals, and to transduce this information into changes in cellular physiology. Most receptors use one of two mechanisms to mediate signal transduction. First, many receptors have ion channels within their protein structures, and binding of transmitter directly influences the flow of ions across the plasma membrane. Because of this direct linkage of transmitter binding and ion flow, these receptors usually mediate rapid physiological responses. Second, other receptors bind to proteins located on the cytoplasmic face of the membrane, which in turn influence both ion channels and enzymes (generating second messengers). Recent studies have revealed that these signal transduction proteins require the binding of guanine nucleotides for their activation (by neurotransmitter-bound receptor); thus the term "G-protein." Because of the involvement of these intermediate molecules, these receptors influence slower physiological responses.

Many, if not all, neurotransmitters are able to influence cells by binding to both types of receptors. In addition to functional differences, these receptor subtypes differ pharmacologically. For example, acetylcholine receptors can be divided into nicotinic (acetylcholine is mimicked by nicotine, ion channel–containing receptor protein) and muscarinic (acetylcholine is mimicked by muscarine, G-protein involved) subtypes. Other examples include ion channel–containing and G-protein–linked forms of γ-aminobutyric acid, serotonin, and glutamate receptors. Molecular approaches have revealed that the two prototypical types of receptors have different structural motifs. G-protein–coupled receptors consist of a single protein subunit that crosses the plasma membrane seven times, while ion channel–containing receptors consist of multiple protein subunits, each of which crosses the membrane several times. Consistent with this division, all G-protein–coupled receptors have significant sequence homology with respect to each other, as do all ion channel–containing receptors. On the other hand, G-protein–coupled and ion channel–linked receptors have little or no sequence homology with members of the other group, even in cases where they bind the same neurotransmitter (e.g., nicotinic and muscarinic acetylcholine receptors).

Even for the G-protein–coupled forms of a given neurotransmitter a heterogeneity of receptor subtypes exists.

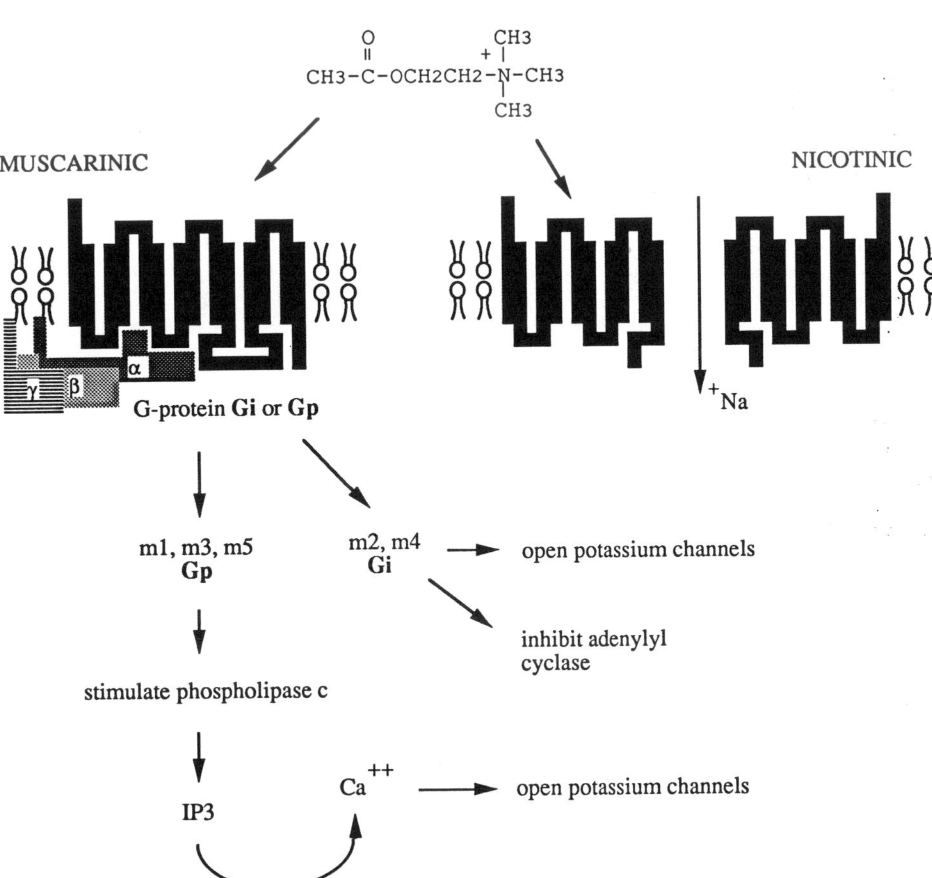

Figure 1. Signal transduction mechanisms for acetylcholine receptors. Acetylcholine binds to one of two types of receptor. Muscarinic receptors consist of a single protein subunit that crosses the membrane seven times. Signal transduction by muscarinic receptors is mediated by G-proteins. Five subtypes of muscarinic receptors (m1–m5) have been identified: m1, m3, and m5 receptors selectively couple to the G-protein Gp, and the m2 and m4 receptors selectively couple to Gi. Nicotinic receptors consist of multiple protein subunits, each of which crosses the membrane several times. Binding of acetylcholine leads to opening sodium channels that are intrinsic to the receptor proteins.

Perhaps the best examples are subtypes of adrenergic and muscarinic receptors. Although these receptors all couple to G-proteins, they have been divided into different subtypes based on both functional and pharmacological differences (adrenergic α1, α2, β1, and β2 receptors and M1–M3 muscarinic receptors). The functional differences of these receptor subtypes are due to their intrinsic selectivities for distinct G-proteins. For example, β1 and β2 receptors stimulate adenylyl cyclase by selectively coupling to Gs, G-protein; α2 receptors and certain muscarinic receptors inhibit adenylyl cyclase by selectively coupling to Gi; and α1 and other muscarinic receptors stimulate phospholipase C by coupling to Gp.

In addition to the classical neurotransmitters, a diversity of molecules modulate neuronal signal transduction via G-proteins. Examples include the initial steps in olfaction and vision. Olfaction is started by an oderant binding to receptors located in the olfactory mucosa; these receptors in turn activate a specific G-protein called Golf (G olfaction). Activated Golf stimulates adenylyl cyclase. Similarly, when photons of light enter the retina, they activate receptors that stimulate a specific G-protein called transducin. Transducin stimulates the enzyme cGMP-dependent phosphodiesterase (increases cGMP levels).

In addition to regulating the activity of those enzymes controlling second messenger systems, G-proteins also modulate ion channels. This can be achieved in two ways: (1) Receptor-activated G-proteins can indirectly alter ion channel activity by virtue of changes in second messengers. For example, muscarinic receptors/Gp are able to open potassium channels by phospholipase C stimulation. That is, stimulation of phospholipase C increases inositol trisphosphate levels, which, in turn, stimulates release of calcium from cytosolic stores. Calcium directly binds to and opens certain potassium channels. In light-sensitive cells, photoreceptor/transducin–elevated cGMP directly opens sodium channels. Similarly in olfactory cells, olfactory receptor/Golf–elevated cAMP opens sodium channels. A variation on this general mechanism is the ability of second messengers to influence the posttranslational modification of ion channels and thus indirectly influence their activity. For example, β receptor/Gs–induced cAMP stimulates certain protein kinases that phosphorylate calcium channels. Phosphorylation of calcium channels increases their stability. (2) Another mechanism involves direct coupling of G-proteins to ion channels.

In addition to inhibiting adenylyl cyclase, muscarinic receptor/Gi directly inhibits potassium channels. Similarly, β receptors/Gs directly stimulate both adenylyl cyclase and calcium channels.

The structures of the signal-transducing G-proteins have been extensively examined using protein chemistry and molecular genetic techniques. All these G-proteins consist of α, β, and γ subunits. The α subunits bind the guanine nucleotides and are thought to confer the selectivity of coupling to distinct receptors and functional responses. The most likely model of G-protein activation suggests that binding of a neurotransmitter to its receptor catalyzes the exchange of GDP for GTP. The GTP-bound α subunit then dissociates from βγ and interacts with its effector (ion channel, phospholipase C, adenylyl cyclase, etc.). The signal is turned off by the hydrolysis of bound GTP to GDP (the α subunit is a GTPase) and the reassociation of α with βγ. A complex between α, β, and γ is required for association of G-protein with receptor. The β and γ subunits do not dissociate under physiological conditions.

Each of the functionally distinct G-proteins have different α subunits. For Gs, the α subunits have two sized species (MW 45 and 52 kDa) which are encoded by alternatively spliced versions of the same gene. These two forms of Gsα are functionally indistinguishable. For Giα, a heterogeneity of related proteins exist with sizes of 40–41 kDa, and three closely related genes have been cloned and designated Gi1–Gi3. The physiological significance of this structural heterogeneity is unknown. For example, it is not known whether the Gi proteins that inhibit adenylyl cyclase and stimulate potassium channels are genetically identical. Transducin α is closely related to the Giα proteins, and Golfα is closely related to Gsα. Gpα has yet to be cloned, or characterized as a protein in great detail; its size is ~40 kDa. Goα is an abundant G-protein of 39 kDa that is selectively expressed by neuronal tissue. Its function remains a mystery; some experiments have suggested an association with calcium channels. Several β and γ subunits have now been isolated and cloned. Their heterogeneity is unexpected, as βγ subunits were thought to be functionally interchangeable.

G-proteins can be distinguished by their sensitivities to the bacterial toxins that are pathogenic in whooping cough and cholera. Gs and Golf are stimulated by cholera toxin, and are insensitive to pertussis toxin. Go and Gi are not sensitive to cholera toxin but are inactivated by pertussis toxin. Gp is

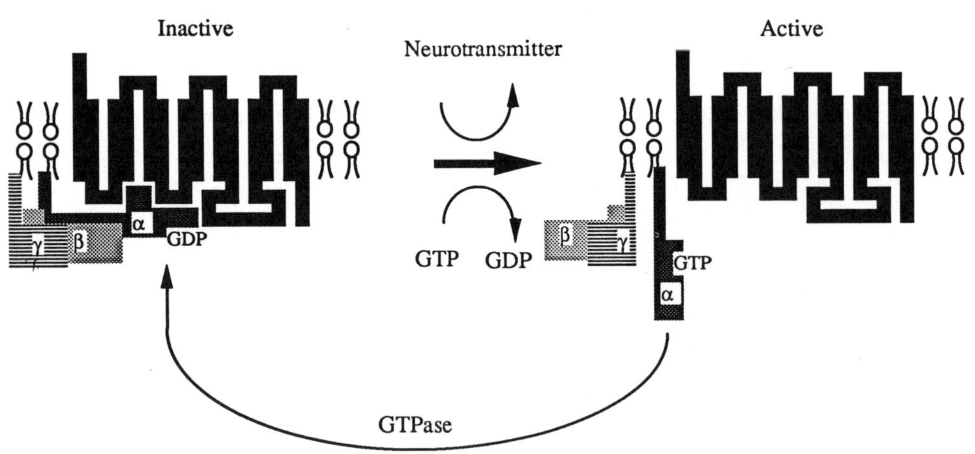

Figure 2. Mechanism of G-protein activation. Neurotransmitter binding catalyzes the exchange of GDP for GTP on the G-protein α subunit. Active α subunit dissociates from receptor and G-protein βγ subunits. GTPase activity intrinsic to the α subunit deactivates the G-protein, and allows reassociation first with βγ and then the receptor.

insensitive to cholera toxin, and both pertussis toxin–sensitive and –insensitive forms have been identified. Transducin is sensitive to both cholera and pertussis toxins.

The present data base clearly indicates the prominent role that G-proteins play in neuronal signal transduction. A number of questions remain to be answered by future experiments: What is the physiological significance of the G-proteins that have recently been discovered using genetic techniques? Are specific G-proteins involved in other neurotransmitter receptor functions (e.g., control of neurotransmitter release, neuronal differentiation)? Do different receptors share the same population of G-proteins, that is, to what extent are G-proteins promiscuous? And do G-proteins play a major role in controlling of the sensitivity of cells to neurotransmitters?

Further reading

Brann MR ed (1991): *Molecular Biology of Receptors Coupled to G-Proteins*. Boston: Birkhäuser

Gilman A (1987): G proteins: transducers of receptor-generated signals. *Ann Rev Biochem* 56:615–649

Nathanson NM, Harden TK, eds (1990): *G-Proteins and Signal Transduction*. New York: The Rockefeller University Press

H

Hair Cells, Sensory Transduction

David P. Corey

The organs of the vertebrate inner ear respond to a variety of mechanical stimuli: *semicircular canals* are sensitive to angular velocity, the *saccule* and *utricle* respond to linear acceleration including gravity, and the *cochlea* is sensitive to airborne vibration, or sound. The ontogenically related *lateral line* organs, spaced along the side of aquatic vertebrates, sense fluid flow. All these organs have a common receptor cell type, which is named the *hair cell* for the bundle of enlarged microvilli protruding from its apical surface (Fig. 1A). In different organs, specialized accessory structures serve to collect and filter these different physical stimuli and to couple them to the hair bundles. Hair cells respond only to the deflection of their bundles.

Structure of sensory epithelia

In each of these organs, hair cells occur in a specialized plaque or strip of sensory epithelium, which generally contains just hair cells and supporting cells. Because the organs are formed during development by the infolding of an otic cyst—derived from surface ectoderm—the apical surfaces of hair cells and supporting cells, as well as cells of the surrounding epithelium, face a closed compartment. Specialized epithelial cells supply this chamber with a unique fluid, termed *endolymph,* which has a high potassium concentration and low sodium and calcium concentrations. Tight junctions ringing the apical surfaces of hair cells and supporting cells separate the endolymph from *perilymph,* a fluid that bathes the basolateral surfaces of hair cells and that has ionic concentrations much like normal saline. In the mammalian cochlea, proton pumps in epithelial cells of the *stria vascularis* raise the potential of endolymph to about $+80$ mV; elsewhere, however, the endolymph potential is within a few millivolts of zero.

Structure of the hair bundle

The mechanosensitive hair bundle comprises 30–300 stereocilia, which have an internal structure similar to microvilli: they are bundles of several hundred actin filaments, tightly cross-linked by fimbrin and bounded by the cell membrane. While ranging in length from about 1 μm (in mammalian cochlea) to over 60 μm (in semicircular canals), stereocilia have such stiff actin cores that they do not normally bend. The cores taper at their bases to a few actin filaments, which anchor the stereocilia into a *cuticular plate* just under the apical surface, and which allow the stereocilia to pivot at the insertion point. Stereocilia adhere to each other, perhaps by means of lateral links that connect their upper ends, so that when the bundles are deflected the stereocilia do not separate, but slide past one another.

The bundle also contains a single, eccentrically placed kinocilium, which has the "9+2" arrangement of microtubules characteristic of motile cilia. The kinocilium may serve to organize the bundle during development (it is lost in mature hair bundles of mammalian cochlea), and it couples the stereocilia to accessory structures in most other organs, but it is not itself mechanically sensitive.

Axis of polarity

The staircaselike variation in heights of stereocilia (Fig. 1A) and the eccentric placement of the kinocilium at one edge of the bundle confer a morphological axis of polarity to the bundle. The positive direction is defined as toward the kinocilium or long stereocilia; negative is toward the short stereocilia. This also describes the physiological axis of the bundle: a positive deflection of the stereocilia is an excitatory stimulus. It increases the membrane conductance, allowing cations to flow into the cell and thereby depolarizing the cell. By opening voltage-sensitive calcium channels in the basolateral surface of the hair cell, the depolarization increases the release of an excitatory neurotransmitter, and increases the firing in the postsynaptic fibers of the eighth nerve. The hair cell itself does not generate action potentials; rather, calcium channels and synaptic release are active at the resting potential of the cell, and receptor potentials of a few millivolts modulate the steady release of transmitter.

The sensitivity of the hair bundle to mechanical stimuli is remarkable: a deflection of a few tenths of a micrometer is sufficient to activate the conductance completely (Fig. 2A), and the deflections at auditory threshold are estimated to be a few tenths of a nanometer. Side-to-side deflections of the bundle, even several micrometers, cause no conductance change, so that the cell has a vector sensitivity and senses the component of stimulus along the bundle axis.

Mechanism of transduction

The last 10–12 years have provided a good understanding of the mechanoelectrical transduction process, whereby deflections of the stereocilia cause the opening of ion channels. The process is extremely fast: at mammalian temperatures, the channels begin to open within about 10 μs after deflection of the bundle. This can explain a sensitivity for frequencies as high as 200 kHz in some animals, but the speed also appears to rule out transduction mechanisms involving an enzyme cascade or second messenger molecule. It was proposed, instead, that deflections exert a force directly on the transduction channels to open them, much as transmembrane voltage exerts a force directly on charged domains of voltage-gated channels. Recent experiments have succeeded in measuring a tiny mechanical deflection of the hair bundle correlated with the opening of transduction channels, which strongly supports the idea of directly gated channels. These experiments also indicated that the movement of each channel upon opening is 2–4 nm, which is large compared with

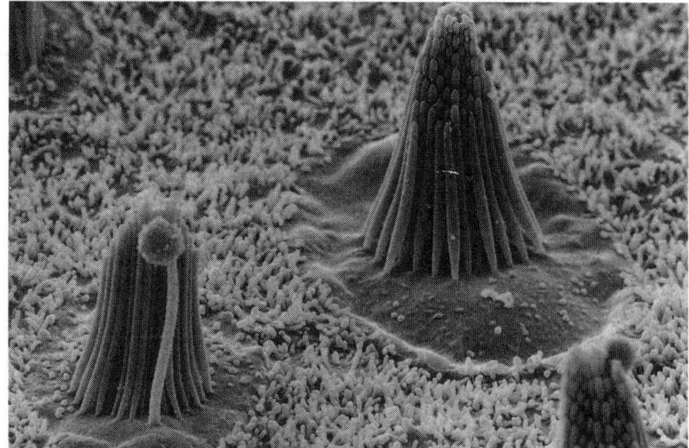

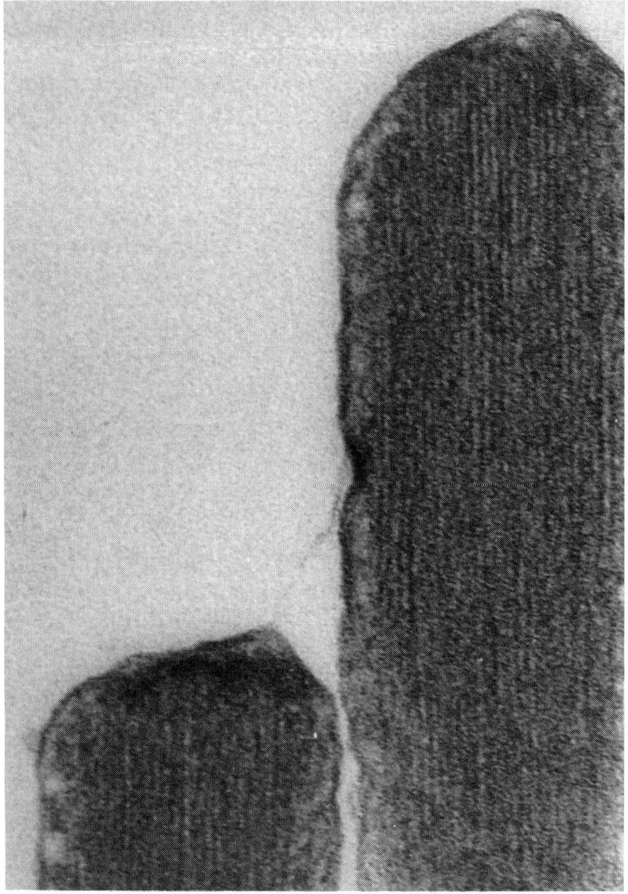

voltage-gated channels but which would provide greater sensitivity to mechanical stimuli. Transduction channels thus represent a third class of ion channel, in addition to those gated by voltage and ligand binding. Other mechanically gated channels have since been found in a wide variety of cell types.

Hair-cell transduction channels are not particularly selective among cations, passing all the alkali cations, many divalent cations including calcium, and small organic cations up to about 0.7 nm diameter. The high potassium concentration in endolymph means that potassium carries most of the receptor current *in vivo*. Several different estimates put the number of channels at 50–100 per cell in lower vertebrates, or 1–2 per stereocilium. While few in number, the channels have a single-channel conductance of 100 picosiemens, sufficiently large to cause receptor potentials of 15 mV or more.

The structural basis of transduction has also been elucidated in recent years. First, the channels were found to be located at or near the tips of the stereocilia, a surprising result that has been confirmed by several independent methods. Second, fine, filamentous "tip links" were discovered extending between the tip of each stereocilium and the tallest adjacent stereocilium. These occur only along the morphological axis of the cell (Fig. 1B), and not side to side. Tip links are extracellular structures but connect osmophilic densities just under the cell membrane at each end (Fig. 1C). Third, the mechanical sensitivity of hair cells is abruptly abolished by a treatment (removing extracellular calcium) that disrupts the tip links. A beautifully simple model for transduction that explains all existing observations is illustrated in Figure 2B. Positive deflection of the bundle causes the tips of stereocilia to shear relative to one another, increasing the tension on tip links. Tip links pull directly on transduction channels, located in the membrane at one or both ends of the link, causing them to open. Conversely, a negative deflection relaxes tension in the tip links, allowing channels normally open at rest to close. Special features of the transduction, such as the remarkable speed and the vector sensitivity of the response, are nicely explained by this model.

Process of adaptation

The very high sensitivity of transduction, as indicated by a range of a few hundred nanometers (Fig. 2A), poses certain problems for the cell. Deflection of 100 nm at the tip of the bundle stretches tip links by just 10 nm, so that processes during the development of the bundle must align transduction structures with this accuracy. Also, relatively modest physical stimuli can exceed the sensitive range and saturate the response, rendering the cell unresponsive to small addi-

Figure 1. Hair cells from the bullfrog inner ear. (A) The hair bundle on the right illustrates the graded heights of stereocilia; the bundle on the left shows the single kinocilium adjacent to the tallest stereocilia, which terminates in a bulb in some organs. Between the hair cells are supporting cells with normal microvilli. (B) At higher magnification, tip links are observed extending between adjacent stereocilia along the morphological axis. (C) A transmission electron micrograph of a pair of stereocilia shows the tip link between them, and its connection to osmophilic densities that lie between the membrane and the actin cores. (Panel B reprinted with permission of Cell Press from Assad JA et al (1991): Tip-link integrity and mechanical transduction in vertebrate hair cells. *Neuron* 7:cover.)

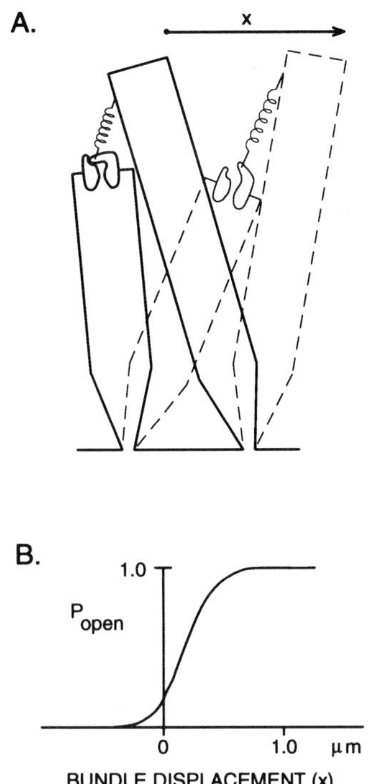

Figure 2. A model for transduction by hair cells. (A) Deflection of the hair bundle toward the tallest stereocilia stretches tip links to pull open the transduction channels (model not to scale). (B) The relationship between deflection (x) and channel opening (Po) is very narrow: the sensitive range of 0.4 µm is about the width of one stereocilium. Reprinted with permission of Elsevier from Pickles JO, Corey DP (1992): Mechanoelectrical transduction by hair cells. *Trends Neurosci* 15:254–259

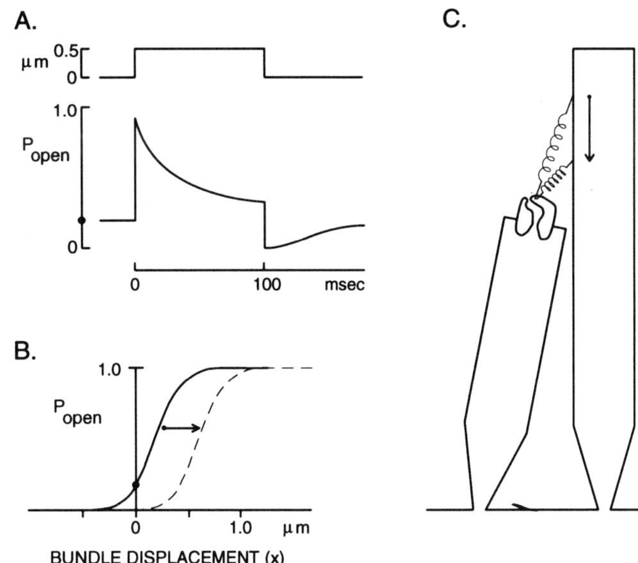

Figure 3. A model for adaptation by hair cells. (A) A maintained deflection of 0.5 µm opens most channels, but they then spontaneously close. (B) Concurrent with the closure is a shift of the sensitive range, suggesting relaxation of tension on the channels. (C) A structural model for adaptation supposes that the upper tip-link attachment points can slip under tension to relax each tip link. Reprinted with permission of Elsevier from Pickles JO, Corey DP (1992): Mechanoelectrical transduction by hair cells. *Trends Neurosci* 15:254–259

tional stimuli. A solution for both problems is an adaptation mechanism that serves to reset the sensitive range of the cell. As shown in Figure 3, a steady deflection of the bundle causes an initial activation of transduction channels, followed—in tens of milliseconds—by a decline in the response toward the resting level. The sensitivity curve measured after 100 ms or so is shifted to the right, relative to its position at rest (Fig. 3B). This suggests that the tension on transduction channels was relaxed during the deflection, since an additional deflection is required to reactivate the channels. Similarly, a negative deflection that closes channels is followed by an adaptation that reopens them, and that shifts the sensitivity curve to the left. This process extends the sensitive range for slower deflections, while retaining a high sensitivity for more rapid stimuli.

The notion that adaptation is a tension adjustment is only consistent with the tip-links model if one imagines that the tip-link attachment point can move. Such a model has been proposed, in which the upper attachment point can climb along the side of the stereocilium (Fig. 3C). In climbing, it exerts tension on tip links, to open some channels at rest. When tension is relaxed by negative deflections, the attachment climbs further to restore tension: when tension is increased the attachment slips to relax the tip link. A number of observations support this model: First, a mechanical relaxation of hair bundles has been observed following deflections, which has the same time course as adaptation. Second, processes that increase or decrease the rate of adaptation—and so change the resting tension on channels—

also cause a freestanding hair bundle to move. A quantitative model for an attachment-point motor can account for the adaptation and the movement. Third, physiological experiments indicate that the adaptation mechanism is within 1–2 µm of the transduction channels, placing it within the tips of the stereocilia. Confirmation of the model awaits electron microscopy of deflected bundles, to see if the attachment-point densities occur higher or lower than normal for negative or positive deflections.

The motor molecule is not yet identified, although circumstantial evidence suggests a myosinlike molecule. The actin in the cores of stereocilia is polarized so that myosin would climb up, and members of the myosin family are the only molecules known to move on actin. The climbing phase of adaptation occurs at 1–2 µm/s, similar to myosin in other systems. Lastly, beads coated with muscle myosin climb on the actin cores of stereocilia, at 1–2 µm/s.

It should be noted that adaptation has been most fully characterized in hair cells of bullfrog and turtle. While occurring in other species and organs, it may be slower or less complete. The adaptation mechanism would nevertheless seem to be important in the development of bundles in all species.

Synaptic connections

The afferent synapse from hair cell to eighth nerve fiber is different from most neuronal synapses, in that neurotransmitter is continuously released and the rate of release is modulated by the receptor potential of the cell. A dense presynaptic ball at each release site, around which vesicles cluster, is reminiscent of the synaptic ribbon in photoreceptors and may serve a similar role in mediating continuous release.

The excitatory afferent transmitter is not known but is probably an amino acid.

The efferent synapse, carrying feedback control from the CNS, uses acetylcholine as the transmitter. The synapse may be excitatory in some hair cells, but in most organs efferent stimulation causes a slow (inhibitory) hyperpolarization. Recent work indicates that the hyperpolarization is a consequence of the activation of potassium channels by intracellular calcium. One possibility is that the acetylcholine receptor in hair cells is a muscarinic type that raises calcium by phosphoinositide metabolism; another is that calcium entry through a brain-type nicotinic receptor causes release of calcium from intracellular stores. A submembranous cisterna that occurs within the hair cell at each efferent synapse may represent the calcium store.

Further reading

Assad JA, Shepherd GMG, Corey DP (1991): Tip-link integrity and mechanical transduction in vertebrate hair cells. *Neuron* 7:985–994

Howard J, Hudspeth AJ (1987): Mechanical relaxation of the hair bundle mediates adaptation in mechanoelectrical transduction by the bullfrog's saccular hair cell. *Proc Natl Acad Sci* (USA) 84:3064–3068

Howard J, Roberts WM, Hudspeth AJ (1988): Mechanoelectrical transduction by hair cells. *Ann Rev Biophys Biophys Chem* 17:99–124

Pickles JO, Corey DP (1992): Mechanoelectrical transduction by hair cells. *Trends Neurosci* 15:254–259

Hepatic Encephalopathy

Anthony S. Basile

Hepatic encephalopathy (HE) is a metabolic neuropsychiatric syndrome that accompanies hepatocellular failure. The manifestations of the syndrome include global depression of the central nervous system (CNS), with the gradual deterioration of higher mental and motor functions (Table 1). HE is a significant public health problem in the United States, where it is associated with 20,000 to 30,000 deaths per year due to liver failure. However, this figure is conservative, as HE is usually recurrent and episodic, resulting in far more hospital admissions than this statistic indicates.

HE is a metabolic encephalopathy of multifactorial origin and as such has the potential to be completely reversible. The underlying liver failure may be either acute or chronic. Few anatomical abnormalities are found in the CNS of patients who die of either acute or chronic liver failure. Those changes that are observed (e.g., astrocytosis) do not involve neuronal structures. The most common precipitant of chronic liver failure is alcoholic cirrhosis, although drug overdose or viral hepatitis commonly cause acute liver failure.

The pathogenetic mechanisms of HE are unclear, although the syndrome has been recognized since the beginning of recorded medical history. As a result, the modalities for treating HE are still evolving. HE arises from the accumulation of potentially neurotoxic substances in the systemic circulation (Fig. 1). Normally, these substances are absorbed from the gut and detoxified by the liver. In hepatocellular failure, these agents may avoid catabolism in the liver by passing through the diseased liver unmetabolized into the systemic circulation. Alternatively, these toxins may bypass the liver altogether through portal-venous shunts that may be surgically constructed, or form spontaneously in cirrhosis. Once these gut-derived substances reach the brain, they may modify CNS function either by altering metabolic processes such as oxidative metabolism or neurotransmitter synthesis, or by interacting directly with neurotransmitter receptors.

Over the years, the involvement of many potential neurotoxins has been proposed in the pathogenesis of HE. Ammonia was the first agent recognized as playing a role in the development of HE, based on its accumulation in liver failure and its ability to cross the blood-brain barrier. Ammonia is produced in the gut by the bacterial degradation of amines, amino acids, and urea. The impaired conversion of ammonia and glutamine to urea that occurs in liver failure also increases plasma ammonia levels.

Ammonia is clearly neurotoxic and can alter normal CNS function in several ways. At relatively low concentrations (approximately 1 mM), it inactivates Cl^- extrusion pumps in the neuron, increasing intracellular Cl^- concentrations. This blocks the formation of hyperpolarizing inhibitory postsynaptic potentials, impairing inhibitory processes throughout the brain.

Moderately elevated concentrations of ammonia (2 mM) block the presynaptic conduction of action potentials, decrease the amplitude of excitatory postsynaptic potentials, and can cause cerebral edema. Finally, high ammonia concentrations (>5 mM) decrease oxidative metabolism in the brain, which may result in convulsions. Present therapeutic regimens for the management of HE attempt to reduce ammonia production and absorption from the gut by administering antibiotics and cathartics together with a low-protein diet.

Despite the clear neurotoxicity of ammonia, plasma ammonia levels do not correlate well with the expression of HE or its severity. For example, hyperammonemia does not cause the sleep inversion and the subtle changes in personality and mentation that are significant components of HE. Finally, standard therapeutic regimens for the management of HE require 4 to 8 days to show an effect, during which time patients may recover without treatment. These observations are not surprising given the multifactorial nature of HE and indicate a role for other agents in its pathogenesis.

Mercaptans, phenol, and medium-chain-length fatty acids have also been reported to be present in elevated concentrations in patients with liver disease (Table 2). However, there is no apparent correlation between the concentrations of these substances and the severity of HE (note the mercaptan concentrations, Table 2). Furthermore, the neurotoxic concentrations of these substances are 8–15 times higher than the levels commonly observed in patients with HE. While it has been proposed that the toxicity of these agents is en-

Table 1. Clinical Stages of Hepatic Encephalopathy

Stage	Mental State	Neuromuscular State
I	Mild confusion, euphoria, depression, decreased attention, slowed analytical ability, irritability, sleep inversion	Mild incoordination, impaired handwriting
II	Drowsiness, lethargy, gross deficits in analytical ability, obvious personality changes, inappropriate behavior, intermittent disorientation; EEG abnormalities: high-amplitude, low frequency waves with nonfocal changes	Asterixis, ataxia, dysarthria, paratonia, apraxia
III	Somnolent but rousable, unable to perform analytical tasks, disorientation with respect to time and/ or place, amnesia, rage, slurred speech	Hyperreflexia, muscle rigidity, fasciculations, abnormal Babinski's sign, seizures (rare)
IV_A	Coma	Oculovestibular responses lost Response to painful stimuli lost
IV_B	Deep coma	Decerebrate posture, no response to painful stimuli

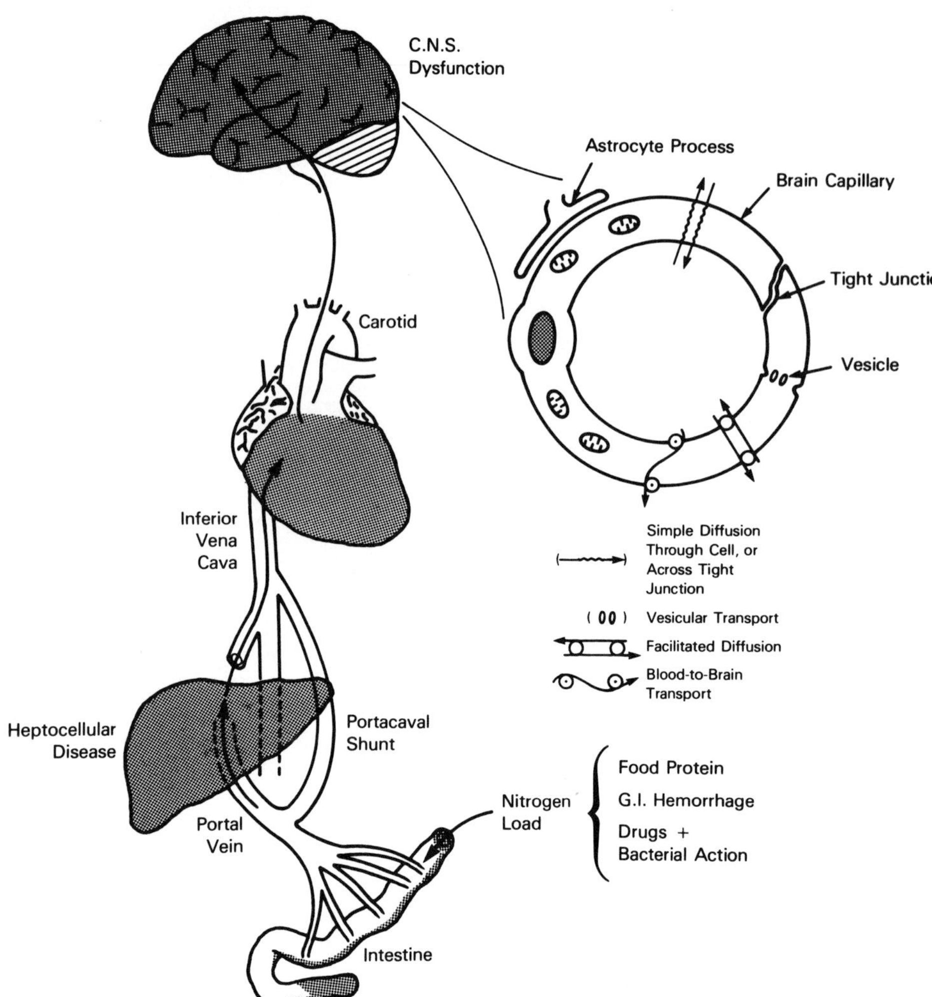

Figure 1. The concept of portal-systemic shunting as it contributes to the development of hepatic encephalopathy. Nitrogenous substances arising from the gut are absorbed into the portal venous system and delivered to the liver where they are extracted and metabolized. However, in liver disease these substances may pass directly into the systemic circulation or bypass the liver completely via portal-venous shunts. Once these compounds enter the systemic circulation, they may gain entry to the CNS by a variety of mechanisms including high-capacity active transport or facilitated diffusion, resulting in abnormal brain function. (Reprinted with permission of the American Society for Pharmacology and Experimental Therapeutics from Basile AS et al (1991): The pathogenesis and treatment of hepatic encephalopathy: Evidence for the involvement of benzodiazepine receptor ligands. *Pharmacol Rev* 43:27–71.)

Table 2. Levels of Mercaptans, Phenol, and Fatty Acids in Hepatic Encephalopathy

Toxin	Control	Liver Failure	Liver Failure with Advanced HE	Comagenic Dose in Rats
Mercaptans				
Methanethiol[a]	5.7 nM/ml	7.7[g]	13[g]	200
Ethanethiol[a]	0.30 nM/ml	0.19	0.35	
Dimethyldisulfide[a]	0.42 nM/ml	0.57	0.62	
Phenol[b]	0.65 μM/ml	1.12	4.20[g]	1.8 mM/kg[c]
Fatty acids				
C_2-C_7[d]	1.11 M/l		1.475	
Octanoic[e]	0.6 μM/l		15	1.6 mM/kg[f]

Values represent mean levels detected in human serum, except for comagenic doses. The comagenic value of methanethiol was measured in rat serum, while the comagenic doses of phenol and octanoic acid represent CD_{50} values in 300 g rats.
[a]Samples obtained from normal subjects and patients with cirrhosis. (From Al Mardini et al., *Gut* 25:284, 1983.)
[b]Samples obtained from normal subjects, patients with cirrhosis (liver failure), and patients with fulminant hepatic failure (Liver Failure with advanced HE). (From Brunner et al., *Artificial Liver Support*, G. Brunner and F.W. Schmidt, eds., p. 25, 1981.)
[c]From Windus-Podehl et al., *J. Lab. Clin. Med.* 101:586, 1983.
[d]Total of all short-chain amino acids present in initial sample of patients with fulminant liver failure and HE. (From Lai et al., *Clin. Chimica. Acta.* 78:305, 1977.)
[e]Samples taken from normal subjects and patients with cirrhosis and HE. (From Rabinowitz et al., *J. Lab. Clin. Med.* 91:223, 1978.)
[f]From Zieve et al., *J. Lab. Clin. Med.* 83:16, 1974.)
[g]$P < .05$.

Table 3. CNS Levels of Amino Acids in Hepatic Encephalopathy

Amino Acid	Control	Liver Failure with HE
Tryptophan[a]	567 ng/ml CSF	2200[c]
Phenylalanine[b]	0.08 μM/g	0.19[d]
Tyrosine[b]	0.21 μM/g	0.36[d]

Values represent mean concentrations per unit of brain or cerebrospinal fluid obtained during autopsy or by lumbar puncture from cirrhotic patients with HE, or from patients with diseases other than liver failure (controls).
[a]From Young et al., *J. Neurol. Neurosurg. Psych.* 38:322, 1975.
[b]From Bergeron et al., *Hepatic Encephalopathy: Pathophysiology and Treatment,* Butterworth, R.F., and Layrargues, G.-P., eds., pp. 389–407, 1988. Humana Press, Clifton, NJ
[c]$P < .001$.
[d]$P < .01$.

hanced in the presence of ammonia, the concept of "synergistic neurotoxicity" clearly applies to the effects of all of the metabolic abnormalities and toxic compounds that arise in liver failure and is not unique to mercaptan, phenol, or fatty-acid toxicity.

CNS levels of aromatic amino acids (e.g., tryptophan, phenylalanine, and tyrosine) are increased between 70 and 300% in patients with cirrhosis (Table 3). These amino acids are precursors to several monoaminergic neurotransmitters, including serotonin. The CNS concentration of serotonin may increase owing to the elevated level of tryptophan observed in liver failure (Table 3). While evidence for increased rates of serotonin metabolism has been observed in patients who succumbed to liver failure, serotonin appears to play a minor role in the development of HE since there is no correlation between the severity of HE and the concentration of serotonin or its metabolites.

Elevated tyrosine and phenylalanine concentrations could also enhance the synthesis of "false" neurotransmitters such as octopamine or tyramine (Table 4). One of the common actions of false neurotransmitters is the depletion of catecholamine neurotransmitters such as dopamine and norepinephrine in the CNS. However, the increase in false neurotransmitters in HE, followed by the depletion of catecholamine neurotransmitters, has not been supported by experimental evidence. Not only are octopamine levels decreased in the brains of some patients with HE (Table 4), noradrenaline and dopamine levels are generally unchanged. Furthermore, the decreases of catecholamine levels reported in animal models of HE (40–50%) are usually not of sufficient magnitude to account for the development of overt behavioral or neurological changes, which requires greater than 90% depletion. Moreover, treatment protocols that either enhance catecholamine neurotransmission (such as L-Dopa or bromocriptine administration) or correct the underlying amino acid imbalances are clinically ineffective in reversing the manifestations of HE.

The syndrome of HE is primarily manifested as a depression of CNS function that may result from decreased excitatory neurotransmission. Thus, significant reductions in glutamate (the primary excitatory neurotransmitter) concentrations might be expected in HE and are observed in various regions of the brain in humans with liver failure (Table 4). However, this decrease may result from CNS accommodation to the hyperammonemic state associated with liver failure. The increased conversion of glutamate plus ammonia to glutamine may account for the overall decline in CNS glutamate concentrations that is primarily associated with metabolic, not neurotransmitter, functions. More recent investigations of specific, neurotransmitter pools of glutamate in animal models of HE indicate that glutamate concentrations in the synapse may increase twofold owing to a decrease in glial reuptake of glutamate. Furthermore, compensatory decreases of 20–40% in the density of NMDA-type glutamate receptors is observed in animal models of liver failure. Although the functional purpose of this enhancement of glutamatergic neurotransmission is presently unclear, it may serve to balance the increase in inhibitory neurotransmission observed in HE.

It is clearly apparent that enhanced GABA-ergic neurotransmission (the primary inhibitory neurotransmitter) plays a significant role in the pathogenesis of HE. No consistent

Table 4. CNS Levels of Neurotransmitters and False Neurotransmitters in Hepatic Encephalopathy

False Neurotransmitter/ Neurotransmitter	Control	Liver Failure	Liver Failure with HE
Octopamine[a]	59 ng/g brain		36
Dopamine[a]	214 ng/g brain	470	158
Noradrenaline[a]	55 ng/g brain	146	83
Serotonin[b]	0.281 ng/mg prot.		1.397
Glutamate[c]	9.54 μM/g brain		7.59[f]
GABA	1.46 μM/g brain[c]		1.21
	0.38 μM/l plasma[d]		3.37[g]
Benzodiazepine receptor ligands[e]	78 ng DZ equiv./g brain		405[h]

Values represent mean concentrations per unit of brain or cerebrospinal fluid obtained during autopsy or by lumbar puncture from cirrhotic patients with or without HE, or from patients with diseases other than liver failure (controls).
DZ equiv.: Diazepam equivalents.
[a]Levels measured in samples of frontal cortex. (From Cuilleret et al., *Gut* 21:565, 1980.)
[b]Levels measured in samples of caudate nucleus. (From Bergeron et al., *Neurochem. Res.* 14:853, 1989.)
[c]Levels measured in samples of prefrontal cortex. (From Lavoie et al., *J. Neurochem.* 49:692, 1987.)
[d]Levels measured in plasma samples from patients with acute liver failure, stage IV HE. (From Levy and Losowsky, *Hepato-gastroenterol.* 36:494, 1989.)
[e]Levels measured in samples of frontal cortex from patients with acute liver failure. (From Basile et al., *New Engl. J. Med.* 325:473, 1991.)
[f]$P < .01$.
[g]$P < .002$.
[h]$P < .05$.

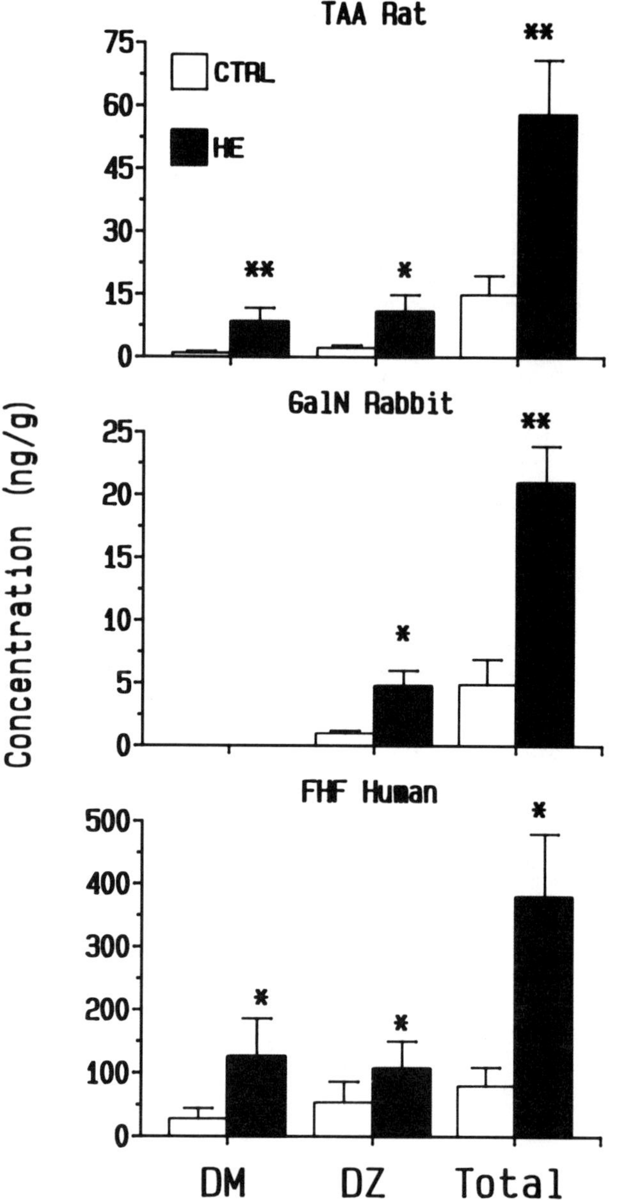

Figure 2. Summary of the levels of N-desmethyldiazepam (DM), diazepam (DZ), and total benzodiazepine receptor ligand activity in the brains of rats (top panel), rabbits (center panel), and humans (bottom panel) with HE due to acute liver failure. The concentrations of diazepam, N-desmethyldiazepam, and total benzodiazepine receptor ligands are significantly increased in all cases of HE compared with controls. However, the levels of these substances are considerably higher in humans with HE than in the animal models. (Reprinted with permission of the American Society for Pharmacology and Experimental Therapeutics from Basile AS et al (1991): The pathogenesis and treatment of hepatic encephalopathy: Evidence for the involvement of benzodiazepine receptor ligands. *Pharmacol Rev* 43:27–71.)

changes in the $GABA_A$ receptor complex or the concentrations of GABA receptor agonists (e.g., GABA, taurine) are observed in animal models of HE or humans with liver failure (Table 4). However, the concentrations of compounds that act as agonists at the benzodiazepine receptor component of the $GABA_A$ receptor complex are increased in several animal models of the syndrome and in humans suffering from HE due to either acute or chronic liver failure. This increase in benzodiazepine receptor agonist concentrations would have the same depressant effect on CNS function as an elevation of GABA levels. Although tissues and body fluids from humans and animals with HE contain numerous benzodiazepine receptor ligands, only two have been chemically identified: the 1,4-benzodiazepines, diazepam and N-desmethyldiazepam. These comprise approximately 20% of the total concentration of benzodiazepine receptor ligands present in HE in both animal models and humans (Fig. 2).

The origin of both the 1,4-benzodiazepines and the other benzodiazepine receptor ligands observed in HE is presently not known. The diazepam and N-desmethyldiazepam present in HE are not of pharmaceutical origin, as they occur in patients who have not received these drugs for at least several months before the occurrence of the encephalopathy, and in animals that were never exposed to these compounds. Alternatively, diazepam and N-desmethyldiazepam are found "naturally" in a variety of foodstuffs and in human tissues preserved before 1,4-benzodiazepines were commercially available. Although these 1,4-benzodiazepines may be synthesized endogenously, it is more probable that they arise from dietary sources, or are synthesized in situ by gut bacteria. Interestingly, present therapeutic protocols aimed at reducing ammonia levels by restricting dietary intake and modifying colonies of enteric bacteria also may reduce the synthesis or uptake of benzodiazepine receptor ligands.

Further evidence implicating benzodiazepine receptor ligands in the pathogenesis of HE is provided by the results of anecdotal trials of benzodiazepine receptor antagonists in the management of HE. Low doses (1–5 mg) of the benzodiazepine receptor antagonist flumazenil are effective in rapidly (0.5–5 min) reversing many of the clinical manifestations of HE in approximately two thirds of the patients tested, regardless of the nature of the underlying liver disease. These preliminary results suggest that benzodiazepine receptor antagonists may be an important adjunct to conventional therapies for HE.

In summary, multiple factors appear to contribute to the pathogenesis of HE. These involve agents that derange CNS oxidative metabolism or neurotransmitter regulation, or interact directly with neurotransmitter receptors. The renewed emphasis on the role of abnormal CNS function in the manifestations of HE suggests that future therapeutic modalities for this disorder will focus upon the brain as well as the gut.

Further reading

Basile AS, Jones EA, Skolnick P (1991): The pathogenesis and treatment of hepatic encephalopathy: Evidence for the involvement of benzodiazepine receptor ligands. *Pharmacol Rev* 43:27–71

Butterworth RF, Pomier-Layrargues G (1989): *Hepatic Encephalopathy: Pathophysiology and Treatment.* Clifton, NJ: Humana Press

Conn HO, Bircher J (1988): *Hepatic Encephalopathy: Management with Lactulose and Related Carbohydrates.* East Lansing, Mich: Medi-Ed Press

Sherlock S (1989): Hepatic encephalopathy. In: *Diseases of the Liver and Biliary System,* Sherlock S, ed. Boston: Blackwell Scientific, pp 91–97

Herpes Simplex Virus and Its Use in Neuroscience Research

Julie K. Andersen and Xandra O. Breakefield

Herpes simplex virus (HSV) is a double-stranded DNA virus that is neurotropic by virtue of the fact that it is selectively transported through the nervous system and can be maintained in a latent state within neurons (Fig. 1). Both type 1 and type 2 HSV have been used as vectors to deliver genes to neurons and as tracers to elucidate neuronal pathways. HSV is well suited for these purposes, as (1) it is readily taken up at nerve terminals; (2) it can infect virtually all cell types in vertebrate animals but spreads preferentially in the nervous system (where the virus is passed by retrograde and anterograde transport, as well as by selective transfer across synapses, thus allowing virus infection to spread through nerve networks); (3) it is able to enter a latent state in some neurons, where it exists as a benign, episomal element within the cell nucleus and, in this state, retains transcriptional activity, allowing stable expression of genes without compromising the cell's neuronal functions; and (4) up to 30 kb of foreign DNA can be replaced at various sites within the virus's large (150 kb), fully sequenced genome without interfering with the ability of the virus to propagate, that is, replicate and package. Disruption of some of the virus's 70 or so genes can reduce or eliminate its ability to replicate within neurons, thus making the virus less neurovirulent.

HSV vectors are the only means to date by which foreign genes have been introduced into postmitotic neurons (Table 1). A major goal in the development of HSV vector technology is to achieve stable delivery of foreign genes into the adult nervous system as a means of modulating nerve cell function, including altering neurotransmission and promoting neuronal survival and regeneration, and of gene replacement therapy for hereditary deficiency states. Two types of HSV vectors have been used for gene transfer: plasmid-derived vectors, termed "amplicons," and recombinant virus vectors. Amplicons consist of a plasmid (10 kb) containing an HSV origin of replication and packaging signal; the plasmid is packaged as tandem repeats in herpes virions with the aid of an HSV helper virus. The advantages of this system are that construction only requires the manipulation of plasmic DNA and multiple copies of the gene of interest are delivered to neurons. However, the amount of foreign DNA that can be introduced into this system is limited by the size of the plasmid vector, and these vectors may not be able to enter a true latent state *in vivo,* although gene expression for up to several weeks has been reported.

Recombinant virus vectors are created by direct alteration of the viral genome. A small portion of the herpes genome cloned into a plasmid is first modified *in vitro,* then introduced back into the virus genome by homologous recombination with infectious herpes virus DNA within cells to create recombinant viruses containing the desired modifications. These vectors can carry multiple large segments of foreign DNA, can enter a true latent state in neurons, and do not need helper virus. Depending on where foreign DNA is placed in the viral genome and what DNA sequences are inserted, successful recombinants can be made that are easily distinguishable from wild-type virus. For example, if foreign DNA is inserted so that it disrupts the HSV thymidine kinase (*tk*) gene, the resulting *tk*-minus virus can be selected on the basis of resistance to nucleoside analogues, such as acyclovir. The bacterial gene *lacZ* can be included in the viral construct as a marker; it encodes an enzyme, beta-galactosidase, that is readily detected by histochemical means and can be distinguished from the mammalian lysosomal enzyme of the same name. *LacZ* has been inserted in vectors downstream of several different viral and mammalian promoters to evaluate their activity *in vivo* during viral replication and latency. The main disadvantage of recombinant virus HSV vectors is that manipulation of a large genome can be difficult from a technical standpoint.

Both helper viruses and recombinant virus vectors contain viral genes that can alter cell metabolism and lead to cell death even without viral replication. Existing HSV mutants can be used either as helper virus or as "backbones" for the

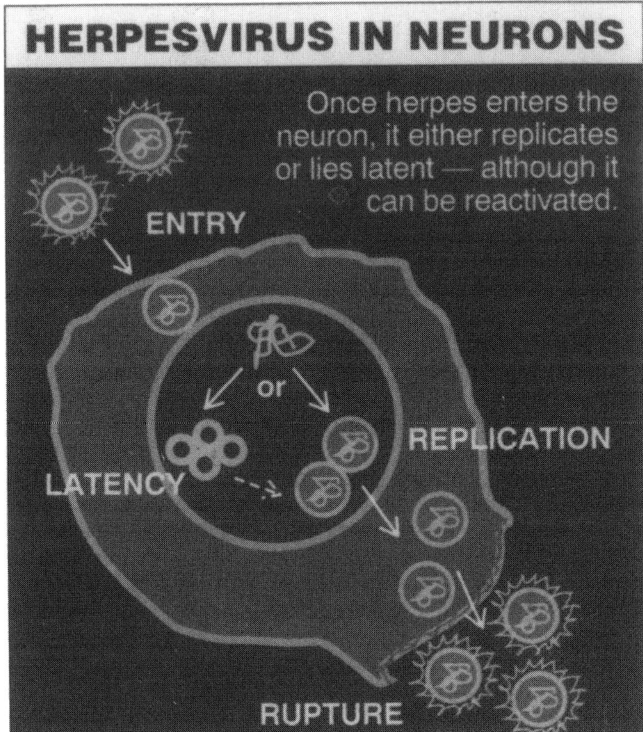

Figure 1. Herpes virus in neurons. Herpes virus enters cells by fusion with the cell membrane. The capsid is transported to the nucleus when the DNA is released. Viral DNA can either begin replication, resulting in a productive infection and death of the neuron, or enter latency as episomal elements. The virus can reactivate from latency at a later time and produce infectious viral particles.

Table 1. HSV as Vectors

Genome can carry multiple inserts, at least 30 kb.
Mutants available that are less pathogenic.
Virus travels by retrograde transport and crosses synapses.
Virus can enter latency in some neurons (episomal, benign state).
Promoters in virus are active in latency and can confer stable gene expression.

creation of recombinant virus vectors. Viruses that have reduced toxicity to neurons as compared with wild-type virus include those that are (1) replication defective or compromised; (2) missing viral genes involved in disruption of cellular macromolecular synthesis; or (3) derived from nonneurovirulent or nonneuroinvasive strains. For example, *tk*-minus HSV is replication-defective in neurons and has been shown to be less toxic than wild-type herpes when injected into the rodent CNS. Viral replication is not required for entry into latency, and replication-defective viruses cannot reactivate from latency to a productive infection.

Herpes-derived virus vectors have been used for delivery of genes both in culture and *in vivo*. In culture, herpes-derived vectors have been used to introduce foreign DNA efficiently into a variety of cell types, including neuronal cells, which are difficult to transfect by traditional means, such as Ca^{2+} phosphate precipitation or electroporation. The ability of herpes virus vectors to deliver and stably express foreign genes in cultured cells depends on the permissiveness of the cell to viral infection and the toxicity of the virus, as well as the activity of the promoter driving foreign gene expression in the context of the herpes vector. Entry is dependent on the availability of viral attachment sites on the cell surface and viral uptake by the cell. Viral replication is governed by the interaction between virus transcription factors and host cellular factors. Toxicity appears to be a function of viral replication and the "shutdown" of host macromolecular synthesis by the virus. Toxicity can be controlled somewhat in culture by using low multiplicity of infection and supplementation with nerve growth factor (NGF).

Amplicon vectors have been used to deliver genes to a variety of neuronal and nonneuronal cells in culture. Transient gene expression has been observed for *lacZ*, nerve growth factor, NGF-receptors, tyrosine hydroxylase, adenylate cyclase, parvalbumin, and protein kinase 2 using either the immediate-early HSV-1 promoter, ICP4, or mammalian promoters like that for neurofilament. Recombinant virus vectors have also been used in culture to express several genes of interest including *lacZ*, hypoxanthine phosphoribosyltransferase, bovine growth hormone, beta-globin, and ovalbumin. Most of these vectors gave transient expression following infection. It is noteworthy that a vector containing the promoter for the adenine phosphoriborsyl transferase gene (a "housekeeping" gene) did not give detectable gene expression during productive infection, and that a vector containing the beta-globin promoter in the context of the herpes vector was active even in human monkey kidney (Vero) cells, in which the endogenous beta-globin promoter is not expressed. These results support the idea that mammalian promoters can behave differently in the context of the herpes genome. A herpes vector containing the promoter for neuron-specific enolase, however, gave neuronal-specific expression of *lacZ*, in various cell lines and primary cultures, and yielded long-term expression in primary dorsal root ganglion neurons.

The most exciting developments in the herpes-derived vector technology are findings *in vivo*, where these vectors appear to mediate both short- and long-term expression of foreign DNA in neurons, with little pathogenicity to animals. Genes delivered *in vivo* include *lacZ*, tyrosine hydroxylase, parvalbumin, adenylate cyclase, protein kinase C, and NGF. In order to optimize stable gene delivery to adult animals using recombinant HSV vectors, work needs to be directed toward (1) defining the optimal promoters and site of gene insertion into the herpes genome to achieve high-level and long-term gene expression; and (2) reducing the toxicity of

the virus. The HSV-1 promoter driving expression of the viral latency-associated transcripts (LATs), viral genes of unknown function expressed in latency, as well as other promoters, appear to be moderately active during latency of the virus and may be used to drive long-term gene expression from a number of positions within the viral genome.

HSV has also been used as an effective method for visualizing transsynaptic connections in the mammalian nervous system. To be effective as a neuronal "tracer," a substance must be (1) transferred selectively across synapses and (2) present at high enough levels in the recipient cell to allow detection. Other compounds used as tracer molecules, such as horseradish peroxidase (HRP), Fast Blue, and wheat germ agglutinin (WGA), while transferring preferentially at synapses, become diluted upon successive transfer between connecting series of neurons. Wild-type HSV, on the other hand, replicates in recipient neurons, resulting in amplification of the tracer signal (Fig. 2). The virus is easily detectable by immunocytochemical techniques using antibodies directed against viral antigens or by monitoring viral DNA

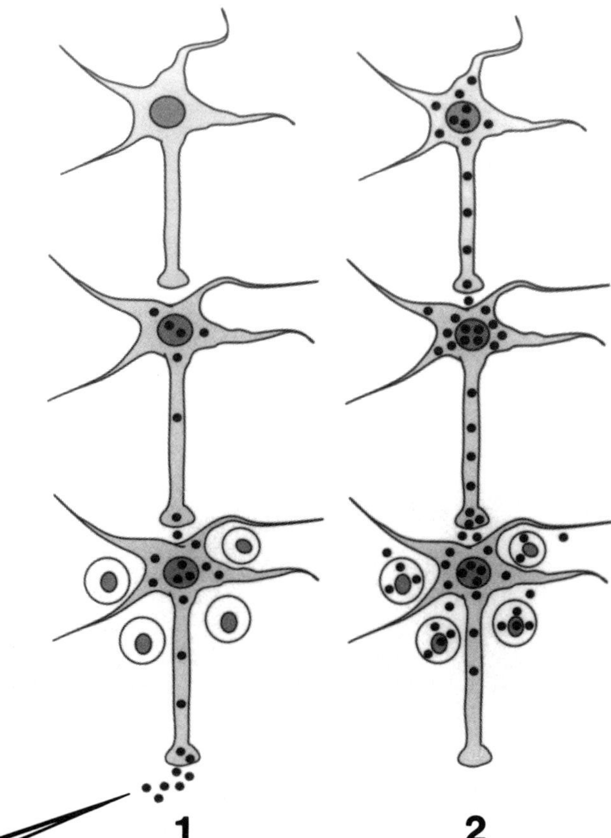

Figure 2. Herpes virus as a transneuronal tracer. (1) Virus inoculated at nerve terminals is taken by rapid retrograde transport to the cell nucleus where viral replication begins. New viral particles are transported initially by anterograde transport to nerve terminals where they cross synapses to connecting neurons. (2) Virus replication proceeds in the secondary neuron and particles are passed to the tertiary neuron in the pathway. Subsequently the primary neuron releases virus to other adjacent cells and dies. (Modified with permission from Kuypers HGJM and Ugolini G (1990): Viruses as transneuronal tracers. *Trends Neurosci* 13:71–75.)

synthesis. Transsynaptic transport of HSV is rapid and can occur over long distances within the nervous system during a period of days. HSV has been shown to give effective labeling of transsynaptic neuronal connections between the periphery and the CNS, for example between trigeminal ganglion sensory neurons and the contralateral thalamus, and hypoglossal motoneurons and the brainstem, as well as between neurons within the brain itself, such as the hypothalamus and the retinal ganglion cell layer.

Further reading

Breakefield XO, DeLuca NA (1991): Herpes simplex virus for gene delivery to neurons. *New Biologist* 3:203–218

Geller AI, During MJ, Neve RL (1991): Molecular analysis of neuronal physiology by gene transfer into neurons with herpes simplex virus vectors. *TINS* II, 428–432

Kuypers HGJM, Ugolini G (1990): Viruses as transneuronal tracers. *Trends Neurosci* 13:71–75

Roizman B, Jenkins FJ (1985): Genetic engineering of novel genomes of large DNA viruses. *Science* 229:1208–1214

Stevens JG (1989): Human herpes viruses: A consideration of the latent state. *Microbiol Rev* 53:318–332

Hyperkalemic Periodic Paralysis

Bertrand Fontaine, James F. Gusella, and Robert H. Brown, Jr.

Periodic paralyses are characterized by acute attacks of muscle weakness accompanied by slight modifications in the blood level of potassium: elevated in the hyperkalemic, and reduced in the hypokalemic form. In each form of the disorder, the cause of muscle paralysis is muscle membrane inexcitability due to depolarization. Periodic paralyses may be either disorders secondary to changes in the blood level of potassium of endocrine or renal origin, or primary genetic affections transmitted as an autosomal dominant trait with a penetrance close to 100%. Primary periodic paralyses are rare disorders: the incidence of the hypokalemic form has been estimated to be 1 in 100,000, the hyperkalemic form being less frequent.

Hyperkalemic periodic paralysis (HYPP), described by Gamstorp in 1956, also known as adynamia episodica hereditaria, typically begins in the first decade with attacks of muscle weakness precipitated by cold, rest after exercise, fasting, or ingestion of potassium. Acute attacks are generally accompanied by myotonia (muscle stiffness following contraction), lasting from 10 minutes to an hour, occurring from daily to weekly, in the day, or early in the evening, and affecting mostly the lower limbs. Rare cases have been associated with cardiac arrhythmia. Recurrent attacks may be accompanied later in life by permanent muscle weakness due to the development of an irreversible vacuolar myopathy.

The diagnosis of HYPP is mostly clinical. Performance of a muscle biopsy in one affected member of the family, or in case of a sporadic case, discloses a vacuolar myopathy and excludes other myotonic syndromes. Provocation can be achieved by the ingestion of chloride potassium but the safety of such a procedure is discussed.

Severe attacks are treated by infusion of glucose (10%) and 20 units of subcutaneously injected insulin, with cardiac monitoring and of the blood level of glucose and potassium. Preventive therapy is most frequently achieved by frequent meals of high carbohydrate content and avoidance of fasting, exposure to cold, or overexertion. If these precautions are not sufficient, diuretic agents such as acetazolamide or thiazides are usually effective in preventing attacks of muscle weakness. They can be recommended on a daily basis over the age of 5.

Advances in the understanding of the pathophysiology of HYPP have recently been achieved. In striated muscle, generation of action potentials along the membrane is due to the sequential opening and closure of voltage-sensitive sodium and potassium channels. In 1987, Lehmann-Horn and Rüdel demonstrated, using muscle fibers from HYPP patients maintained alive *in vitro,* that a tetrodotoxin-sensitive sodium current was triggered in patients with HYPP by a slight increase in extracellular potassium that would normally not be sufficient to open the voltage-gated sodium channels. Different forms of voltage-gated sodium channels are expressed in brain and muscle. In striated muscle, two species of voltage-gated sodium channels are present: the fetal voltage-gated sodium channel, which is insensitive to tetrodotoxin and expressed in immature and denervated muscle, and the adult voltage-gated sodium channel, which is blocked by tetrodotoxin and expressed in mature innervated and denervated muscle. Muscle-specific sodium channels are

composed of two subunits, the α- and β-subunits, but the α-subunit is responsible for most of the voltage-gated sodium channel activity. Indeed, the α-subunit of the adult sodium channel can itself form a voltage-gated sodium channel that can be blocked by tetrodotoxin in vitro. Ionic channels are highly conserved throughout different species. Therefore, using the rat sequence of the adult form of the α-subunit of the muscle sodium channel to design primers, parts of the human counterpart were isolated by the polymerase chain reaction from complementary DNAs prepared from human muscle. One probe (HNa2) defined a restriction fragment length polymorphism in humans and was therefore used for genetic linkage analysis in one big family displaying HYPP. A tight linkage with no recombinants was found between HYPP and the locus defined by HNa2 (SCN4A). The locus SCN4A was subsequently found tightly linked with no recombinants to the locus of growth hormone (GH1), localized on the long arm of chromosome 17 (17q22–24). Accordingly, the locus GH1 was also linked to the locus HYPP with no recombinants. The most tempting interpretation of these results is that the genetic defect of HYPP resides in the gene coding for the α-subunit of the adult muscle-specific sodium channel.

Definite proofs of the implication of a defect in the muscle sodium channel gene in HYPP will require definition of the molecular lesions. Transgenic animals bearing the abnormal sodium gene should help to understand the pathophysiology of HYPP and the basis of the development of a debilitating myopathy in some patients. Allelic mutations of the sodium gene have already been excluded as the causative defect in the hypokalemic form of primary periodic paralysis. It may also be possible to classify both the periodic paralyses and other myotonic syndromes such as paramyotonia congenita or Schwartz-Jampell syndrome according to whether or not they arise from a primary sodium channel gene defect and, ultimately, by specific categories of allelic mutations in the sodium channel gene.

Note: Since the writing of this chapter, mutations in the sodium channel gene were demonstrated. (See Barchi, 1992.)

Further reading

Barchi RL (1992): Sodium channel gene defecs in the periodic paralyses. *Current Opinion in Neurobiology* 2:631–637

Fontaine B, Khurana TS, Hoffman EP, Bruns GAP, Haines JL, Trofatter JA, Hanson MP, Rich J, McFarlane H, McKenna-Yasek D, Romano D, Gusella JF, Brown RH Jr (1990): Hyperkalemic periodic paralysis and the adult muscle sodium channel α-subunit gene. *Science* 250:1000–1002

Gamstorp I (1956): Adynamia episodica hereditaria. *Acta Paediatr* 45 (suppl.) 108:1–126

Lehmann-Horn F, Küther G, Ricker K, Grafe P, Ballanyi K, Rüdel R (1987): Adynamia episodica hereditaria with myotonia: A non-inactivating sodium current and the effect of extra-cellular pH. *Muscle and Nerve* 10:363–374

Trimmer JS, Cooperman SS, Tomiko SA, Zhou J, Crean SM, Boyle MB, Kallen RG, Sheng Z, Barchi RL, Sigworth FJ, Goodman RH, Agnew WS, Mandel G (1989): Primary structure and expression of a mammalian skeletal muscle sodium channel. *Neuron* 3:33–49

Induced Rhythms, Oscillations, in the Brain

Theodore H. Bullock

Induced rhythms in the brain are special and unusual though widespread phenomena, requiring particular conditions in each situation where they have been found. This class of phenomena has been defined (Başar and Bullock 1992) as oscillations caused or modulated by stimuli that do not directly drive successive cycles. Hence they are a class of endogenous rhythms distinguished from purely spontaneous and from entrained (driven) rhythms. The class overlaps with evoked potentials since some induced rhythms are evoked by discrete stimuli and may last a few cycles, whereas others are maintained modulations of ongoing rhythms induced by a slow change of state.

"Induced rhythms" as a term, and as a category of brain oscillations, has hardly been recognized until recently, although many examples have been known for more than 50 years. New attention has been attracted to them by the discovery of several interesting cases. One, for instance, occurs in the visual cortex when a visible stripe, of the preferred orientation for that electrode locus, moves at right angles to its axis (Gray, Singer et al., and Eckhorn et al. in Başar and Bullock 1992). This was long overlooked but includes both spike bursts and slow waves, usually between 35 and 60 Hz, lasting for a second or so. Neighboring columns may show no rhythm if their preferred stripe orientation is different; distant columns may show similar but independent rhythms if stimulated by stripes moving out of phase. If the stimuli are stripes moving in phase, the rhythms in columns of the same orientation tuning even several millimeters apart are well correlated.

Induced rhythms have been recorded in many places, preparations, and conditions, from invertebrates and vertebrates, from peripheral and central neurons, from spinal to cortical levels. They apparently do not comprise a homogeneous class but a diverse set of responses sharing only the defining feature. The greater number and variety of reports from higher cortical levels of mammals may or may not be significant; it may be a sampling artifact or it may reflect a greater tendency to induced rhythms in more advanced neural tissue.

Diversity also seems to apply to the mechanisms, although these are principally unknown in most cases. Isolated axons, receptors, or neuron somata possess a degree of iteration in response to membrane depolarization, characteristic for that type of unit and level of depolarization, from a few cycles per second or less up to over 100 Hz. Some receptors oscillate after a pulse or step stimulus at a characteristic frequency of many hundreds of hertz. Some rhythms are triggered by an impinging event or state, perhaps a particular transmitter. Some "ring" for a number of cycles and cease as though damped. Some appear to be circuit rhythms, dependent on the connectivity and the circuit time constants.

A major variable is the degree of cycle-by-cycle cross-correlation between the oscillations in separate populations of cells. In cat visual cortex this can be high when the two populations have similar preferred stimuli (moving stripes at the same orientation) and the respective stimuli are stripe segments moving in phase to appear as one object. The correlation is low when the two striped segments are moving out of phase, like two objects, suggesting a role of the induced rhythm in "binding" images together that share signs of belonging to one object. Such a suggestion, however, deserves some search for counter examples as well as additional examples. A similar wide range of values of coherence (a measure like cross-correlation but computed for each frequency) is known for the ongoing micro-EEG. Values typically decline rapidly with distance between recording electrodes; in a few millimeters of cortex, tangentially, but in fractions of a millimeter, radially. In these cases a role for the coherence is more difficult to demonstrate.

Induced rhythms are sometimes well time-locked to the inducing stimulus and hence appear large in averaged repetitions of the stimulation; sometimes, however, successive responses are not well time-locked and disappear in averages. For this reason, one suspects that many event-related responses actually include an induced rhythm as a component, undetected because it is too small to see on single trials. Others may have been missed, as was true for the visual cortex for decades, because attention was concentrated on spikes, whose burstiness may not be obvious in single units that misfire in many cycles.

Induced rhythms may be more or less regular, and vary from less than 10 Hz up to at least 80 Hz. Some spontaneous oscillations of 1 kHz and more, in electric fish central pacemakers for electric organ discharges, are characteristically modulated by ethologically significant stimuli, making these examples of induced rhythms.

Proposals that such brain rhythms play some role in information processing go back at least to Bremer and Titeca in 1940, Gerard in 1941, and Bullock in 1945 but became more explicit with Freeman (1975). Some have associated gamma band (ca. 40 Hz) rhythms with focused arousal and cognitive performance. One experiment (see Başar and Bullock, 1992) has led to the proposal that the gamma rhythm contributes to the "binding" of parts of the pattern of excitation on the retina into the subjective experience of a single object. Such correlations are suggestive but do not show a causal role or explain the many other situations exhibiting such rhythms.

Systematic study of this class of oscillations has barely begun. It is too early to tell whether there will be a few distinct subclasses or many, or a continuum graded by properties or by mechanisms. Many citations to literature, going back for decades, as well as a number of recent studies are treated in a recent book (Başar and Bullock, 1992).

Further reading

Başar E, Bullock TH (1992): *Induced Rhythms in the Brain*. Boston: Birkhäuser

Freeman WJ (1975): *Mass Action in the Nervous System*. New York: Academy Press

Long-term Depression (LTD)

Masao Ito

Long-term depression (LTD) is activity-dependent persistent reduction in the efficacy of signal transmission across a synapse. LTD occurs in the cerebellar cortex through the association of two distinct inputs from parallel fibers and climbing fibers, and apparently plays a role in motor learning. Another form of LTD occurs in the hippocampus and cerebral neocortex by the antiassociative interaction of two inputs and probably contributes to the process of cognitive learning together with the long-term potentiation (LTP) prevailing in the hippocampus and neocortex. In recent years, the occurrence of LTD has been confirmed by using various recording and stimulating techniques, and knowledge of molecular and cellular mechanisms of LTD has accumulated.

Occurrence

Cerebellar LTD. In the cerebellar cortex, each Purkinje cell receives two distinct types of excitatory synapses, one from numerous (some 100,000) parallel fibers (axons of granule cells in the cerebellar cortex) and the other from a (normally only one) climbing fiber (axon of a neuron located in the inferior olive of the medulla oblongata; Fig. 1). LTD takes place when these two types of synapses are activated repeatedly in approximate synchrony, and leads to an enduring decrease of synaptic strength of the parallel fibers. Typically, LTD is induced by conjunctive stimulation of parallel fibers and climbing fibers at 4 Hz for 25 s (100 stimuli). LTD so induced has been followed for 1–3 h without sign of recovery. LTD is input specific; it occurs only in those parallel fiber synapses activated conjunctively with climbing fibers. LTD is associative; it takes place only when two types of synapses are conjunctively activated. Activation of parallel fibers alone induces potentiation instead of LTD, and activation of climbing fibers alone causes neither of these effects.

Hippocampal and neocortical LTD. Pyramidal cells in the hippocampus receive excitatory input from numerous (several thousand) presynaptic fibers (Fig. 2). When bundle A of presynaptic fibers is stimulated with high-frequency stimulus trains (for example, bursts with 5 stimuli at 100 Hz repeated at 200 ms intervals for 2 s), it induces long-term potentiation (LTP) in A as well as in another bundle B if B is stimulated in phase with A. However, when B is stimulated out of phase with A, LTD occurs in B. LTD so induced lasts for several hours.

In the visual cortex, tetanic stimulation of an optic nerve induces potentiation in those synapses supplied by the stimulated optic nerve to cortical cells, but the same stimulation induces LTD in those synapses derived from another optic nerve. The LTD lasts for up to 9 h. The occurrence of LTD is less frequent when the nontetanized optic nerve is deprived of spontaneous activity by intraocular injection of tetrodotoxin. Hence, spontaneous activity out of phase with

tetanus stimuli seems to facilitate LTD. LTD has been observed also in the sensorimotor cortex and in the prefrontal cortex.

Underlying cellular and molecular processes

Involvement of Ca²⁺ ions. Impulses of climbing fibers evoke entry of Ca^{2+} ions into Purkinje cell dendrites through voltage-sensitive Ca^{2+} channels. This Ca^{2+} entry is an essential step in the induction of LTD, since LTD is abolished by injection of Ca^{2+}-chelator, EGTA, into Purkinje cell dendrites. It is an interesting contrast that whereas cerebellar LTD needs an enhanced Ca^{2+} level, hippocampal and neocortical LTD is presumed to require a lowered Ca^{2+} level, as has been deduced from the following observations.

In the hippocampus, membrane hyperpolarization follows afferent stimuli with a long delay so that membrane hyperpolarization prevails in the out-phase period. Membrane hyperpolarization directly induced in a hippocampal cell by current injection induces LTD when paired with afferent

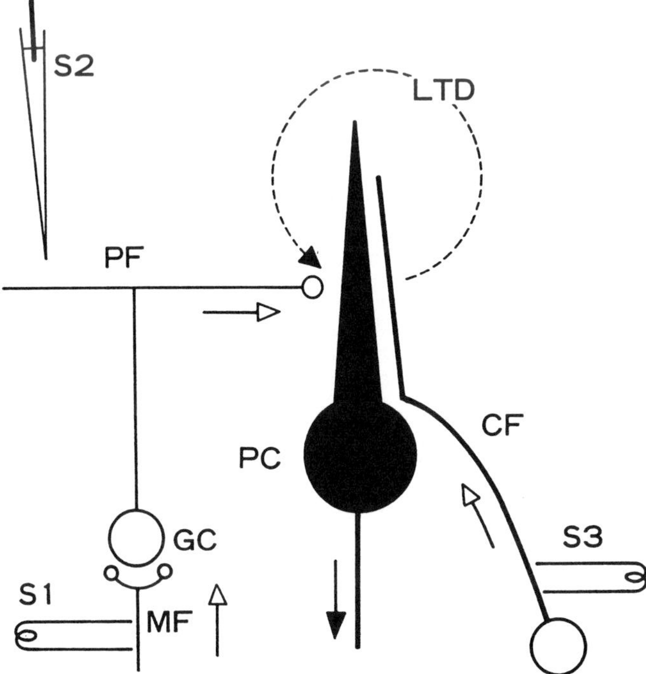

Figure 1. Occurrence of LTD in a cerebellar Purkinje cell. PC, Purkinje cell. GC, granule cell. MF, mossy fiber. PF, parallel fiber. CF, climbing fiber originating from cell body in the inferior olive. S_1, S_2, stimulating electrode. Arrows indicate the direction of impulse propagation.

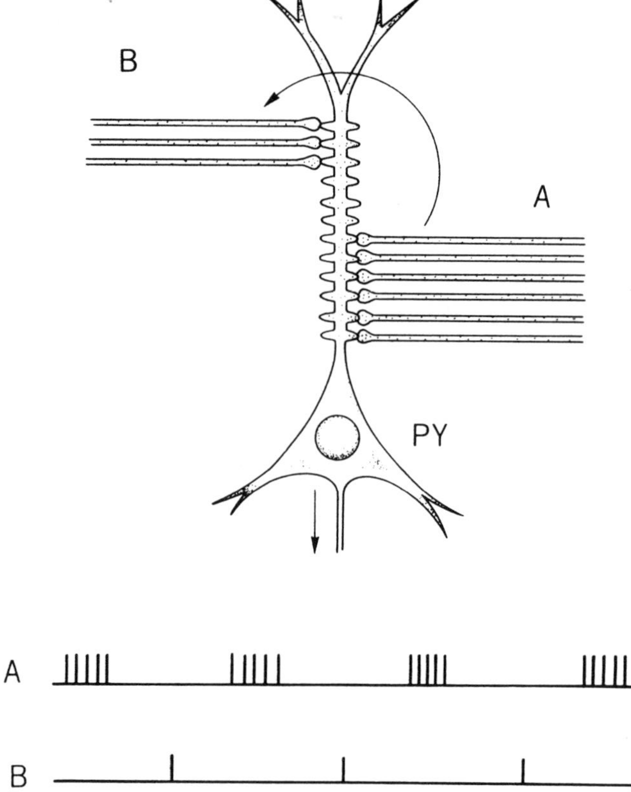

Figure 2. Occurrence of LTD in a cerebral pyramidal cell. PY, pyramidal cell. A, B, bundles of presynaptic fibers for conditioning and testing, respectively. An out-phase stimulus paradigm effective in inducing LTD is indicated.

stimuli. It is probable that LTD occurs when Ca ions are driven out of cells by membrane hyperpolarization. A similar situation also applies to visual cortical neurons in which LTP is converted to LTD by injection of EGTA. However, the membrane potential dependence of LTD in the visual cortex is slightly more complex. LTD occurs when the membrane is depolarized over a critical level but remains below a threshold above which LTP is induced. The neocortical LTD seems to require Ca^{2+} levels to be set at an appropriate level by membrane potential in a critical range.

Ca^{2+} ions enter hippocampal and neocortical neurons through channels associated with a particular (N-methyl D-aspartate–selective) subtype of glutamate receptors under the influence of membrane potential. These channels are normally closed by Mg^{2+} ions, which are removed at a depolarized membrane potential level. At a less depolarized level, these channels are closed, so that intracellular Ca^{2+} concentration is reduced by the action of Ca^{2+} binding proteins or accumulation into intracellular stores.

Desensitization of glutamate receptors. Cerebellar LTD is ultimately due to reduced sensitivity of glutamate receptors that mediate parallel fiber–Purkinje cell transmission. When a Purkinje cell is activated by ionotophoretic application of L-glutamate through a micropipette in conjunction with stimulation of climbing fibers or with induction of Ca^{2+} spikes by passage of depolarizing currents, a long-lasting reduction

occurs in the glutamate sensitivity of that Purkinje cell. At the same time, transmission from parallel fibers to the Purkinje cell is persistently depressed. These observations disclose that glutamate receptors mediating parallel fiber–Purkinje cell synapses are desensitized when exposed to glutamate released from parallel fibers in conjunction with entry of Ca^{2+} ions into Purkinje cell dendrites. Glutamate receptors involved in the cerebellar LTD are a subtype selective to AMPA (α-amino-3-hydroxy-5-methyl-4-isoxazole-propionic acid).

Whether the hippocampal or neocortical LTD is due to desensitization of AMPA-selective glutamate receptors, which also mediate excitatory synaptic transmission in these cells, or to reduced release of transmitter from presynaptic fibers is yet an unanswered question.

Nitric oxide and cyclic GMR. In brain tissues, Ca^{2+} ions activate calmodulin, which is required for activation of synthase of nitric oxide (NO). NO in turn accelerates guanylate cyclase to enhance the level of cyclic GMP, which activates cGMP-dependent protein kinase. These chain reactions may underlie the cerebellar LTD, since the LTD is blocked by an inhibitor of NO synthase and hemoglobin, which absorbs NO. An inhibitor of protein kinase G blocks desensitization of AMPA-selective glutamate receptors in Purkinje cells.

Roles of LTD in cerebellar function

In the cerebellum, climbing fibers convey signals representing control errors involved in performance of a motor system. Hence, those parallel fiber–Purkinje cell synapses activated during misperformance of the system will be depressed by LTD. Since these synapses should be responsible for the misperformance, their functional removal by LTD will lead to improved performance. This view matches the major principle of motor learning in which skills are acquired by repeated trials of error-correcting practice.

VOR adaptation. The vestibuloocular reflex (VOR) exhibits remarkable adaptability. Under the conditions where VOR fails to compensate for head movement to stabilize retinal images of the environment, the dynamics of VOR is gradually modified to minimize retinal errors. This adaptive mechanism is due to action of the cerebellar flocculus, to which climbing fibers convey retinal error signals and which send Purkinje cell axons to relay cells of VOR. LTD triggered by retinal errors would alter responsiveness of flocculus Purkinje cells to head rotation, and thereby lead to adaptive modification of VOR dynamics. This proposition is supported by a recent finding that subdural application of hemoglobin to the flocculus in rabbits and a monkey abolished adaptive modification of VOR, which normally is induced by sustained imposition with artificially amplified retinal errors.

Classical conditioning of eye blinking. Conditioning or airpuff–induced eye blinking with auditory or visual stimuli in rabbits has its central mechanisms in the interpositus nucleus of the cerebellum. Unconditioned corneal stimuli evoke climbing fiber signals that induce LTD in those parallel fiber–Purkinje cell synapses conveying conditioned stimuli. If the eye blinking to be evoked originally via an interpositus pathway is inhibited by Purkinje cells driven by the conditioned stimuli, LTD releases the reflex from inhibition. There is evidence supporting this view, but more information is needed to account for details of the mechanisms of classical conditioning. Involvement of LTD in other forms of cerebellar learning is a likely candidate for testing in future investigations.

Further reading

Artola A, Brocher S, Singer W (1990): Different voltage-dependent thresholds for inducing long-term depression and long-term potentiation in slices of rat visual cortex. *Nature* 347: 69–72

Crepel F, Audinat E (1991): Excitatory amino acid receptors of cerebellar Purkinje cells. *Progr Biophys Mol Biol* 55:31–46

Ito M (1989): Long-term depression. *Ann Rev Neurosci* 12:85–102

Ito M (1991): The cellular basis of cerebellar plasticity. *Curr Opin Neurobiol* 1:616–620

Stanton PK, Seznowsky TJ (1989): Associative long-term depression in the hippocampus induced by Hebbian covariance. *Nature* 339:215–217

Thompson RF (1986): The neurophysiology of learning and memory. *Science* 233:941–947

M

Malignant Tumors of Brain and Head, Boron Neutron Capture Treatment

William H. Sweet

The isotope ^{10}B has a number of extraordinary features, leading to the use indicated by the title. It has a remarkable avidity to capture neutrons when the velocities with which they are emitted at atomic disintegration are reduced to the "thermal" neutron range of circa 2200 m/s. This property makes it the agent of choice to control the evolution of power in a nuclear reactor. Of equal interest to biologists is the consequence of that neutron capture, namely the immediate disintegration of the unstable compound nucleus [^{11}B] to an atom of ^{7}lithium and a He^{++} (alpha particle) plus 2.4 MeV of energy divided between the two particles. At this energy even the alpha, the smaller of the two particles, can travel in tissue only about 14 μ. Hence the cell that takes up a boron-labeled molecule bears the brunt of the destructive force. In 1936 Locher proposed in general terms that the neutron capture reaction of an isotope taken up by a cancer might be helpful in its treatment. However, it was the report of Conger and Giles in 1950 that the radiation damage to lily bulbs exposed to slow neutrons was largely due to the trace amounts of boron they contain that attracted the attention of Sweet. The quantitative measure of an isotope's tendency to capture a slow neutron, its "capture cross section," in the unit called barns, is 3900 for ^{10}B. This contrasts sharply with the figures for 10 of the 11 common elements in the body, which range from 0.0009 for oxygen to 2.05 for potassium. Only the relatively sparse chlorine at 32.5 is much higher. The largest contribution to radiation injury from neutron capture by the chemical elements normally found in the brain and brain tumors comes from hydrogens with nitrogen in second place.

The relevant data may be summarized as follows:

σa capture or absorption cross section	Atom + Thermal neutron	Unstable compound nucleus Product	Immediate Final Product
0.33	^{1}H + n$_{th}$ →	[^{3}H] →	^{2}H + gamma ray (2.23 MeV)
1.7	^{14}N + n$_{th}$ →	[^{15}N] →	^{14}C + proton p$^+$ (0.6 MeV)
3900	^{10}B + n$_{th}$ →	[^{11}B] →	^{7}Li + alpha particle^{++} (2.4 MeV)

The normal brain has a mechanism called the blood–brain barrier that impedes entry into it of many substances circulating in the blood. This barrier is much less in malignant brain tumors. Clearly what was needed was an innocuous substance that entered brain tumors but not normal brain and was then made destructive by some kind of maneuver directed only at the brain. Hence the large amounts of the substance in the rest of the body would be unaltered. The isotope ^{10}B appeared to Sweet and colleagues in 1950 to have this potential.

Clinical therapeutic trials were carried out in the fifties and sixties at the two nuclear reactors at Brookhaven National Laboratory (BNL), initially under the aegis of Farr and Sweet. The latter shifted in 1960–61 to using a special operating room constructed when the research reactor at the Massachusetts Institute of Technology (MIT) was built. At all three reactors the treating neutron portion of the beam was essentially confined to slow neutrons, which penetrate tissue poorly, the flux decreasing roughly 50% with each 2 cm increment of depth in the brain. Hence within a few centimeters the dose drops to levels sublethal for tumors. All of the reactor-treated patients died at varying intervals, nearly all from further tumor growth and/or radiation injury.

Postmortem studies on them showed that the radiation from the boron still in the bloodstream caused "eosinophilic" degeneration, narrowing or occluding the blood vessels to the normal brain in 10 of the 14 patients treated at MIT. In order to stop the radiation in time to prevent this cerebral ischemia, we had to know the precise amount of boron in the circulating blood of the patient at the time of radiation—not possible then because the available method of determination required 4 hours. This problem was solved by Fairchild and colleagues at BNL in 1986, who measured the amount of 478 keV gamma radiation accompanying each disintegration of a boron atom at neutron capture. The method, requiring only minutes, gives adequate accuracy in specimens as small as 1 g.

Fairchild and many other physicists after him have also steadily improved the complex beam of fast neutrons and gamma rays emerging from the specifically designed portal of a nuclear reactor. None of these radiations is selectively destructive to tumor cells, so the physicist's objective, now achieved, is to filter out as much of the gamma component as possible while leaving enough intermediate speed (epithermal) neutrons at energies above that of the slow neutrons so that they will penetrate deeply enough into the tumor-infested brain.

The first major biological step forward after the early fifties was achieved by Albert Soloway at the Massachusetts General Hospital when he synthesized the substance B$_{12}$H$_{11}$SH, more of which was taken up by brain tumor and retained longer than any of the numerous compounds he and others had developed earlier. Our scientific colleagues refused to approve a clinical trial with this compound until cure of glioma-bearing animals by the method. However, Hiroshi Hatanaka, having worked with us for 3 years on the project, returned to Japan determined to pursue this approach. When the medical science community there would not recommend support, he persuaded the then prime minister of the nation to do so. Despite the steep uphill effort required from him, and with the cooperation of the appropriate Japanese physicists, he treated his first patient in Japan in 1968, using for this and all of his over 100 subsequent patients Dr. Soloway's $Na_2B_{12}H_{11}SH$. He has kept the whole concept alive by his indefatigable persistence and achievement of a few spectacular successes in patients. The most convincing of these is a man 50 years old when diagnosed and operated on in 1972. He presented the 7 most unfavorable prognostic features delineated by the analyses of Walker and colleagues from the experience of 17 neurosurgical services. Not a single other such case report of survival as long as 5 years appears in this gloomy congeries of findings. This man has no neurological deficit over 19 years later and is an active, industrious farmer. The last CT scans continue to show no sign of recurrent tumor. Hatanaka also has survivals 5 years in 7 of 12 patients with high-grade malignant brain tumors lying within 6 cm in the cerebral cortex. There are 3 survivors for over 10 years; one of them at work at age 73 is 13 years post BNCT.

A further impetus for neurooncologists to converge on this approach has been the dismal result after more than 25 years of well-conceived and well-supported research on combinations of surgery, x-radiation, and a huge variety of chemicals. The status of these efforts at the University of California at San Francisco has been tersely summarized by Charles B. Wilson, organizer of the group, as follows: "In the near future the development of a better chemical does not appear to be likely; possibly one will emerge within 10 years. The current group of drugs do not qualify. An entirely new agent must be identified."

The malignant gliomas spread rarely out of the cranial cavity and only 6% are multiple in that cavity. Moreover, their almost invariable recurrences are in the great majority at or near the original site. Hence this group's growth patterns are nicely tailored to be vulnerable to the characteristics of BNCT. Sweet has also advanced evidence for this form of treatment for three other groups of tumors in the head. The commonest of these are the cancers of the deep face such as those of the throat, sinuses, or ear. Of the order of 85 to 90% of these tumors are diagnosed before they metastasize but can rarely be totally removed at the primary site by radical surgery. The dose of standard radiation therapy is most critically limited by the risk of injury to the neighboring frontal and temporal lobes of the cerebrum and brain stem. These regions are never invaded by tumor, so have a completely intact blood–brain barrier and are hence ideal from the standpoint of sparing of brain by BNCT. The same considerations apply to the treatment of the usually benign tumors of the membranes covering the brain, the meningiomas, when they invade the base of the skull and cranial nerves, or become malignant. Similar features apply to a few pituitary tumors.

An extremely important unknown is the extent to which any boron compound localizes in and around the small streamers of tumor cells invading the largely normal brain. These are the cells the capture radiation from the boron must kill. Nearly all of our data at present have been obtained from the main tumor mass, that is, from tissue the neurosurgeon removes. The best evidence to date was gleaned from our post mortem on a patient whose highly unusual malig-

nant glioblastoma had seeded the whole length of the spinal cord down into the nerve rootlets of the cauda equina. $B_{12}H_{11}SH$ had been infused about a day before death; these seedlings were filled with it as indicated by the alpha particle autoradiography of Slatkin et al. (1989).

The most encouraging of the recent data are those from Drs. Joel, Coderre, and Slatkin and colleagues at BNL. Their rat animal model of a brain tumor consistently grows at a pace that kills the rat in 22–24 days. They found about twice as much boron was retained in the tumor and persisted therein for a much longer period when the dimer of the original monomer $Na_2B_{12}H_{11}SH$ was used. There was about 13 times as much boron in the tumor as in the normal brain. Of the 10 radiated rats 6 lived on to die of some other cause than their gliosarcoma. Similar numbers of control animals in each of three other groups died in a month or two. This is the most successful, well-controlled analysis of treatment yet reported of an animal glioma with any modality or any combination of the modalities surgery, radiation, and/or chemicals. Since then two additional highly favorable facts have been discovered. In the first of these Dr. Coderre used boronophenylalanine as the boron carrier in a series of radiations to rats with gliosarcomas like those of Dr. Joel and with similar survival figures. This result has been confirmed by the Tufts-MIT group of Zamenhof and Harling in gliomabearing mice and they have proposed a protocol for a clinical trial in humans. Finally, in unpublished studies Joel and colleagues found that administration of a tripeptide to the gliosarcoma-bearing rats preceding that of the B compound increased the total tumor uptake of both the boron SH monomer and dimer and brought the uptake of the less toxic monomer up to the higher level of the dimer.

At present the upper limit of the radiation dose is determined by the tolerance of the vessels in the normal brain to the amount of boron in the blood passing through them. Slatkin and colleagues have shown that the great majority of both the boron SH monomer and dimer in the blood are bound to albumin. In the safe standard procedure of plasmapheresis the blood is partially depleted of its plasma protein. The radiation dose to the brain can then be correspondingly increased.

In summary, the wealth of recent discoveries markedly brightens the hope for successful treatment of a number of malignant tumors of the brain itself and those near it.

Further reading

Coderre JA, Joel DD, Micca DL, Nawrocky MM, Slatkin DN (1992): Control of intracerebral gliosarcomas in rats by neutron capture therapy with p-boronphenylalanine. *Radiat Res* 129: 290–296

Hatanaka H, ed (1986): *Boron-Neutron Capture Therapy*. Nigata, Japan: Nishimura, p 463

Joel DD, Fairchild RG, Laissue JA, Saraf SK, Kalef-Ezro JA, Slatkin DN (1990): Boron neutron capture therapy of intracerebral rat gliosarcomas. *Proc Natl Acad Sci (USA)* 87:9808–9812

Slatkin DN (1991): A history of boron neutron capture therapy of brain tumors. *Brain* 114:1609–1629

Marijuana Receptor Gene, Cloning and Characterization

Lisa A. Matsuda

For centuries, the flowering tops of hemp, also known as cannabis or marijuana, have been used for medicinal purposes and for their mood- and mind-altering effects. Marijuana's low toxicity coupled with its euphorigenic effects have resulted in widespread hedonistic use of this plant throughout the world. Although generally viewed as a substance of abuse, the therapeutic potential of marijuana stems from its ability to provide relief from pain, loss of appetite, insomnia, asthma, nausea, and vomiting. Elucidating the mechanisms responsible for marijuana's many actions has been difficult. The cannabinoid constituents of cannabis, including Δ9-tetrahydrocannabinol (Δ9THC), are very hydrophobic compounds and thus marijuana-induced effects were often thought to result from interactions between the cannabinoids and lipids, which constitute most of a cell's plasma membrane. While some of marijuana's effects may result from such nonspecific interactions, a receptor-mediated mechanism has also been established.

The receptor for cannabinoids belongs to a class of membrane-located proteins that interact with guanine nucleotide-binding proteins (G-proteins) as part of their second messenger transduction system. The structural characteristics of receptors of this type (G-protein–coupled receptors) include the presence of seven hydrophobic domains in a single polypeptide subunit. In addition, a limited yet characteristic amount of primary structure identity is apparent between a majority of the members of this receptor class. These shared structural features reflect a substantial amount of conserved nucleic acid sequence in the genes for these receptors. These conserved nucleic acids provide a means for cloning the genes that encode many different G-protein–coupled receptors, one of which was ultimately found to be a marijuana or cannabinoid receptor.

Information on the marijuana or cannabinoid receptor gene comes from a complementary DNA (cDNA), SKR6, which was obtained by using a strategy based on the structural features that are common to different types of G-protein–coupled receptors. This strategy involved the use of an oligonucleotide probe the sequence of which complements a portion of the bovine substance K receptor gene. The portion of this gene from which the probe's sequence was derived encodes the second hydrophobic domain of the bovine substance K receptor protein. This domain contains a substantial number of amino acids that are conserved between many different types of G-protein–coupled receptors (adrenergic, muscarinic, serotonergic, dopaminergic); thus the probe does not selectively target one specific receptor subfamily but contains a sequence that should hybridize with genes encoding numerous members of this receptor class.

The SKR6 cDNA was one of several clones isolated from a rat cerebral cortex cDNA library. Potential receptor clones were identified within the library by screening pools of cDNAs using Southern blotting techniques; candidate clones were those that hybridized to the oligonucleotide probe. By using low stringency conditions, which allowed the probe to hybridize with cDNAs containing sequences that are similar to, yet different from, the substance K receptor sequence, cDNAs encoding novel receptors could be obtained. Once candidate cDNAs were identified, the relevant pools of cDNAs were sequencially subdivided until the clones could be isolated as individual colonies. Plasmid DNA from these colonies was analyzed for appropriate cDNA length and correct positioning of the hybridizing region within the cDNA. The SKR6 cDNA was 5.7 kilobases (kb) in length and its predicted protein product contained seven hydrophobic domains and numerous amino acid residues that are highly conserved among other G-protein–coupled receptors (Fig. 1).

The SKR6 clone clearly differed from all other previously cloned G-protein–coupled receptors, but its amino acid sequence also did not resemble one receptor any more than another. The identity of the SKR6 receptor, therefore, could not be surmised from its similarity to any one specific receptor or receptor family. To gain clues regarding the identity of the SKR6 receptor and its ligand we determined in which tissues the SKR6 gene was active. RNAs isolated from various peripheral tissues or cell lines were screened on Northern blots by using a probe designed to hybridize specifically with the RNA for the SKR6 receptor. The expression of the SKR6 gene in the central nervous system was also determined by *in situ* hybridization techniques using this same probe on rat brain sections. While little information was obtained from the Northern blot analysis of RNA obtained from peripheral tissues, a regionally distinct and unique localization pattern was observed for the SKR6 RNA in rat brain. The RNA for this clone was notably abundant in the cerebral cortex, caudate nucleus, hippocampus, and cerebellum; these findings, at minimum, suggested an important role for the SKR6 receptor in numerous brain systems.

Northern blot analysis of RNA obtained from cell lines provided the first indication that the SKR6 clone encoded a cannabinoid receptor. A single hybridizing band (~6 kb) in RNAs isolated from N18TG-2 neuroblastomas and NG108-15 neuroblastomaglioma cells revealed that the SKR6 gene was active in two cell lines known to contain cannabinoid receptors. Consequently, the localization pattern of the SKR6 RNA and the autoradiographic map of cannabinoid receptors in rat brain were compared and, in many instances, the SKR6 RNAs were present within the same regions as the cannabinoid binding sites. This generally similar distribution of cannabinoid receptors and the SKR6 RNA throughout the brain was consistent with the possibility that the SKR6 protein product was a cannabinoid receptor. Consequently, cannabinoid compounds were tested as ligands for the SKR6 clone.

In N18TG-2 and NG108-15 cells, the naturally occurring cannabinoid receptor mediates a cannabinoid-induced inhibition of adenylate cyclase activity resulting in decreased accumulation of cyclic AMP. In Chinese hamster ovary (CHO) cells transfected with the SKR6 cDNA, cAMP production decreased in a dose-dependent and stereoselective manner in response to either Δ9THC or CP 55940, a potent, synthetic nonclassical cannabinoid analog (Fig. 2). This cannabinoid-induced effect was not observed in nontransfected CHO cells or in CHO cells expressing other cloned G-protein–coupled receptors. Furthermore, the cloned receptor also appeared to couple with the appropriate G-protein in the CHO cell line, since the effect on cAMP production was completely prevented in cells pretreated with pertussis toxin.

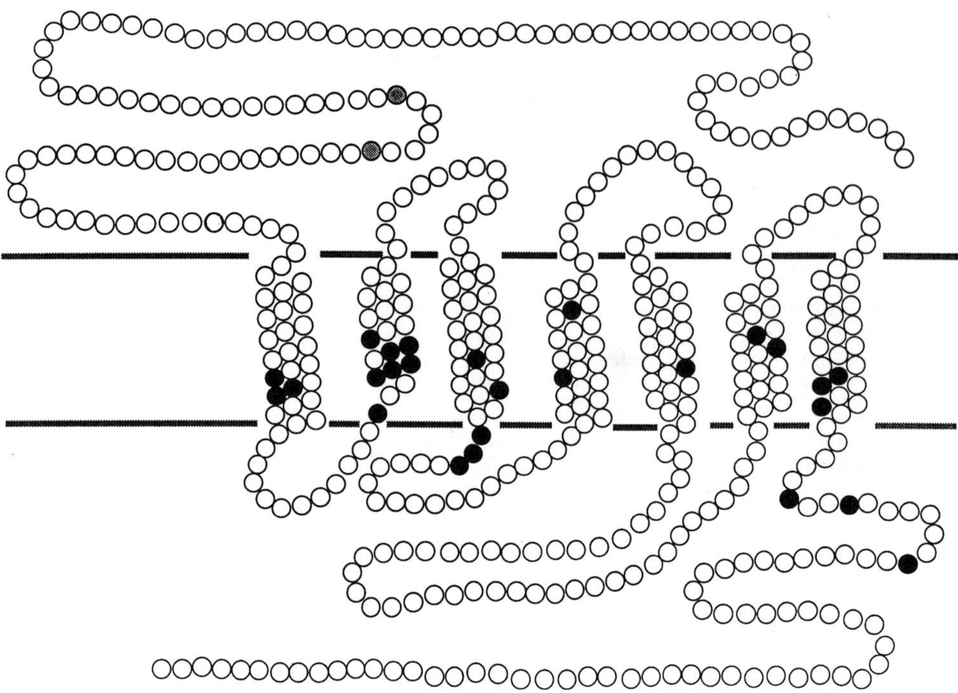

Figure 1. Cartoon depicting predicted structural features of the SKR6 protein product. Circles represent individual amino acid residues, which have been arranged to illustrate seven hydrophobic domains. These domains are presumed to reside within plasma membranes (boundaries indicated by shaded lines), thereby positioning the protein on a cell's surface. Lightly shaded circles indicate potential sites for N-linked glycosylation of this protein. Darkly shaded circles feature amino acid residues that are commonly found in other membranes of the G protein–coupled receptor class.

Thus, the SKR6 cDNA encodes a protein that responds to cannabinoids in a manner identical to that of the cannabinoid receptor found in neural cell lines.

While the original cloning and characterization of the cannabinoid receptor was accomplished with a cDNA obtained from rat, the corresponding human gene has also been cloned and expressed. The human gene encodes a protein that is extremely similar to the rat cannabinoid receptor with more than 97% of the amino acid residues being identical between the receptors from these two species. The human receptor also mediates a cannabinoid-induced reduction in cyclic AMP in cells transfected with the cloned gene. The ligand specificity of this response indicates that this receptor is likely involved in the psychoactive effects that are experienced by humans intoxicated with marijuana. Moreover, saturable and specific binding of ^{3}H CP 55940, a ligand that binds specifically to a cannabinoid receptor in brain tissues, can be demonstrated in membranes prepared from transfected cells. This binding to the cloned receptor is competitive with Δ9THC and another cannabimimetic compound,

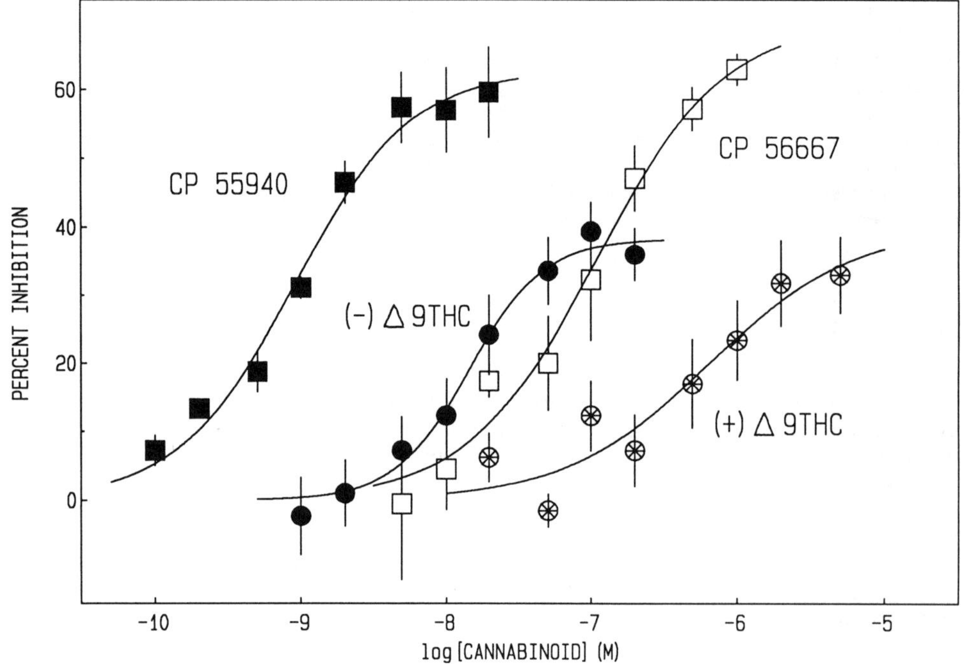

Figure 2. Cannabinoid-induced inhibition of cAMP production in cells transfected with the SKR6 cDNA. Data represent the average per cent inhibition of forskolin-stimulated cAMP production ± SEM. (Reprinted from Matsuda et al. (1990). Reprinted by permission from *Nature,* vol. 346, p. 564. Copyright © 1990 Macmillan Magazines, Ltd.)

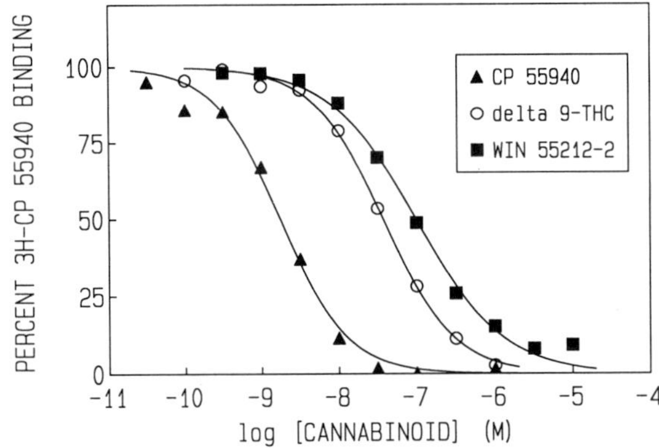

Figure 3. Competition binding of synthetic (CP 55940 and WIN 55212-2) and naturally occurring (Δ9THC) cannabinoids with ^{3}H CP 55940 to crude membranes prepared from mouse L cells expressing the cloned human cannabinoid receptor gene.

WIN 55212-2 (Fig. 3) and further confirms the identity of the cloned receptor as that of the cannabinoid receptor naturally expressed in brain tissue.

The cloning and characterization of a cannabinoid receptor gene provides conclusive evidence for a receptor-mediated mechanism of action for marijuana. The predicted structural features of this protein further qualify the marijuana or cannabinoid receptor as a member of the G-protein–coupled receptor class. These data and the cloned materials themselves will facilitate research efforts aimed at increasing our understanding of receptor biochemistry and the consequences of marijuana and/or cannabinoid exposure. As with other classical neurotransmitter and hormone receptors, it follows that an endogenous ligand should exist for the cannabinoid receptor. Although the identity of this substance is unknown, the predominant localization of the receptor in the brain suggests that this ligand influences a variety of brain functions including sensory, cognitive, "limbic," motor, and autonomic systems. While the determination of the pathological and physiological significance of this receptor will increase our understanding of the central nervous system, the unknown endogenous ligand will continue to intrigue scientists until it is found.

The development of specific drugs that either mimic or antagonize the effects of natural agonists is another event that should follow the characterization of the cannabinoid receptor. While history documents the medicinal use of marijuana, current therapeutic use of cannabimimetics is very limited. The approved indications for marijuana include the relief of nausea and vomiting associated with cancer chemotherapy or the stimulation of appetite in patients infected by HIV. Thus far, beneficial therapeutic effects of the cannabinoids have not been successfully dissociated from the psychoactive effects of these compounds. It is possible that cannabimimetic agents will not fulfill their therapeutic potential until they are formulated in a manner that limits their access to the central nervous system.

Further reading

Gérard CM, Mollereau C, Vassart G, Parmentier M (1991): Molecular cloning of a human cannabinoid receptor which is also expressed in testis. *Biochem J* 279:129–134

Herkenham M, Lynn AB, Johnson MR, Melvin LS, deCosta BR, Rice KC (1991): Characterization and localization of cannabinoid receptors in rat brain: A quantitative *in vitro* autoradiographic study. *J Neurosci* 11:563–583

Howlett AC, Bidaut-Russell M, Devane WA, Melvin LS, Johnson MR, Herkenham M (1990): The cannabinoid receptor: Biochemical, anatomical and behavioral characterization. *Trends Neurosci* 13:420–423

Matsuda LA, Lolait SJ, Brownstein MJ, Young AC, Bonner TI (1990): Structure of a cannabinoid receptor and functional expression of the cloned cDNA. *Nature* 346:561–564

Mental Imagery

Stephen M. Kosslyn and Lisa M. Shin

Mental images are brain states that give rise to the experience of perceiving in the absence of the appropriate sensory input; for example, one "sees with the mind's eye" or "hears with the mind's ear." Research on mental imagery has progressed during the past five years, increasing our understanding of key aspects of the phenomena and the neural mechanisms that produce them. This research has tended to focus on three aspects of imagery: image generation, image inspection, and image transformation.

Image generation

Mental images are transient, only coming to mind in specific circumstances. The information necessary to form an image is stored in long-term memory, and an image generation process uses this information to produce the image itself. Recent research on image generation has focused on the claim, originally made by Martha Farah, that this process depends on the posterior region of the left cerebral hemisphere. Farah and her collaborators provided additional support for this hypothesis by testing the image generation ability of a patient (R.M.) with a left-posterior lesion. R.M. was asked to evaluate both statements that had previously been shown to require imagery (e.g., "A grapefruit is larger than an orange") and statements that do not require imagery (e.g., "The U.S. government functions under a three-party system"). R.M. had more trouble evaluating the imagery sentences than the nonimagery sentences. He also performed poorly in two nonverbal imagery tasks (a coloring task and a drawing completion task), which ruled out the possibility that his problem was language related. Farah and her colleagues suggest that R.M. had a deficit in the image generation process that assembles separately stored parts; according to Kosslyn and his colleagues, one form of this process may be lateralized to the left hemisphere.

Georg Goldenberg and his collaborators have used SPECT brain-imaging to show that the left occipital lobe is in fact active during mental imagery. In subsequent work on this topic, these researchers studied patients with lesions in the posterior left hemisphere, posterior right hemisphere, and normal subjects. The subjects were asked to evaluate low-imagery statements or high-imagery statements (concerning shape or color), with or without pictures of the objects that should have been visualized to evaluate the statements. The results were mixed: the left-hemisphere patients did make more errors than the two other groups on the high-imagery statements, but they did not show any improvement when the pictures were available—which would have been expected if the deficit were due to impaired ability to form (as opposed to interpret) images. The normal control subjects and the right-hemisphere patients did show improvement when pictures were presented.

Michael Corballis and Justine Sergent tested image generation in the isolated cerebral hemispheres of a split-brain patient, L.B. In the first image generation task, L.B. was presented with uppercase letters in either the left or right visual half-field for a brief period of time (and so the image was seen only by the right or left hemisphere, respectively), and was instructed to decide whether the corresponding lowercase letter was "short" or "tall." In the second image generation task, L.B. was presented with digital clock times in either the right or the left half-field and was asked to decide whether the hands on an analog clock would subtend more or less than 90 degrees at that particular time of day.

On both image generation tasks, L.B. made significantly fewer errors when the stimuli were presented to the left hemisphere, but he responded significantly faster when stimuli were presented to the right hemisphere. This speed–accuracy tradeoff makes the results difficult to interpret: a hemisphere could have made more errors because it was responding too quickly, before it was ready.

In a subsequent study, Sergent administered the letter-height task to normal subjects and found that they responded faster when the cues were presented initially to the right cerebral hemisphere; and there was no speed–accuracy trade-off in these results. This finding contradicts those reported by Farah. Sergent subsequently reviewed the case studies of brain-damaged patients who had image generation deficits and concluded that the evidence favoring a left-hemisphere locus for image generation is not compelling.

Kosslyn provided evidence that both hemispheres can generate visual mental images but do so in different ways. The left hemisphere assembles parts using mental descriptions of their spatial relations (e.g., "connected to" or "above"), whereas the right hemisphere assembles parts using metric coordinate specifications of location. These findings are consistent with those of other researchers, who have failed to find (as reported earlier by Farah) selective left-hemisphere evoked potentials during image generation, and instead find evidence that both hemispheres can generate images.

In conclusion, it seems likely that there is more than one way to form visual mental images, and the cerebral hemispheres may differ in their relative efficacies at using the different methods.

Image inspection

Imagery is useful in part because imaged patterns can be "inspected," allowing one to "perceive" previously unconsidered characteristics of an object or scene. Kosslyn and his collaborators used PET scanning to provide evidence that primary visual cortex (area 17, in the medial occipital lobe) is activated during imagery. This area was activated more during an imagery task than during a corresponding perceptual task, which may reflect the fact that imaging a pattern requires more effort than perceiving it. Thus, it is not surprising that imaged patterns can be operated on by higher-level processes that allow one to inspect objects during perception.

In 1985 Deborah Chambers and Daniel Reisberg reported that subjects were unable to "reverse" mental images of classical ambiguous figures, such as the Necker cube, the duck/rabbit, and the Schroeder staircase. This finding suggested to them that images are interpretations, which cannot be reorganized.

The figures used as stimuli in this research were rather complex, however, and subjects' difficulties may have reflected an inability to retain the image instead of an inability to reorganize figures. If objects in mental images can be reorganized, then images cannot be just interpretations—they must contain the same kinds of geometric information that perceptual representations do.

Ronald Finke, Steven Pinker, and Martha Farah asked subjects to juxtapose or superimpose mental images of familiar patterns, describe the result with their eyes closed, then draw their image. For example, subjects were asked to form an image of an uppercase "D," rotate it mentally 90 degrees counterclockwise, and then place an uppercase "J" below it. Subjects could identify the result (as an umbrella), which clearly requires reinterpreting the image. The authors concluded that it is possible to assign novel interpretations to ambiguous images and that these interpretations are not merely the results of guessing.

Image transformation

Imaged patterns can be transformed in various ways, being rotated, expanded, truncated, and so forth. Michael Corballis and Justine Sergent tested a split-brain patient, L.B., to study mental image rotation in the cerebral hemispheres. L.B. was presented with single letters at different degrees of rotation and was asked to mentally rotate each one to its upright position and to decide whether it was "normal" or "mirror-reversed." L.B. responded faster and more accurately when the rotated letter was presented to the right hemisphere than to the left hemisphere. However, by the final three sessions L.B. could mentally rotate stimuli presented to his left hemisphere—but he was much slower and made more errors than when stimuli were presented to his right hemisphere. Thus, we again find that imagery is not strictly lateralized to one or the other hemisphere.

Daniel Voyer and Phillip Bryden recently argued that inconsistencies in the literature regarding the locus of mental rotation in the brain may be a result of methodological problems. They tested the prediction that the cerebral lateralization of mental rotation varies with the subject's level of spatial ability. Subjects were presented with two figures at different degrees of rotation and were asked to decide whether they were "same" or "different." Males showed a significant right-hemisphere advantage and females showed a marginally significant left-hemisphere advantage. Furthermore, subjects with low spatial ability showed a right-hemisphere advantage, subjects with medium spatial ability showed no particular advantage, and subjects with high spatial ability showed a left-hemisphere advantage. Thus, individuals may differ in the lateralization of processes used in image transformation.

In conclusion, the recent work on imagery reinforces the inference that it is not a unitary phenomenon but rather is accomplished by a collection of distinct processes. In addition, these processes appear to be implemented in different parts of the brain, they can vary in efficacy in different people, and numerous sets of processes (with each corresponding to a distinct method or strategy) may allow one to accomplish any given imagery task.

Further reading

Farah MJ (1988): Is visual imagery really visual? Overlooked evidence from neuropsychology. *Psychol Rev* 95:307–317

Hampson PJ, Marks DF, Richardson JTE (1990): *Imagery: Current Developments*. London: Routledge

Kosslyn SM, Shin LM (1992): Visual mental images in the brain: Current issues. In: MJ Farah, G Ratcliff, eds. *The Neural Bases of Mental Imagery*. Hillsdale: Lawrence Erlbaum

Logie RH, Denis M (1991): *Mental Images in Human Cognition*. Amsterdam: North Holland

Neuropsychiatry

Richard M. Restak

Neuropsychiatry is the practical application of neuroscience for the purpose of understanding and treating disorders of mood, thought, and behavior. As the name implies, it is a hybrid discipline combining elements of both psychiatry and neurology. It differs from more traditional psychiatric approaches (psychodynamic, social, etc.) by reason of the neuropsychiatrist's working belief that mental illness is the result of brain dysfunction. It also differs from traditional neurology, in which neuroscientific findings are applied to all disturbances of the brain and central nervous system, whether or not the disturbances have behavioral consequences.

The roots of neuropsychiatry are found in humankind's earliest efforts to understand the influence of physical forces on human behavior. The Hippocratic doctrine of the humors was the first to attempt a systematic, consistent explanation for the phenomena of mental illness. It emphasized physical rather than supernatural forces and could be tested by observation of physical phenomena. Hippocrates wrote: "As long as the brain is at rest, the man enjoys his reason, but the depravement of the brain arises from phlegm and bile." This was the first recognition of the importance of the brain, rather than the heart, as the mediator of psychological and emotional experience.

The next great advance occurred in the seventeenth century. The great neuroanatomist Thomas Willis, discoverer of the circle of blood vessels at the base of the brain that bear his name, began the practice of correlating physical signs in dying patients with postmortem changes in the brain. Willis was the first to employ the term "Neurologie," although neurology, as a discipline distinct from the much older psychiatry, did not emerge as a separate specialty until 1817. In that year, James Parkinson wrote his essay "The Shaking Palsy."

Progress during the remainder of the nineteenth century was rapid. In the early years, Franz Joseph Gall formulated what we now see as a pseudoscience, phrenology. According to his theory, personality and character traits were associated with "bumps" and other variations of the skull. Although it failed to explain character and personality, it did stimulate critical clinicopathologic correlations and thus hastened the development of neurology. It was the foundation from which neuroscience has developed.

During the mid and late 1800s the neurologist John Hughlings Jackson formulated a hierarchical explanation of brain function. In this, higher, more evolved areas of the brain suppressed the actions of the "lower," more primitive centers. Impairment of the higher centers produced signs and symptoms of mental function by, negatively, removing the normal function and, positively, releasing the lower functions from suppression. This concept Freud would later retain in modified form in such concepts as repression, primary process thinking, and some other defense mechanisms.

Other contributions to the emerging neuropsychiatric orientation included those of Broca and of Wernicke, each of whom discovered areas within the brain responsible for speech and language. In this period, a German physician, Theodore Fritsch, discovered while dressing a head wound of a soldier injured in the Prusso-Danish War of 1864 that touching the cerebral hemisphere produced twitching of the opposite side of the body. He discussed this finding with Eduard Hitzig, a medical practitioner in Berlin. While carrying out experiments to further study this phenomenon, they observed that electrical stimulation of the cortical surface in dogs produced muscular contractions in the muscles on the opposite side of the body. This crude experiment inspired subsequent more sophisticated demonstrations of cerebral localization.

As always, the greatest advances in neuropsychiatry were made through the study of illness and disease. In 1826 Calmeil had identified what may be called the first neuropsychiatric illnesses: general paralysis of the insane, secondary to syphilis. Over the next century it was the model for neuropsychiatric illnesses. At times it accounted for as much as 50% of all hospitalized psychiatric patients. Its manifestations and courses were the instructors of generations of physicians. It was a central theme in Meynert's comprehensive compendium *Diseases of the Forebrain*.

Thus instructed by syphilis, prominent teachers of psychiatry and of neurology, such as Wilhelm Greisinger in Germany and Henry Maudsley in England, proclaimed that mental illness can be correlated with brain disease. In 1868 Greisinger wrote: "Psychiatry and neuropathology are not merely two closely related fields; they are but one field in which only one language is spoken and the same laws rule."

However, the clinicoanatomical method, so successful in neurology, proved far less useful in understanding psychiatric illnesses. For instance, neuroanatomical studies of schizophrenia failed to reveal correlations of brain and behavior along the lines observed in Alzheimer's disease. We recognize now that this failure stemmed principally from the absence of technological developments capable of revealing function instead of simply structure.

In the meanwhile, such students of mind-brain problems, as was Freud, turned attention to thought, emotion, beliefs, and interpersonal relationships. Freud was a prominent and influential neurologist, author of authoritative works on aphasias and infantile paralysis. Looking over the centuries, it is ironic that the failure of pathologists to discern a disordered pattern of brain structure in mental illnesses returned behavioral investigation to the ineffable, unmeasurable, immaterial concepts of two millennia previous. A large field of neuropsychiatry went into a shadow for some 70 years, well into the 1950s. A substantial part still is.

During the Second World War, practical considerations dictated a limited revival of neuropsychiatry. A shortage of psychiatrists in the military prompted the employment of neurologists in the care of psychiatric patients. Following the war, upon return to civilian life, some of these neurologists continued their newfound interest in psychiatry via the maintenance of small psychiatric practices in addition to their regular neurological practice. As another spin-off of the war experience, some of the psychiatrists associated with academic and training institutions reorganized their departments to include either combined psychiatry-neurology programs, or psychiatric training programs with emphasis on the brain. These efforts to apply neuroscience to the care of the mentally ill failed to have a substantial influence on the welfare of patients, and the field of psychiatry remained essentially that of psychoanalysis.

In 1954, chlorpromazine came into wide use, shortly followed by imipramine. Their effectiveness in the two major psychoses, schizophrenia and depression, again led to new and progressive developments in understanding the function of the brain. Chlorpromazine's usefulness correlated with its ability to induce at higher doses almost all of the symptoms of lethargic encephalitis. Based on these neurological effects, chlorpromazine, along with other antipsychotics of the same class, were termed "neuroleptic": that which takes the neuron. Imipramine's effectiveness also correlated with biological variables described by the drug's discoverer Roland Kuhn as "a vital depressive disturbance": a slowing down of thinking, action, and decision. Over the period of the next two decades, three other factors stimulated the development of neuropsychiatry.

First, new observations about epilepsy, particularly the description in 1963 by Slater and Beard of the schizophrenia-like psychoses of epilepsy, have been very provocative. They observed that symptoms of some forms of epilepsy are very like those of some schizophrenias. This is particularly so when the seizure discharge is in the temporal lobe and associated amygdala and hippocampus of the dominant hemisphere. As is now evident, the frontal lobe also plays a significant role in both conditions.

Second, the Canadian neurosurgeon Wilder Penfield showed that electrical stimulation of the temporal lobe and its environs in conscious and cooperative patients undergoing neurosurgical procedures could induce altered perceptions. Penfield called some of these "experiential illusions." His work further buttressed the view that mental illness is the result of brain disease.

Third is the evidence drawn from patients who sustained "minor" head injuries. They regularly undergo changes in mood, cognition, and personality. They are subject to rage attacks and temper outbursts, rapid shifts of feeling, difficulty in concentrating and remembering, change in sexual drive, and disruptions of the sleep-waking cycle. These disturbances occur irrespective of issues of financial or legal gain and may happen to persons who have no previous history of psychiatric disturbance. Furthermore, electrophysiological changes such as diffuse slowing or paroxysmal discharges can often be recorded from the brain, corresponding to the behavioral changes.

The major factor giving rise to a revival of neuropsychiatry has been technical advances. The development of the EEG in 1929 provided an instrument of proof that psychological processes involve detectable and measurable electrical activity in the brain. Subsequent work, beginning in the 1940s and 1950s, using depth electrodes placed in the brains of schizophrenics, revealed spike and sharp wave discharges below the cortex. These abnormalities could often be recorded from patients who had normal records by standard EEG using scalp electrodes. These observations prepared the way, three decades later, for the introduction into psy-

chiatry of concepts and treatment approaches borrowed from epileptology (i.e., anticonvulsants in bipolar disorder and selected instances of rage and aggression).

Additional technological developments in the 1980s included computerized axial tomography of the brain (CAT), computerized 32 channel EEG, magnetic resonance imaging of the brain (MRI), and brain electrical activity mapping (BEAM). Now neuroscientists have a succession of ways to view the brain in whole or in part, continuously, safely, as an anatomical study, as a metabolizing organ, or as an electrophysiological marvel. We have moved from the static brain to the dynamic, from the passive to the interactive. No longer relatively restricted to the study of schizophrenia, these instruments are revealing distinctive physiological "signatures" for neuropsychiatric illnesses. Panic attacks, depression, mania, and obsessive-compulsive disorder are marked by PET scan changes that suggest a functional pathophysiological localization, usually within either the frontal or temporal areas. In addition, research is currently underway aimed at defining normal PET scan patterns accompanying sensory stimulation, memory, thinking, and other cognitive activities in normals.

Stimulated by these discoveries, clinicians are now altering their approaches to certain patients who can be helped only by a combined neuropsychiatric approach. An example is provided by a patient who had been attacked and suffered concussion. This woman, thereafter, suffered recurrent and treatment-resistant nightmares of a hand approaching her from the rear. She also experienced hypnogogic hallucinations of a threatening figure standing at the foot of her bed. These symptoms, together with headaches and cognitive disturbances, improved only when attention was directed to the anguish and terror she continued to experience as a result of her inability to remember, or to identify her attacker.

Another patient suffered from complex partial seizures with frequent generalization followed, two or three days later, by the onset of a psychosis in which he believed himself to be Jesus Christ. Improved seizure control eliminated the psychotic episodes altogether. Additional improvement in mood and social functioning occurred when the patient learned that he did not suffer from two illnesses (schizophrenia and epilepsy) as his former physician had told him, but from the single condition, that is, the schizophrenia-like psychosis of epilepsy.

While neuropsychiatry combines elements of both psychiatry and neurology, it remains a subspecialty all its own. From neurology, the neuropsychiatrist borrows a concern for clinicopathologic correlations and the detailed and painstaking analysis and exploration of the patient's symptoms. From psychiatry the neuropsychiatrist takes an interest in the patient's subjective experiences, qualities of thought, abnormalities of mood—in fact, everything that will help in understanding what it must be like to be the person with emotional illness. The following quotation is from William Alwyn Lishman, Professor of Neuropsychiatry at the Institute of Psychiatry, London:

> The neurologist's decisions and inferences hinge on matters that are common to all men and women: their neural physiology and anatomy. The raw material for his science is "the species in aggregate." The psychiatrist's conclusions, by contrast, hinge very often on matters peculiar to individuals: knowledge of persons in particular settings, of the "characters which make them unique."

Few neurologists or psychiatrists possess this particular blend of talents, training, and interests. As a result, only a small number of physicians, drawn from the neurological and psychiatric disciplines, practice the specialty. In 1986, less than 100 practicing neuropsychiatrists existed in the United States. In the ensuing five years, the field underwent a surge of growth: two textbooks, two journals, and the formation of the American Neuropsychiatric Association with over 200 members.

What will be neuropsychiatry's future? Will it once again cycle back into oblivion? This seems unlikely. In the light of what we have learned in the last two decades, a return to strictly psychological approaches to understanding and treating emotional disorders is as difficult to imagine as a world without computers or television. With additional knowledge about the brain and its relation to behavior, neuropsychiatry is likely to exert even more influence. It will do so as long as its advocates do not make the mistake of claiming that an emphasis on the brain precludes the central humanistic mission shared throughout history by all schools and methods of psychiatry: providing sympathetic understanding, treatment, and support for those suffering from mental illnesses.

Further reading

Cummings JL (1985): *Clinical Neuropsychiatry*. Orlando, FL: Grune and Stratton

Gualtieri C (1991): *Neuropsychiatry and Behavioral Pharmacology*. New York: Springer Verlag

Restak R (1986): Neuropsychiatry. *Psychiat Clin N Amer* 9:2

Reynolds EH, Trimble M (1989): *The Bridge Between Neurology and Psychiatry*. Edinburgh: Churchill Livingstone

Trimble M (1991): *The Psychoses of Epilepsy*. New York: Raven Press

The NO Hypothesis

P. Read Montague

The NO hypothesis, put forward by J. A. Gally, P. R. Montague, G. N. Reeke, and G. M. Edelman in 1990, proposed three physiological and developmental functions of nitric oxide (NO) in the vertebrate central nervous system (CNS). These are (1) regulation of blood flow in accordance with local neural activity at excitatory synapses; (2) modulation of the efficacy of synaptic contacts within a small tissue volume, such that simultaneously active synapses are strengthened and others are weakened; and (3) shaping and segregation of axonal arbors during synaptogenesis, so that neurons that tend to fire together innervate common volumes of target tissue.

The second and third parts of the NO hypothesis involve what might be considered learning at individual synaptic contacts. These hypotheses suggest how NO could mediate changes in the efficacy of transmission at individual synapses in a manner consistent with experimental observations.

Nitric oxide (NO)

Nitric oxide is an oxygen free radical that is produced in cells of the immune system as well as in neurons in the CNS. The recent excitement about the possible roles of NO in the vertebrate nervous system derives from an experiment done by J. Garthwaite in 1987. He showed that cerebellar granule cells release NO in response to glutamate application and that this effect is blocked by N-methyl-D-aspartate (NMDA) receptor antagonists. This finding was confirmed by Bredt and Snyder, who subsequently isolated and characterized a synthetic enzyme for NO called NO synthase. NO synthase produces NO from the basic amino acid L-arginine and is activated by Ca^{2+}-calmodulin (Fig. 1). These findings suggest that activation of NO synthase and subsequent NO production would be controlled by fluctuating excitatory neural activity in local volumes of neural tissue through postsynaptic calcium levels. NO readily diffuses through cell membranes, is produced in a calcium-dependent fashion upon activation of NMDA receptors, is highly reactive with a biological half-life probably less than 1 second, and has soluble intracellular enzymes as targets. Moreover, NO is a potent vasodilator and thus could link local changes in neural activity with local changes in blood flow. Since NO is destroyed by hemoglobin, this blood flow effect would provide one mechanism by which NO levels would be self-limiting in an active volume of tissue.

Nitric oxide release and cerebral blood flow

In 1890, Sherrington and Roy demonstrated that blood flow in the brain is under local and endogenous control. Although it is known that local neural activity is linked to local changes in cerebral blood flow, the nature of this link is unknown. The proposed role of *neurally derived* NO in the regulation of the blood flow in the brain parallels the previously established role of NO produced in the endothelial cells of blood vessels. The NO produced in these endothelial cells, called endothelial derived relaxing factor (EDRF), causes relaxation of the smooth muscle cells that control the diameter of blood vessels. NO produced in the endothelium thus causes local increases in blood flow. Likewise, the NO hy-

pothesis suggests that NO produced in neurons due to excitatory activity contributes to local changes in cerebral blood flow. This proposal is an obvious one, but it provides a simple explanation for the link between transmission at glutamate synapses and local changes in cerebral blood flow.

Spatial signaling and the modulation of synaptic efficacy

The NO hypothesis posits the existence of a new kind of signaling mechanism (Fig. 2B), called a *spatial signal,* to explain how temporal correlations in the firing of axonal terminals could be communicated throughout local volumes of neural tissue to control changes in synaptic efficacy. In par-

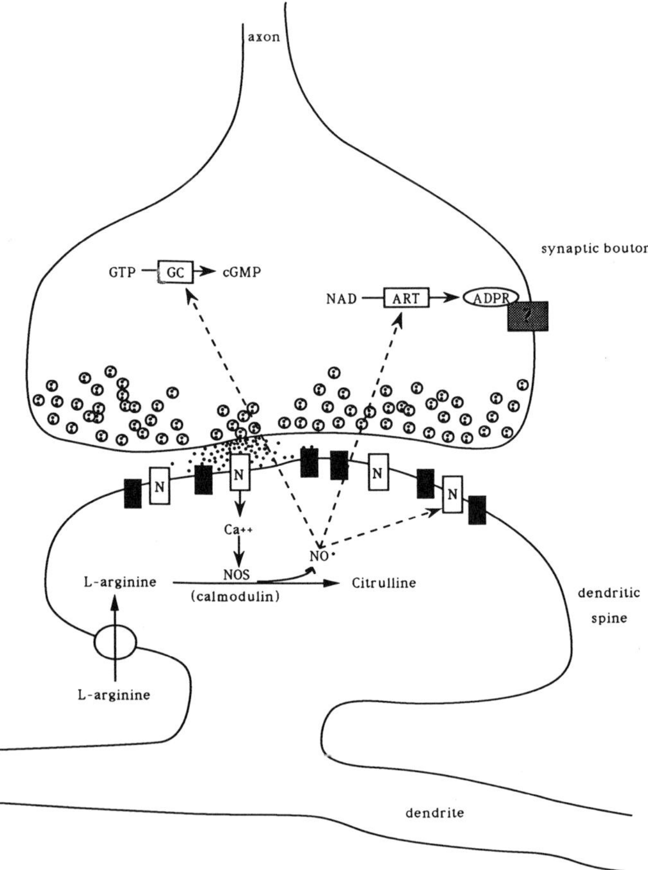

Figure 1. Production of nitric oxide from L-arginine. Glutamate is released from presynaptic terminals in response to depolarization; it binds to glutamate receptors including the NMDA receptor; calcium enters the postsynaptic spine through activated NMDA receptors; in the spine, the elevated calcium level activates NO synthase, which produces NO from the basic amino acid L-arginine. The two known target enzymes for NO are GC (soluble guanylate cyclase) and ART (ADP-ribosyltransferase); however, the function of these enzymes in the brain remains obscure. ADPR = ADP-ribose, N = NMDA receptors; cGMP = cyclic guanosine monophosphate; GTP = guanosine triphosphate, solid rectangles = other receptors including other glutamate receptors.

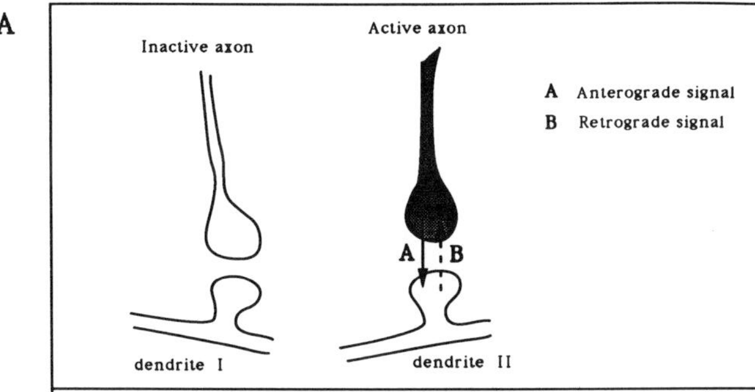

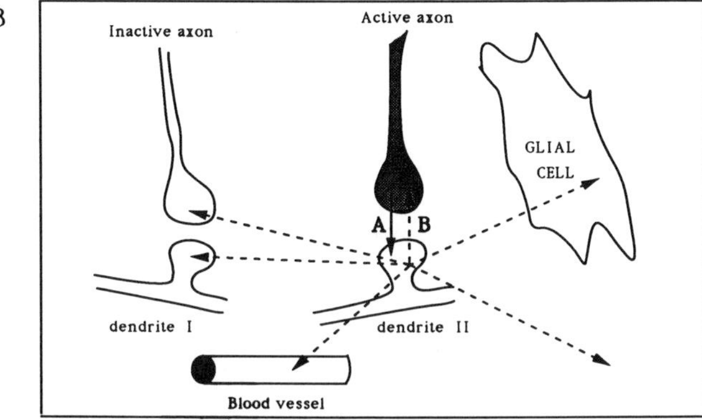

Figure 2. (A) Anterograde and retrograde signals in synaptic transmission. Evidence for the retrograde signal currently is found only in the hippocampus. (B) Spatial signaling via a rapidly diffusing substance. In this scheme, anterograde neurotransmission is followed by the production of a diffusible spatial signal in the postsynaptic dendritic spine. The diffusible spatial signal would extend to all elements in a local volume. In this scheme, the retrograde signal (B) is a special case of the volume effect.

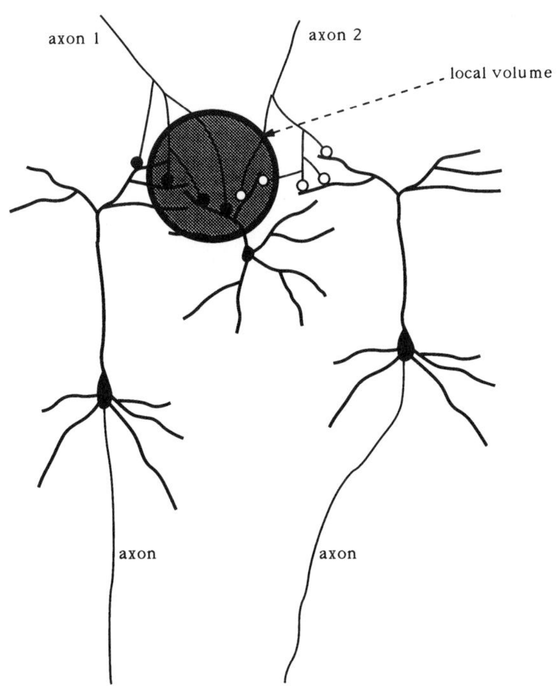

Figure 3. Volume learning. As described in the text, local volumes of neural tissue are the elements that learn associations between convergent afferent pathways. In this diagram, synaptic terminals from axon 1 can influence changes in synaptic efficacy at the terminals from axon 2 and vice versa *whether or not* the axons innervate the same postsynaptic cell. The *size* of the local volume in which the interaction occurs depends on the temporal correlations in the activity of axon 1 and axon 2, the production rate of the substance in response to the activity, and the destruction rate of the substance. These dependencies suggest that the size of these local volumes could change depending on the patterns of activity they experience.

ticular, the NO hypothesis suggests that the oxygen free radical nitric oxide (NO) mediates this communication both during the development of neural architecture and subsequently in an adult nervous system.

The existence of a spatial signal is suggested by evidence from the developing vertebrate brain. During development, axon terminals segregate into spatial domains based on their firing patterns. Those axons that tend to fire together make stabilized synaptic contacts in common volumes of neural tissue, and those that tend not to fire together segregate into separate volumes. This segregation of axon terminals has been called *activity-dependent segregation* or refinement. The activity-dependent segregation effect suggests that some kind of signal must communicate these patterns of activity throughout a volume of tissue to all those presynaptic terminals that innervate that volume (Figs. 2B and 3). If this were not the case, axon terminals could not be included or excluded from the volume of neural tissue based on their firing patterns. Since this signal transforms temporal correlations among synaptic terminals into a *signal which extends throughout a local volume of tissue,* it has been termed a "spatial signal." The NO hypothesis suggests that this spatial signaling is mediated by nitric oxide.

The synaptic learning mechanism suggested by the NO hypothesis is as follows: Activity at axon terminals causes the release of a rapidly diffusible, short-lived compound from postsynaptic dendrites (Fig. 2B). This substance moves directly through the tissue to influence all synaptic contacts

in a local volume. Synapses that are active when the surrounding volume contains high levels of the substance are strengthened. Silent synapses in those regions with high substance levels or active synapses in regions with low substance levels are weakened. In this scheme, the fluctuating substance levels reflect the activity of many synapses within a local volume.

This scheme stores associations among coactive terminals in the distribution of synaptic efficacies in a local volume. The spatial signaling mechanism, therefore, permits local *volumes* of neural tissue to learn associations (see Fig. 3). Neurons contributing dendrites into a number of such volumes would develop specific response properties by integrating input from both the local volumes as well as the direct synaptic input onto their dendrites.

How would spatial signaling through NO work in a volume of tissue?

It is likely that production of detectable levels of NO would require the nearly simultaneous firing of many synapses within a local volume. In such a local volume, the neuronal compartments that contained NO synthase and experienced a rapid elevation in Ca^{2+} concentration would become a composite source of a rapidly expanding sphere of NO that could affect presynaptic, postsynaptic, and glial elements throughout the volume. After the return of intracellular Ca^{2+} to its normal level, NO synthase would inactivate and the production of NO would cease. The concentration of NO present in that volume would quickly diminish as a result of dilution via diffusion, destruction by hemoglobin, and chemical degradation through oxidation to nitrites and nitrates. During its brief existence, however, NO could cause physiological responses by modulating the activity of target enzymes. The best characterized of these is the soluble form of guanylate cyclase, which is induced to initiate the synthesis of cGMP from GTP when NO binds to its heme group. There are many possible functional consequences of elevated intracellular levels of cGMP, but none of these have been well delineated in the cerebral cortex and thalamus. NO can also activate an ADP-ribosyltransferase present in brain tissue, but the consequences of this activation are likewise unknown.

Spatial signaling in LTP and activity-dependent segregation

In 1991, Montague, Gally, and Edelman showed that this kind of spatial signaling mechanism, acting in concert with axonal growth and retraction during development, can account for the self-organization of various neuroanatomical structures in the thalamus and cerebral cortex. Through large-scale computer simulations, they illustrated that an activity-dependent spatial signal could operate to self-organize ocular dominance stripes, somatotopic maps, and the refinement of topographic mappings between sensory sheets and CNS structures. To date, no direct experimental tests of the role of NO in activity-dependent segregation have been performed. In an adult brain, when large-scale changes in neuroanatomy are no longer possible, this spatial signaling mechanism and its attendant rules for synaptic change are consistent with experimental results from work on long-term potentiation (LTP). LTP is a persistent change in the efficacy of transmission between pre- and postsynaptic elements following epochs of sufficient correlation between presynaptic activity and postsynaptic depolarization. Because of this "experience-dependent" plasticity, LTP has long had an appeal as a neural basis for learning in the vertebrate CNS.

Recent experimental work on NO and synaptic plasticity

Experimental evidence suggests that the induction of LTP requires postsynaptic activation of NMDA receptors followed by the production of a retrograde signal (Fig. 2A) that moves to the presynaptic terminal to affect subsequent transmitter release. In this context, NO produced at active glutamatergic synapses would serve as the retrograde signal to the presynaptic terminal, yet it would also control synaptic plasticity at synaptic sites in the vicinity, *whether or not* they innervate the same postsynaptic cell (Fig. 2B). In 1991, four laboratories reported, almost simultaneously, that inhibitors of NO synthase block the induction of LTP in hippocampal slices. These observations provide direct evidence that NO may play an essential role in LTP induction and thus act as a spatial signal in the vertebrate CNS.

A conundrum for the NO hypothesis in synaptic learning

Recent experimental work has shown that the synthetic enzyme for NO in the vertebrate brain, NO synthase, is equivalent to a previously described enzyme called NADPH diaphorase. This enzyme is found diffusely throughout cortical and subcortical regions and is distributed throughout the soma and dendrites of the neurons in which it is found: smooth stellate cells that stain positively for both GABA and somatostatin. These cells represent only 1–2% of the neurons in the CNS. This suggests that a special subset of cells, those containing NADPH diaphorase, would direct learning at synapses throughout the CNS. Contradicting these observations that NO synthase is located only in local circuit neurons, E. Shuman and D. Madison at Stanford University and O'Dell, Hawkins, Kandel, and Arancio at Columbia University, have provided functional evidence that NO is made in hippocampal pyramidal cells. A definitive answer as to the actual distribution of functional NO synthase and the possible role of NO as a spatial signal awaits further work.

Emerging questions for the role of NO and spatial signals in the brain

The LTP work suggests that NO may be critically important in learning and memory or it could simply be one more in a long list of required signals no one of which is sufficient for learning. One central question for the role of NO is the spatial scales over which it can act during normal synaptic transmission. The answer to this question will require definitive evidence of exactly which cells contain a functioning version of NO synthase, how much NO is made during normal synaptic transmission, and catabolic rates for NO.

One important but completely unexplored question involves the possible effects of NO on synaptic terminals containing other neurotransmitters such as norepinephrine, acetylcholine, or peptides. If production of NO at glutamate synapses affects the release of glutamate, then it is reasonable to suggest that this NO production could also affect the release from presynaptic terminals containing other neurotransmitters that act at different time scales and through different mechanisms. For example, in the cerebral cortex and hippocampus, this would imply that transmission and the potentiation of transmission at glutamate synapses could also modulate the efficacy of other synaptic contacts in the vicinity. In this fashion, the local release of a transmitter like norepinephrine could also be learned, depending on how frequently these terminals fired synchronously with the local glutamate terminals.

Further reading

Bredt DS, Snyder SH (1992): Nitric oxide, a novel neuronal messenger. *Neuron* 8:3–11

Gally JA, Montague PR, Reeke GN, Edelman GM (1990): The NO hypothesis: Possible effects of a short-lived rapidly diffusible signal in the development and function of the nervous system. *Proc Nat Acad Sci* (USA) 87:3547–3551

Garthwaite J, Charles SL, Chess-Williams R (1988): Endothelium-derived relaxing factor release on activation of NMDA receptors suggests role as intercellular messenger in the brain. *Nature* (London) 336:385–388

Haley JE, Wilcox GL, Chapman PL (1992): The role of nitric oxide in hippocampal long-term potentiation. *Neuron* 8:211–216

Montague PR, Gally JA, Edelman GM (1991): Spatial signaling in the development and function of neural connections. *Cereb Cort* 1(3):199–220

O'Dell TJ, Hawkins RD, Kandel ER, Arancio O (1991): Tests of the roles of two diffusible substances in LTP: Evidence for nitric oxide as a possible early retrograde messenger. *Proc Nat Acad Sci* (USA) 88:11285–11289

Shuman E, Madison D (1991): A requirement for the intercellular messenger nitric oxide in long-term potentiation. *Science* 254:1503–1506

Nociceptors, Novel Classes

Stephen McMahon and Martin Koltzenburg

There seems to be consensus among scientists and clinicians that the peripheral neural mechanisms of pain have by and large been settled. The conventional wisdom states that specialized primary afferents—the nociceptors—detect actual or imminent tissue injury and by use of specific and fairly inflexible connections to cognitive or motor centers initiate the body's protective responses. This straightforward model originates essentially in the views of researchers working at the turn of the century. Their influence appears to have dominated the conceptual approach to pain and nociception, and even today most of us will probably still regard the nociceptors as the body's alarm system sensing immediate threat. Within this conceptual framework, researchers sought and found primary afferents in animals and humans that fit the predicted properties of nociceptors reacting to acute tissue injury. Indeed, it is now established that these specific nociceptors can encode important aspects of some stimulus-induced pain, including its magnitude and temporal and spatial properties.

Although the conventional view describes one important aspect of nociception, the generality of this concept is now in need of revision with the recent emergence of new data that has relevance for those interested in pain mechanisms. Evidence is emerging that the structural and functional characteristics of nociceptive afferents are constantly changing, indicating that the peripheral nociceptive nervous system adapts dynamically in response to tissue perturbations. In this article we will concentrate on these properties. One important aspect of this primary afferent plasticity, which has been recognized recently, appears to be the recruitment of dormant primary afferents that are not excited in normal tissue but that become active following tissue injury. The recruitment of these novel types of nociceptors is not achieved by acute or transient stimuli but requires more serious tissue damage as might occur when the immediate alarm system has failed. It is evident that these long-lasting changes are particularly important, if we want to understand the neural mechanisms affecting patients in chronic pain. In other words, the mechanisms operating in recuperation of damaged tissue are qualitatively different and not simply a prolongation of acute phenomena.

A striking example of the peripheral plasticity of nociceptors is their morphological malleability and the dynamic fluctuation of receptive field size. It has been determined that the receptive terminals of primary afferents are far from being stable, as might have been anticipated from the elegant reconstruction of nociceptors from electron micrographs. On the contrary, using vital staining and repeated microscopy, a remodeling has been visualized within hours. So far, little is known about the factors that regulate the malleability of the terminal arborization, yet it could be an adaptive regenerative response following minor tissue abrasion or trivial trauma. Moreover, the expansion of receptive fields lasting tens of minutes has been witnessed neurophysiologically in a subpopulation of nociceptors. An example of this phenomenon is illustrated in Figure 1A where, following application of a noxious stimulus, the receptive field enlarged to incorporate the adjacent stimulated area. This suggests that some terminals of nociceptors are normally dormant but may be sensitized and become active in certain conditions. Moreover, research on invertebrates indicates that nociceptors

might even be instigated to grow and expand their receptive area following application of a noxious stimuli.

In addition to the changing morphology, the receptive properties of nociceptors are highly variable and even a brief stimulus can entail marked excitability changes, resulting in considerable change in the functional properties. By definition, the adequate stimuli can produce tissue damage, and the immediate consequences for nociceptors are, surprisingly, more often than not desensitization. Thus, repeated heating, pinching, or application of algesic chemicals, notably bradykinin, usually causes tachyphylaxis with rapid stimulus repetition. This fatigue of the nociceptors' discharge is particularly striking with long-lasting stimuli such as squeezing of skin folds. Thus, it is a paradox that the more damage the stimulus inflicts, the weaker becomes the immediate afferent barrage. However, if some tissue damage

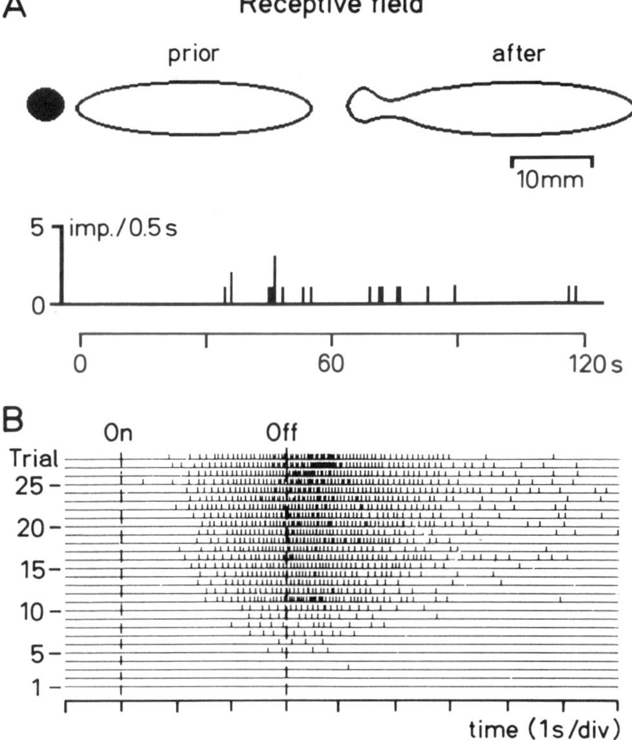

Figure 1. Examples of the dynamic change of receptive field size and coding properties of thin myelinated nociceptors. (A) In the rat tail, noxious pressure of 4 N applied outside the receptive field for 120 seconds (black dot) elicits only some few action potentials with a long time lag. Subsequently, however, the receptive field expands to incorporate the stimulated area. (Modified from Reeh et al. 1987 with permission.) (B) In the monkey glabarous skin a short heat stimulus of 53°C lasting 3 seconds does not initially excite the unit. However, with stimulus repetition and an interval of 30 seconds between trials the units lower their thermal threshold, resulting in a progressive increase of the afferent activity. (Reprinted with permission of author from Campbell JN et al. (1979): Sensitization of myelinated nociceptive afferents that innervate monkey hand. *J Neurophysiol* 42:1669–1679.)

has taken its effect, nociceptors can also become sensitized. Some begin to discharge in the absence of further manipulations (so called spontaneous activity) or lower their threshold so that previously ineffective stimuli become excitatory. It is plain that this temporal increase of afferent activity can result in spontaneous pain, tenderness, and increased pain perception with previously nonpainful stimuli.

The processes of sensitization and desensitization indicate that the excitability of the nociceptive terminal following adequate stimulation is very dynamic and emphasize that the history of the nociceptor is a crucial determinant of its properties. There is yet a further mechanism that can contribute to a change in nociceptive activity. It has been demonstrated that some nociceptors can take on novel receptive and coding properties in the aftermath of tissue damage. In normal monkey glabarous skin a moderately painful heat stimulus excites mainly unmyelinated nociceptors. Although thin myelinated nociceptors have a thermal sensitivity, they are apparently not activated, as most have considerably higher thresholds that are above the withdrawal threshold. However, following stimulus repetition or creation of a burn, the threshold of these afferents decreases considerably, and importantly, the stimulus response function changes so as to encode the increased thermal pain perception in the burned skin (Fig. 1B). In contrast, most unmyelinated afferents become desensitized. This implies a switch between the dominant nociceptor classes that inform the CNS of noxious thermal events.

The importance of the recruitment of nociceptors and the acquisition of a new sensitivity as a sequela of tissue injury has been further emphasized by the discovery of a novel class of primary afferent neurons. These fibers do not respond to transient excessive mechanical or thermal stimuli, but they appear to have a sensitivity that activates them, if the tissue becomes inflamed. This novel type of receptor has been given various names such as "silent afferents," "sleeping nociceptor," and "mechanically insensitive afferents." They are present in many tissues, including skin, join, and viscera, and in several species including rat, cat, and monkey. The exact percentage of this class of receptors has not been fully worked out and it is likely that their percentage varies between tissues and species.

The excitability changes of this novel type of receptors has been directly observed with neurophysiological techniques that have recorded units prior to and at the onset of an acute inflammation. In the feline joint, many thin myelinated or unmyelinated afferents do not respond to even strong stimuli, such as twisting of the joint, that are aversive in the conscious animal. Yet, in the hours after the induction of an experimental arthritis, previously unresponsive afferents exhibit ongoing activity and develop new receptive properties (Fig. 2). They begin to discharge during movements in the normal working range of the joints and some units display mechanosensitive receptive fields in the joint capsule that were absent prior to inflammation.

Essentially similar results have been obtained for visceral organs that are supplied by the pelvic nerve. Few unmyelinated afferents respond to acutely painful stimuli such as overt distension of urinary bladder or distal colon (which are the main organs supplied by this nerve). However, with acute inflammation of the urinary bladder a significant population of the previously unresponsive afferents start to discharge. Most of these afferents are chemosensitive and respond vigorously to intravesical injection of irritant chemicals (Fig. 3). These afferents show an initial burst of activity and continue to discharge as the inflammation progresses. In the inflamed tissue some afferents also start to respond to mechanical stimuli that had previously been ineffective and this novel property can persist for several hours. Chemosensitive afferents with negligible mechanical or thermal sensitivity have also been identified in the skin. Again, some of these units take on new receptive properties and start to respond to heat or mechanical stimuli that are usually used to identify nociceptive afferents.

The important conclusion from these examples is that most tissue contains afferent fibers that require some sort of chronic tissue injury to become active. Estimates on the relative proportion of these fiber classes indicate that in the

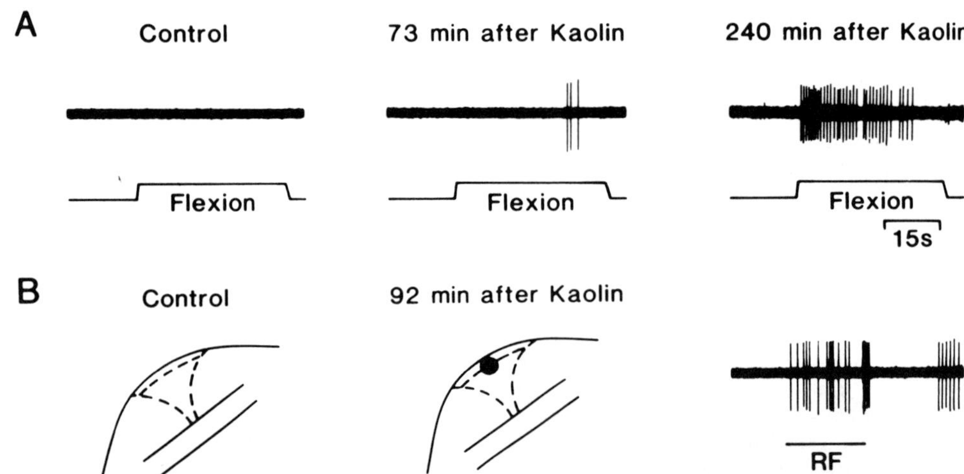

Figure 2. Recruitment of an initially unresponsive unmyelinated afferent fiber innervating the knee joint of the cat during inflammation. In control conditions the unit did not respond to movements (A), nor was there a mechanosensitive receptive field (B). Some time after induction of an experimental inflammation with kaolin the unit started to respond to innocuous movements. It also exhibited a receptive field on the medial side of the joint where pressure with a glass rod excited the unit (RF). (Reprinted with permission of author from Schaible H-G and Schmidt RF (1988): Time course of mechanosensitivity changes in articular afferents during a developing arthritis. *J Neurophysiol* 602180–2195.)

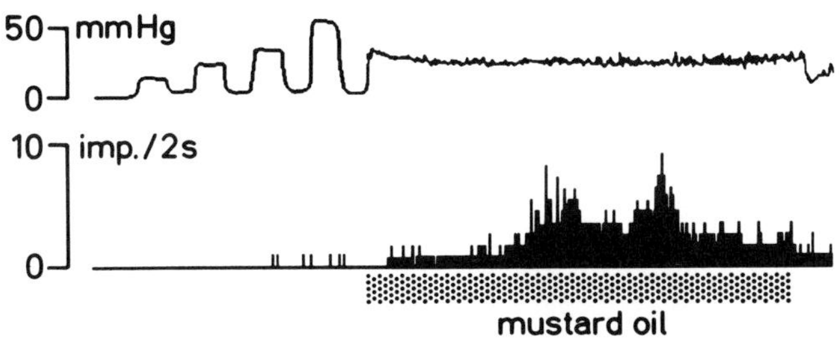

3 hours after mustard oil:

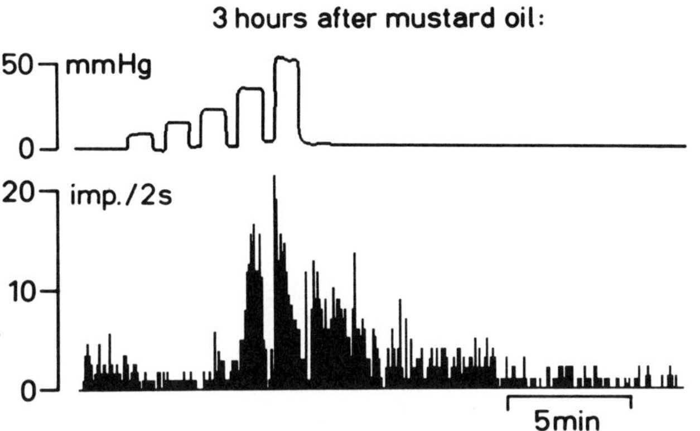

Figure 3. Recruitment of an initially unresponsive unmeylinated afferent fiber innervating the urinary bladder of the cat at the onset of an acute inflammation. In the control state the unit did not respond to distension of the bladder (upper trace) that can be presumed to be noxious. However, the unit discharged vigorously during in intravesical injection of dilute mustard oil. As the chemically induced inflammation progressed the unit developed ongoing activity and started to respond to distension of the bladder. (Reprinted with permission of author from Häbler H-J et al. (1990): Activation of unmyelinated afferents by mechanical stimuli and inflammation of the urinary bladder of the cat. *J Physiol* (London) 425:545–563.)

feline urinary bladder four times as many unmyelinated afferents are responsive during inflammation compared with healthy tissue. In the knee joint of the cat some one third of the nociceptive afferents are "sleeping nociceptors." A somewhat smaller proportion have been reported to exist in the skin of the rat or primate. The gist of these numbers is that most tissues are endowed with a reserve of nociceptors that only come into action following special pathological events, notably inflammation. Taken together with sensitization and desensitization affecting the more classical type of nociceptors, it is obvious that the peripheral neurophysiological correlates of nociception depend critically on the state of the tissue. The transition of normally silent afferents into an active state could significantly increase the afferent activity impinging on second order neurons in the spinal cord. By the same token it is clear that the CNS has to handle information relevant for pain from a constantly changing population of active nociceptive afferents.

Further reading

Campbell JN, Meyer RA, LaMotte RA (1979): Sensitization of myelinated nociceptive afferents that innervate monkey hand. *J Neurophysiol* 42:1669–1679

Campbell JN, Raja SN, Cohen RH, Manning DC, Khan AA, Meyer RA (1989): Peripheral neural mechanisms of nociception. In: PD Wall, R Melzack, eds. *Textbook of Pain,* 2nd ed. Edinburgh: Churchill Livingstone, pp 22–45

Häbler HJ, Jänig W, Koltzenburg M (1990): Activation of unmyelinated afferents by mechanical stimuli and inflammation of the urinary bladder of the cat. *J Physiol* (London) 425:545–563

McMahon S, Koltzenburg M (1990): The changing role of primary afferent neurones in pain. *Pain* 43:269–272

Reeh PW, Bayer J, Kocher L, Handwerker HO (1987): Sensitization of nociceptive cutaneous nerve fibers from the rat's tail by noxious mechanical stimulation. *Exp Brain Res* 65:505–512

Schaible H-G, Schmidt RF (1988): Time course of mechanosensitivity changes in articular afferents during a developing arthritis. *J Neurophysiol* 60:2180–2195

O

Obsessive-Compulsive Disorder, Neurobiology of

Thomas R. Insel

Obsessive-compulsive disorder (OCD) is one of the most common major psychiatric syndromes, affecting over 2% of the U.S. population. There is an equal prevalence of males and females and as many as 30% of cases begin in childhood. The syndrome is characterized by intrusive, reprehensible thoughts and repetitive, ritualized behaviors. A typical OCD patient might report a nagging, irrepressible sense of being contaminated. Although the patient realizes this idea is irrational, this feeling of contamination generates such intense discomfort that the patient washes compulsively, often to the point of abrading the skin on both hands. Other patients will have to check repetitively to assure themselves that they have not committed some harmful or sinful act. Although the cause of this syndrome remains a mystery, psychiatrists for the past century have suggested that such bizarre and "egodystonic" symptoms must have an organic or neurobiological basis. In the past decade, the search for a neurobiological basis for OCD has included genetic, neuroanatomic, and neuropharmacologic studies. The etiology of this disorder (or group of disorders) remains unclear, but several interesting leads have emerged from each of these areas of study.

Genetic studies

The search for a genetic factor has been limited by the absence of large pedigrees positive for OCD. Nevertheless, a genetic etiology has been suggested from various reports of twins concordant for OCD. It is not yet clear, however, if the genetic factor in question codes for OCD or some character trait that may predispose to the development of the syndrome. It should also be noted that these sorts of reports are subject to a reporting bias: discordant twins are less likely to be reported in the literature. Furthermore, even the presence of a very high rate of concordance does not exclude epigenetic factors.

In addition to reports of concordance in twins, several family studies have described an increased prevalence of OCD in the first-degree relatives of OCD patients. Again, this observation does not necessarily imply a genetic as opposed to an experiential factor. Perhaps the most important report in this general area described a high incidence of OCD in the family members of Tourette's syndrome probands. Tourette's patients frequently have obsessional symptoms, but interestingly, in this study, OCD was found in the relatives of both obsessional and nonobsessional Tourette's patients, thus suggesting a genetic link between these two disorders.

Neuroanatomic studies

There has long been a belief in the neurologic basis of OCD, partly because the symptoms seem so contrary to the conscious will of the patient and partly because some OCD patients often have subtle motor tics or clumsiness, which might signify a neurologic disorder. Several frank neurologic disorders have been associated with the symptoms of OCD. Perhaps the best known example occurred following the von Economo's influenza epidemic of 1919. Many of these patients developed a parkinsonian syndrome, and among this subgroup a fraction showed both oculogyric crises and fairly classic OCD. The association of parkinsonism and OCD suggests abnormal dopaminergic innervation of the striatum, an area that is involved in three other neurologic disorders associated with OCD: Sydenham's chorea, bilateral necrosis of the globus pallidus, and Tourette's syndrome. It is unlikely that any of these syndromes represents an etiology for very many of the psychiatric patients presenting with OCD, but the association with striatal pathology may serve as a marker for an anatomic site in which to seek physiologic, if not pathologic changes in OCD.

Historically, another suggestion for neurological involvement in this disorder came from the relative success of psychosurgery. Here, "relative" refers to the consistently higher rate of symptom relief following surgery in OCD compared with the same intervention in several other disorders such as anorexia nervosa, schizophrenia, and depression. Specifically, two surgical lesions have been reported to be especially useful for OCD: cingulectomy and "stereotaxic leukotomy" (i.e., transection of the tracts from frontal cortex to subcortical sites such as striatum and thalamus). Indeed, using modern neurosurgical techniques with stereotaxic placement of both of these lesions, this procedure has been reported successful in as many as 80% of OCD patients refractory to behavior therapy and pharmacotherapy. These methods have not undergone rigorous double-blind testing with sham surgery, but even as extensive case reports, these results are impressive and may provide some insight into neuroanatomic circuits involved in OCD.

The most exciting development toward a neuroanatomic model of OCD is certainly the advent of new imaging techniques. Several studies have used PET scans with ^{18}F-2 deoxyglucose to localize regions of abnormal metabolic activity in the brains of OCD patients. Results across studies describe three areas of abnormally high activity: the orbital region of the frontal cortex, the cingulate cortex, and the head of the caudate nucleus. The changes reported in orbital and cingulate cortex are perhaps most remarkable in view of the many regions such as hippocampus and temporal cortex, previously described as active in anxiety states, that appear normal in OCD. It is also important to note that the striatum was previously implicated by neurologic disorders associated with OCD and that transecting the tracts projecting from orbital cortex to striatum or lesioning the cingulate appears to decrease obsessional symptoms. On the basis of these findings, Rapoport (1991) has suggested that an orbital frontal-striatal-thalamic-cortical circuit is activated in OCD, resulting in the emergence of excessive grooming activity.

Neuropharmacologic studies

From a neuropharmacologic perspective, the most remarkable aspect of OCD is its high selectivity for drug response. Most drugs that reduce anxiety, depression, or psychosis are not effective against obsessions. Thus far the only medications that appear effective in studies of OCD are clomipramine, fluvoxamine, and fluoxetine. These three compounds are all antidepressants, yet they reduce obsessional symptoms in nondepressed OCD patients. For this reason, they have been described as antiobsessional drugs. What is surprising is that so many other excellent antidepressants, many of which are structurally related to these antiobsessional drugs, appear ineffective for the symptoms of OCD.

What distinguishes these few antiobsessional compounds from the antidepressants that are not effective? All three of these drugs are potent inhibitors of the synaptic reuptake of serotonin. Serotonin is cleared from the synaptic space largely by reuptake into the presynaptic nerve terminal, where it is either recycled into storage vesicles or metabolized to 5-hydroxyindoleacetic acid (5-HIAA). Reuptake is thus the first and most important step for inactivating serotonin neurotransmission. Inhibiting reuptake makes more serotonin available to postsynaptic receptors and also reduces the formation of 5-HIAA. Although most antidepressants, including imipramine and amitriptyline, inhibit the synaptic reuptake of serotonin, none are quite as potent as clomipramine (or its metabolite desmethylclomipramine), fluvoxamine, and fluoxetine.

Beyond this unique selectivity of serotonin uptake inhibitors for reducing OCD symptoms, the following results from clinical studies support a role for serotonin in the mediation of antiobsessional drug effects. With clomipramine treatment, the decrease in obsessional symptoms correlates with the decrease in the concentration of the serotonin metabolite 5-HIAA in cerebrospinal fluid and serotonin content in platelets, implicating the inhibition of reuptake in the antiobsessional effect. And in a study of changes after chronic treatment, a serotonergic antagonist, metergoline, was found to induce a partial relapse of symptoms in OCD patients who had improved on clomipramine.

Is serotonin involved in the pathophysiology of OCD? Serotonin cannot be directly measured in the CNS of patients, but most studies have not found significant abnormalities in the concentrations of serotonin or its metabolite 5-HIAA in blood, platelets, or CSF of OCD patients. Various serotonergic agonists have been administered to patients to assess responsiveness in one of the many central serotonin receptors. Here, the main finding is that m-CPP, a mixed serotonin receptor agonist, can induce obsessional symptoms when orally administered to OCD patients but is largely without effects in healthy controls. Several other compounds that affect brain serotonin receptors, such as 1-tryptophan, ipsapirone, and MK-212, have equal behavioral effects in OCD patients and healthy controls. The heightened responsiveness to m-CPP suggests an increase in the sensitivity of that population of 5-HT receptors subserving these behavioral effects. These effects are blocked by pretreatment with the nonselective serotonin antagonist metergoline and they are blunted by chronic treatment with clomipramine, but the exact subtype of receptor and the precise location of m-CPP's effects remain unknown.

Summary

OCD was for many years considered one of the classic forms of neurosis for which psychoanalysis was the recommended treatment. Although some patients undoubtedly improved while undergoing years of psychoanalysis, in most cases nondirective psychological treatments have not been useful. The reduction in OCD symptoms following pharmacologic and surgical interventions has spurred recent interest in a possible neurobiologic basis for OCD. Thus far, the data are inconclusive, although cerebral imaging studies point to focal metabolic abnormalities associated with the symptoms of OCD, and pharmacologic studies have raised the possibility of altered sensitivity to a serotonergic challenge. This disorder is currently the focus of intense investigation as these early leads are followed up by more extensive studies.

Further reading

Jenike MA, Baer L, Minichello WE, eds (1990): *Obsessive Compulsive Disorder: Theory and Management,* 2nd ed. Littleton, Massachusetts: Yearbook Publishers

Rapoport JL (1991): Recent advances in obsessive-compulsive disorder. *Neuropsychopharmacol* 5:1–10

Zohar J, Insel TR, Rasmussen S (1991): *The Psychobiology of Obsessive Compulsive Disorder.* New York: Springer Press

Olivopontocerebellar Atrophy

Sid Gilman

Olivopontocerebellar atrophy (OPCA) is a chronic progressive neurological disease of undetermined cause characterized by degeneration of neurons in the inferior olives, pons, and cerebellum. The disease occurs both sporadically and with hereditary transmission. When sporadic the disease is of undetermined cause. When hereditary, the disease is usually expressed as an autosomal dominant trait, but in some families it is inherited as an autosomal recessive trait. A locus on chromosome 6 has been identified in patients with dominantly inherited OPCA. The age of onset varies widely, but most patients become symptomatic in middle age. The presenting symptoms usually consist of ataxia of gait, followed soon afterward by incoordination of the legs, dysarthria, and incoordination of the arms. Neurological examination in symptomatic patients usually reveals ocular dysmetria, ataxic speech, and ataxia of limb movements and gait. In addition to signs of cerebellar disturbance, many patients show involvement of corticobulbar and corticospinal systems, with spasticity of speech, spasticity of the limbs, hyperactive deep tendon reflexes, and extensor plantar responses. Dementia has been reported in OPCA but is not common.

Neuropathological changes in OPCA

Typical cases of OPCA show at postmortem examination gross shrinkage of the ventral part of the pons, the inferior olives, the middle cerebellar peduncles, and the cerebellar cortex. The pons and inferior olives show loss of neurons with corresponding loss of pontocerebellar (mossy) fibers and olivocerebellar (climbing) fibers, respectively. Purkinje cells are severely reduced but not completely lost, and granule cells are reduced. Golgi cells are probably unaffected, and basket cells are relatively preserved. Dentate neurons also are relatively well preserved. Often the long tracts of the spinal cord are affected, and neurons in the striatum and substantia may degenerate. Neuronal loss in the basal nucleus of Meynert occurs frequently in OPCA.

Classification of the adult-onset cerebellar degenerations

Many attempts have been made to classify the adult-onset cerebellar degenerations. Most of these have been unsuccessful because they do not adequately account for the diversity of clinical presentations and neuropathological changes in patients with degenerative changes within the cerebellum and brainstem. Table 1 presents a classification. In this table, primary cerebellar degeneration denotes cases

Table 1.

Primary Cerebellar Degeneration
Olivopontocerebellar Atrophy
Cerebellar Cortical Atrophy
Olivocerebellar Atrophy
Machado-Joseph Disease

Secondary Cerebellar Degeneration
Remote Effects of Neoplasm
Alcoholic Cerebellar Degeneration
Drug-Related Cerebellar Degeneration

in which the cause of the degeneration is unknown. Olivopontocerebellar atrophy consists of cases in which the pathological changes affect the pons, cerebellum, and inferior olives. Cerebellar cortical atrophy is usually a sporadic disorder involving degeneration of the cerebellar cortex without involvement of the brainstem. This entity probably has a slower rate of progression than OPCA. Olivocerebellar atrophy is a rare hereditary degenerative disease involving the inferior olives and cerebellum, but sparing the pons. Machado-Joseph disease is a separate heredodegenerative disease transmitted as an autosomal dominant trait involving degeneration of the cerebellum, basal ganglia, and corticospinal projections. The secondary degenerations are those in which an immunologic disorder (remote effects of neoplasm) or toxins (alcoholic cerebellar degeneration, phenytoin-induced cerebellar degeneration) can be identified as causal agents.

Multiple system atrophy

In some cases, OPCA occurs as part of generalized multiple system atrophy (MSA). MSA is defined as an adult-onset sporadic neurodegenerative disease presenting clinically with any combination of extrapyramidal, cerebellar, and autonomic symptoms and signs. The neuropathological features of MSA include the changes seen in striatonigral degeneration (SND), Shy-Drager syndrome (SDS), OPCA, and degeneration of the central autonomic pathways associated with progressive autonomic failure (PAF). These neuropathological changes consist of degeneration with neuronal loss and gliosis in the putamen, substantia nigra, locus ceruleus, inferior olives, pontine nuclei, cerebellar Purkinje cells, dorsal vagal nuclei, intermediolateral columns of the spinal cord, Onuf's nuclei, vestibular nuclei, and pyramidal tracts. Other central nervous system structures that can be affected include the cerebral cortex, caudate nuclei, globus pallidus, subthalamic nuclei, red nuclei, dentate nuclei, anterior horn cells, sympathetic nuclei, and optic nerves. In SND the neuropathological changes affect chiefly the basal ganglia, in OPCA the inferior olives, pons and cerebellum, and in PAF and SDS the basal ganglia, brainstem, cerebellum, and spinal cord. In some patients, MSA begins with signs of cerebellar degeneration and features of extrapyramidal involvement and/or autonomic failure occur as the disease evolves. In others, levodopa-unresponsive, or poorly responsive, parkinsonism is the presenting disorder and later autonomic failure or cerebellar degeneration occurs.

Positron emission tomography studies of OPCA

Studies of cerebral metabolism with [^{18}F]fluorodeoxyglucose and positron emission tomography have demonstrated hypometabolism in the cerebellum and brainstem in OPCA. The degree of hypometabolism is directly related to the degree of atrophy of these structures in anatomical imaging studies and to the severity of the clinical neurological disorder. Cerebral metabolic rates in the cerebral cortex, thalamus, and basal ganglia are within normal limits in OPCA. Patients with MSA show marked hypometabolism in the cerebellum and brainstem, but also in the cerebral cortex, basal ganglia, and thalamus.

Neurotransmitter abnormalities in OPCA

Neurochemical studies of postmortem brain tissue in OPCA have demonstrated abnormalities of neurotransmitter levels and of the densities of neurotransmitter receptors. A number of biochemical profiles have been described in the cerebellar tissues of patients with dominantly inherited OPCA. One set of cases is characterized by decreased glutamate, aspartate, and GABA levels in the cerebellar cortex and by decreased GABA levels in the dentate nucleus. Taurine content is increased in the cerebellar cortex. A second group shows decreased levels of glutamate, aspartate, and GABA in the cerebellar cortex and decreased levels of GABA in the dentate nucleus. In this group, however, taurine content in the cerebellar cortex is normal. In a third group, the levels of glutamate, aspartate, and GABA are normal in the cerebellar cortex, but the GABA content of the dentate nucleus is reduced. In a fourth group, glutamate, aspartate, and GABA levels are decreased in the cerebellar cortex and GABA levels are decreased in the dentate nucleus; however, glutamate is decreased in other brain regions, notably the cerebral cortex and striatum.

The foregoing groups may not represent distinct biochemical subtypes of OPCA, but only differences in the degree of degeneration of particular types of neurons among different cases of OPCA. Evidence supporting this idea came from correlative studies of neurotransmitters and neuronal cell densities in OPCA. In these studies, the concentrations of glutamate and aspartate in the cerebellar cortex varied markedly from case to case. Glutamate concentrations in the anterior vermis were correlated with the density of granule cells there, and aspartate concentrations in the anterior vermis were correlated with the density of neurons in the inferior olives. GABA concentrations in the dentate nucleus were decreased in all cases of OPCA and were correlated with the degree of loss of Purkinje cells.

Neurotransmitter receptor abnormalities in OPCA

Only a few postmortem studies of GABA/benzodiazepine receptor binding in the cerebellum have been reported in patients with OPCA, and the results are variable. GABA binding in the cerebellar cortex was reported to be increased in one study and decreased in another. Benzodiazepine binding was within normal limits or elevated in the cerebellar cortex in one study and decreased in another. In preliminary studies of patients with OPCA utilizing positron emission tomography and the ligand [^{11}C]flumazenil, benzodiazepine binding in the cerebellum was found to be within normal limits or decreased.

Other neurochemical abnormalities in OPCA

A deficiency in the specific activity of glutamate dehydrogenase in fibroblasts and leukocytes has been reported in autosomal recessive and autosomal dominant forms of OPCA, but abnormality of GDH activity has not been found in postmortem tissue. Other neurochemical abnormalities have been found in postmortem tissues of OPCA patients, including reduction of noradrenalin levels in the cerebellar cortex, decreased levels of choline acetyltransferase in the cerebral cortex and hippocampus, and changes in the concentration of immunoreactive thyrotropin-releasing hormone in the cerebellar cortex, dentate nuclei, and olivary nuclei.

Therapy of OPCA

No satisfactory treatment has been developed to arrest the progress of OPCA or to treat the symptoms effectively. Elastic stockings and treatment with fluorocortisone acetate for the postural hypotension resulting from autonomic failure can be helpful in MSA. Clonazepam in small doses administered at bedtime lessens the cerebellar symptoms of a small number of patients with OPCA.

Further reading

Duvoisin RC, Plaitakis A, eds (1984): Advances in Neurology. Vol. 41, *The Olivopontocerebellar Atrophies*. New York: Raven
Gilman S, Bloedel J, Lechtenberg R (1981): *Disorders of the Cerebellum*. Philadelphia: FA Davis
Gilman S, Markel DS, Koeppe RA, Junck L, Kluin KJ, Gebarski SS, Hichwa RD (1988): Cerebellar and brainstem hypometabolism in olivopontocerebellar atrophy detected with positron emission tomography. *Ann Neurol* 23:223–230
Harding AE (1984): *The Hereditary Ataxias and Related Disorders*. New York: Churchill Livingstone

P

Pain as Learning

Clifford J. Woolf

The external environment is potentially hostile to all living organisms. Essential for survival is the ability to detect harmful stimuli and, having established their presence, to initiate avoidance responses. This defense strategy is present in all the major animal phyla and escape responses are the evolutionary antecedent of pain. While unicellular organisms can move away from threatening environments, it is with the emergence of a specialized nervous system in multicellular organisms that elaborate and coordinated escape responses become prominent. Simple invertebrates, for example, have sensory neurons that can discriminate low- and high-intensity peripheral stimuli and, having detected the latter, activate motor neurons to produce a movement of that body part in contact with the offending stimulus, away from the stimulus. In vertebrates this type of escape response has become organized as a stereotyped segmental reflex, the flexion withdrawal reflex, whereby noxious stimuli result in a contraction of appropriate flexor muscles and the relaxation of antagonist extensor muscles. The sensory inflow producing the flexion withdrawal reflex also activates ascending pathways to the brain, where the stimulus is perceived as unpleasant, uncomfortable, distressing, or intolerable—what we have come to call pain.

The escape responses and the sensation of pain elicited by noxious stimuli represent an early warning system. In this sense pain is a highly adaptive sensation, preventing excessive stimulation and minimizing tissue damage. We experience such pain in our everyday lives whenever we make contact with objects that are too hard, too hot or cold, or contain irritable chemicals. Absence of such a warning system, as in those patients with congenital analgesia due to a lack of unmyelinated primary afferents, can lead to horrific deformities. This pain is a true physiological sensation, with distinct sensory thresholds, a clear stimulus-response relationship, and a highly specialized neural apparatus devoted to its function.

Physiological pain reflects the high degree of specialization of primary sensory neurons (Fig. 1). There is marked distinction between those primary sensory neurons that are activated by low-intensity or innocuous stimuli and those whose thresholds are such that only intense, damaging, or noxious stimuli activate them and which are consequently called nociceptors. This functional specialization is conserved in the anatomical arrangements of the central terminals of the two different classes of afferents; each terminate in distinct regions of the dorsal horn of the spinal cord. This, together with the somatotopic organization of afferent terminals in the central nervous system, provides the structural framework for the ordered transfer of specific sensory input from primary afferent to dorsal horn neuron. The receptive field properties of dorsal horn neurons are characterized, however, by a convergence of sensory inputs including low- and high-threshold afferents in many cells. This indicates that ascending pathways carry a complex signal and that the

generation of the sensory experience of pain is due to activity generated in central neurons that also respond to innocuous inputs. However, whatever the sensory processing involved, the system is sufficiently distinct in its operation that we can easily predict which stimuli, based on experience, are likely to produce pain. By the process of association we "learn" to recognize that certain types of stimuli are likely to be accompanied by the sensation of pain and therefore attempt to avoid them.

This is the first aspect of pain as learning. The association between certain stimuli and the unpleasant sensations they invoke is sufficiently robust that we tend to confer on a stimulus the properties of the response it elicits, by, for example, talking of a painful stimulus. This really is a prediction that a certain input is likely to produce a particular type of sensory experience. Most of us flinch when shown a needle or even at the mention of a dentist. This form of associative learning reaches its most sophisticated elaboration in humans, where we have learned to control potentially noxious stimuli as part of our capacity to exploit the environment, by developing tools or the use of strategies such as protective clothing.

The second form of pain as learning is far more primitive from an evolutionary perspective and relates to pain, not as a warning signal, but as a signal that frank injury has actually occurred and that recovery or recuperation is necessary. This is the type of pain that presents in the clinical context, such as postoperative pain, and which is qualitatively quite

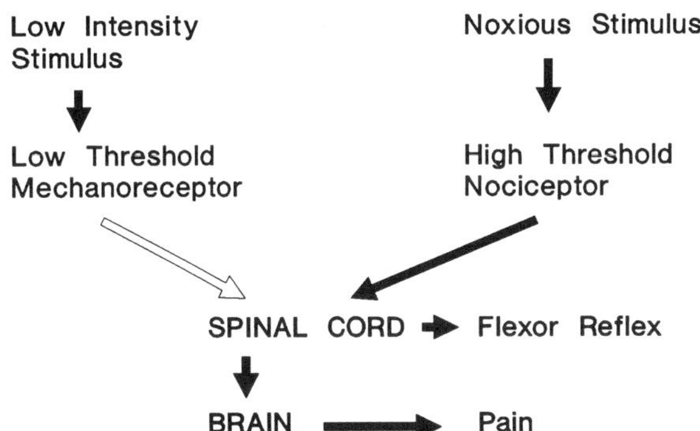

Figure 1. Physiological pain. Under normal circumstances pain is produced only following the activation of high-threshold nociceptors by intense or noxious stimuli (solid arrows). Low-intensity stimuli, which activate low-threshold mechanoreceptors, are ineffective in generating either the flexion withdrawal reflex or the sensation of pain (unfilled arrow). This system acts as a warning of potential danger in the environment.

different from physiological pain. The key differences lie in the reduced intensity of stimuli necessary to elicit the pain. This pain can be produced by stimuli that normally would be considered to be innocuous, the response to noxious stimuli is exaggerated (hyperalgesia), and there is a spread of sensitivity from the site of the injury to noninjured areas. Together these sensory changes constitute the phenomenon postinjury pain hypersensitivity. Teleologically, the role of this type of pain is to assist repair by preventing contact of the injured part not just with a noxious stimulus but with any stimulus until healing has occurred. Clinically this manifests as a limp or sustained flexion following injury to a limb, stabilization of joints with arthritis, and generally, the avoidance of excessive stimuli.

What makes this form of pain so interesting is that it results from an activity- or use-dependent modification of the function of the nervous system—which is a form of learning. Two general types of use-dependent learning can be observed in the somatosensory system among most animals. Both involve changes in response produced by previous inputs; in other words, the present response depends on the history of preceding inputs—the system has a form of memory. The first is a gradual decrement in the response elicited by repeated application of an identical stimulus to the same area on the body surface. This is the phenomenon of habituation. The second is a progressive increase in the amplitude and duration of the response with repeated application, the phenomenon of sensitization. As a generalization, habituation occurs readily to non-tissue-damaging stimuli, while sensitization accompanies intense stimuli. Habituation may enable an animal to remain in a "safe" environment without constantly eliciting a response to an innocuous input, while sensitization enables an animal to detect or react to any stimulus earlier or in a more exaggerated way than normal, having been exposed previously to a harmful stimulus. It is precisely this situation that occurs during the recuperation phase following tissue injury.

Nociceptive sensitization involves the conversion of a high-threshold nociceptor-driven pain system into a low-threshold nociceptor- and non-nociceptor-mediated pain system. Two mechanisms operate. The first is a change in the transduction sensitivity of nociceptor terminals in the periphery, the second is a change in the sensitivity of neurons in the spinal cord, representing peripheral and central sensitization, respectively. Peripheral sensitization results from changes induced in the afferent terminal by chemicals liberated by tissue damage and the inflammatory process (Fig. 2). The actual molecular mechanisms remain unclear but bradykinin, prostaglandins, 5-hydroxytriptamine, histamine, and K^+ and H^+ ions interact synergistically to increase sensitivity. The inflammatory process also recruits or instigates an interaction between sympathetic terminals and primary afferent terminals such that the release of purines, noradrenaline, and eicosanoids may result in a cross-talk between the two types of neuron.

Central sensitization is a state of hyperresponsiveness of dorsal horn neurons that is triggered by nociceptive afferents and which outlasts the initiating input (Fig. 3). Again the detailed cellular and molecular mechanism remains to be unraveled, but central sensitization appears to be the consequence of the summation of slow synaptic potentials generated by small-caliber afferents. These slow synaptic potentials, operating via N-methyl-D-aspartic acid (NMDA) and tachykinin receptors activated as a result of glutamate and neuropetide release from primary afferents, can, with repeated afferent inputs, produce a buildup or cumulative

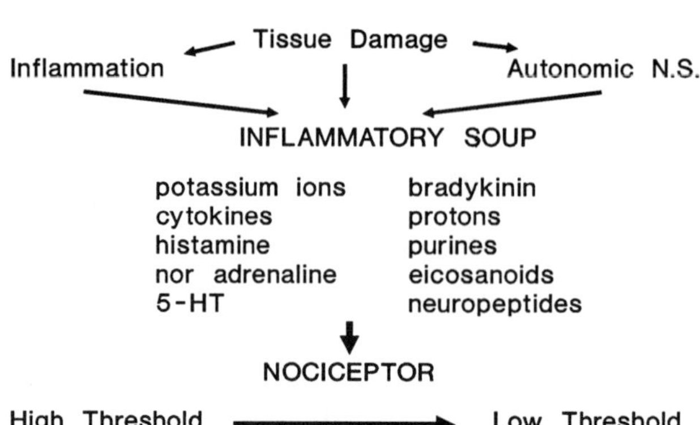

Figure 2. Clinical pain I: Peripheral sensitization. One mechanism for the reduction in threshold for the elicitation of pain under circumstances associated with inflammation is an increase in the sensitivity of the transduction process of nociceptors as a result of an "inflammatory soup" acting on the peripheral terminal. This enables low-intensity stimuli to activate nociceptors at the site of tissue injury.

depolarization. This persists for several minutes after the stimulus is completed. Both ligand and voltage-dependent ion channels activated during this cumulative depolarization may change second messenger levels in the target neurons, by, for example, raising intracellular calcium or altering GTP-binding proteins. These in turn could activate protein kinases directly and indirectly change gene expression via the stimulation of immediate-early gene products.

There is a scenario, therefore, whereby brief activation of certain afferent inputs can produce prolonged changes in the

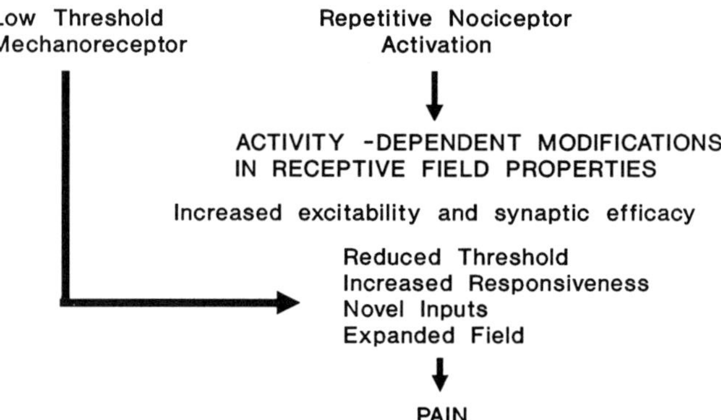

Figure 3. Clinical pain II: Central sensitization. An important feature of clinical pain states is the spread of sensitivity to uninjured tissue and the development of a state of hypersensitivity such that innocuous inputs acting via low-threshold afferents begin to produce pain. This is due to an activity-dependent increase in the excitability of spinal cord neurons triggered by nociceptive afferent input acting on NMDA and tachykinin receptors. The use-dependent change is an example of the functional plasticity of the nervous system and illustrates how dynamic changes in response properties can produce prolonged pathological changes in sensitivity.

response properties of neurons in the dorsal horn. The duration of the changes can vary from minutes (membrane depolarization) to hours (protein kinase activation) to days (gene expression). Precisely what is responsible for the increase in responsiveness in the spinal neurons is not known, whether it be increased receptor sensitivity, modified ion channel kinetics, or modulation of presynaptic transmitter release. What is certain, though, is that following injurious peripheral stimuli, the receptive field properties of dorsal horn neurons change dramatically and these changes—a reduction in threshold, an increase in responsiveness and an expansion in size—parallel clinical changes in pain hypersensitivity. The reason for this is that the normal receptive field of dorsal horn neurons is composed of sub- as well as suprathreshold inputs. Increasing the excitability of a cell or the synaptic efficacy of its inputs results in the recruitment of the subthreshold inputs so that they generate a suprathreshold response, changing the output of the cell. This includes abnormal responses to low-threshold primary afferents, which begin to be able to elicit the sensation pain (Fig. 3).

Central sensitization resembles other examples of activity-dependent changes in synaptic efficacy, although it does differ from long term potentiation in the hippocampus, in being a form of heterosynaptic facilitation and because of its shorter duration. Nevertheless, nociceptive sensitization is present in invertebrates as well as vertebrates and may be considered to represent one of the earliest forms of learn-ing—whose persistence throughout evolutionary change is a measure of the survival advantages it offers.

The extent to which activity-dependent changes in the spinal cord and other centers can lead to permanent changes in function is uncertain. What is sure, however, is that the experience of pain is not simply the cortical representation of ongoing specific peripheral inputs. Our present sensory experiences are colored by their association with previous experiences and reflect the dynamic capacity of the nervous system to change. This modifiability is a form of plasticity or learning that, as far as acute pain is concerned, is an essential basis for its clinical manifestations. Central and peripheral sensitization within nociceptive pathways exaggerates and prolongs pain of diverse causes, by changing the gain of the system.

Further reading

Basbaum AI, Besson J-M, eds (1991): *Towards a New Pharmacotherapy of Pain*. Dahlem Workshop Report. Chichester: John Wiley, 457 pp

Dubner R (1991): Neuronal plasticity and pain following peripheral tissue inflammation or nerve injury. In: *Proceedings of VI World Congress on Pain*, MR Bond, JE Charlton, CJ Woolf, eds. Amsterdam: Elsevier, pp 263–276

Wall PD, Melzack R, eds (1989): *Textbook of Pain*, 2nd ed. Edinburgh: Churchill Livingstone, 1064 pp

Woolf CJ, Walters ET (1991): Common patterns of plasticity contributing to nociceptive sensitization in mammals and aplysia. *Trends Neurosci* 14:74–78

Pain Receptors (Peripheral) and Chronic Pain

Daniel F. B. Bossut and Edward R. Perl

Introduction

Normal human beings ordinarily experience pain as a consequence of injury to body tissue. Usually, such pain is short lived or "acute" and lasts only for the duration of tissue injury and its repair. In medical usage, a pain lasting more than a few months (a period well past usual wound healing) is labeled chronic. While acute and chronic pain can be the result of differing circumstances, in some cases they can have a common basis. When the circumstances evoking acute pain do not regress or even develop further, they produce chronic symptoms. Although pain and healing of injured tissue may be the product of similar events and coexist in the acute phase, they are independent. Pain from peripheral injury is initiated by the activity of particular sense organs. Healing is a complex process combining an inflammatory reaction in the injured tissue and tissue regeneration. Healing can occur without pain and pain without (or after) healing.

The reaction to injury leads to changes in the afferent neural elements (discussed presently) and to central neural processing. Such changes may cause pain to appear in the absence of a normally noxious stimulus and persist for no obvious reason. Only some of the neural processes related to protracted or chronic pain are understood. While central factors are involved in many aspects of the development of chronic pain syndromes, the present commentary will be limited to events occurring peripherally.

It is now broadly accepted that the primary afferent neurons whose activation leads to pain under normal conditions have small-diameter peripheral fibers and are effectively excited by stimuli physically threatening the integrity of the bodily tissues, that is, noxious. Relative to other sensory elements, these "nociceptors" in normal tissues have elevated thresholds for all of the usual events occurring in the immediate vicinity of their terminals. There are several varieties of nociceptors. These differing nociceptors are distinguished by unique combinations of responses to (a) marked mechanical distortion of the tissue, (b) temperatures high enough or low enough to damage cells, and (c) the accumulation of certain chemicals. Nociceptors can be and often are modified as a consequence of past history or repeated exposure. These changes can include the development of new features leading to substantial changes in their response to usual stimuli.

Major peripheral factors that should be considered in relation to chronic pain are (1) changes in the milieu of the terminals of the primary afferent fibers, (2) modifications of the primary afferent terminals themselves, and (3) alterations in the fibers of the nerve trunk. For example, the milieu can be altered so as to provide agents continuously excitatory for the nociceptor terminals. Enhancement of nociceptor responsiveness can foster the signaling of stimuli that previously were below threshold. Injury of a nerve trunk can disrupt the usual arrangements between the peripheral afferents and central neuronal outflow, a situation from which novel responsiveness of some nociceptors might develop (e.g., sensitivity to the neural mediators, norepinephrine).

Sensitization of nociceptors

The discharge from most primary afferent neurons adapts (decreases) to a constant stimulus or upon repeated stimuli,

contributing to a fading sensation. In contrast, most nociceptors "sensitize" when an effective stimulus is repeated. The term *sensitization* implies that a primary afferent neuron develops a lower threshold of activation (responds more readily) and produces an exaggerated response relative to its usual or past characteristics.

Figure 1 illustrates the development of sensitization by responses of a "C-fiber" polymodal receptor (CPM) during and after consecutive noxious heat stimulations. The receptive terminals of these unmyelinated fibers are named polymodal because they vigorously respond to strong mechanical stimulation, to noxious heat, or to a variety of irritant chemicals. In Figure 1, the sensitization in later trials is represented by the development of responses at lower temperatures and by the generation of more discharges at a

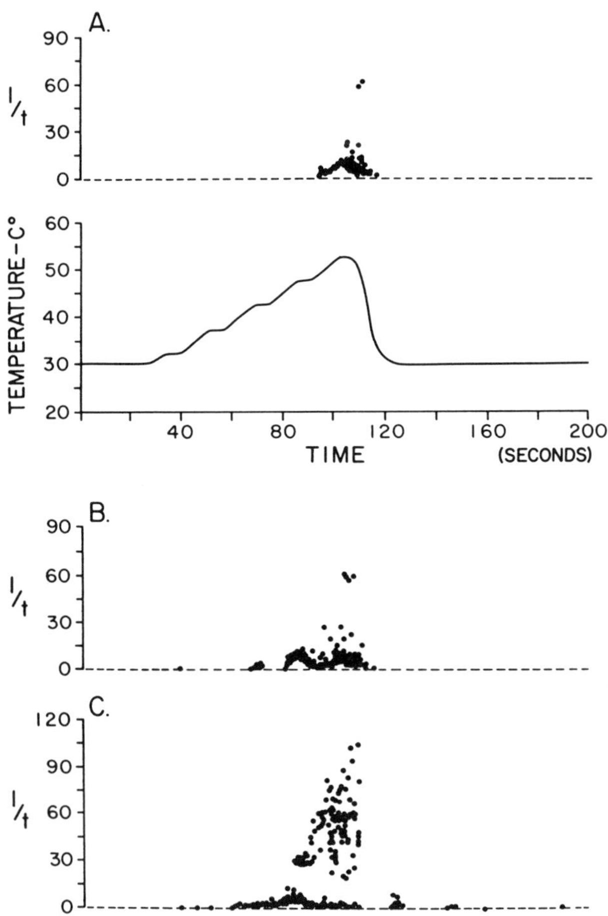

Figure 1. Responses of a C-fiber cutaneous nociceptor (polymodal type) to repeated noxious heating. A thermode contacting the skin generated 5°C step increases in temperature at ~0.5°C/s to about 50°C, followed by active cooling to a neutral holding temperature. Each essentially identical heating cycle was repeated at 200 s intervals. (A) First cycle of heating; (B) third cycle of heating; (C) fifth cycle of heating. Graphs of B and C were aligned to A by superimposing the heating cycle. Reprinted with permission of the American Physiological Society from Kumazawa T and Perl ER (1977): Primate cutaneous sensory units with unmyelinated (C) afferent fibers. *J Neurophysiol* 40(6):1325–1338.

particular temperature than appeared in the initial test. In undamaged tissue, nociceptors that initially respond vigorously only to noxious mechanical stimulation demonstrate sensitization to mechanical stimuli and may develop a response to a repeated noxious heat. An interesting concept of "silent" or "sleeping" nociceptors has emerged from studies on experimental inflammation in animals. This work describes increased avoidance responses and associated activation of otherwise mechanically inexcitable or "silent" nociceptive fibers from inflamed tissues (joints, bladder). Such "awakened" nociceptors could represent a form of inflammation-induced sensitization and has putatively been linked to instances of hyperalgesia (enhanced pain sensation).

Circumstantial evidence strongly suggests that the sensitization of nociceptors correlates in time with hyperalgesia in human beings and with enhanced adversive behavior in animals. Together, these parallels infer that some forms of chronic pain are related to and associated with a sensitization of nociceptors. However, it must be kept in mind that nociceptor sensitization takes more than one form and involves more than a single cause or mechanism.

Sensitization of nociceptors by inflammatory mechanisms

Inflammation is characterized by redness (rubor), swelling (tumor), and pain (dolor). It is the usual sequela of mechanical, thermal, chemical, or immune (e.g., in response to microorganisms) injuries. It is a complex process involving connective tissue cells, circulatory cells, and nerve terminals themselves. The development of inflammation is associated with the liberation or production of a series of chemical messengers that act in part sequentially to facilitate the various vascular and tissue reactions. The combination of substances has been termed, albeit misleadingly, the "inflammatory soup." Serotonin, histamine, substance P (SP), bradykinin, and the metabolites of arachidonic acid are among the substances that induce, independently or in concert, vasodilation, plasma extravasation, alteration of the behavior of inflammatory cells, and pain. In addition, the local environment of inflammation exhibits an increased concentration of potassium ions extracellularly and a drop in pH. Of these substances, only bradykinin, potassium ions, hydrogen ions, and high concentrations of histamine have been found to evoke pain independently.

Arachidonic acid forms part of the plasma membrane of cells and becomes metabolically available in increased quantities in the cell breakdown accompanying tissue injury. Two independent pathways of arachidonic acid metabolism produce prostaglandins (cyclooxygenase path) or leukotriene compounds (lipooxygenase path). Historically, involvement of the cyclooxygenase pathway in inflammation was indicated by the effects of agents that suppress the production of prostaglandins (indomethacin, acetylsalicylic acid, ibuprofen) and were found to relieve some of the inflammatory symptoms. Recently, prostaglandin E_2 (PGE_2) has been shown to sensitize directly cutaneous nociceptors, an effect consistent with its association with inflammatory pain. The evidence concerning the role of leukotrienes in inflammation and their potential contribution to inflammatory pain is still circumstantial and fragmentary.

In the present context, there are many possible changes in the local environment that might make inflammation progress from acute to chronic. In the progression to a chronic state, it is likely that inflammatory and algogenic conditions develop concomitantly. Several factors have been recognized to lead to chronic inflammation. They include (1) the persistence of the inflammatory irritant, inducing a granulomatous inflammation; (2) the formation of novel, endogenous, antigenic substances from modified or catabolized proteins, leading to the persistent activation of the immune response (a process considered important in the progression of adjuvant arthritis); (3) the production of endogenous chemotactic agents altering the cellular participation and production at the site of inflammation. In addition, weakened structures, swollen tissues, and loss of adequate blood supply all have the possibility of enhancing the chances for normally innocuous events to become noxious or to chronically activate nociceptors.

An important lesson from recent experimental work is that the mechanism of sensitization can differ for different types of nociceptors of the same tissue. Hence, a therapeutic agent highly effective in reducing inflammation and inflammatory pain from one set of nociceptors sensitized by PGE_2 process, for instance, will remain ineffective against a sensitization of other nociceptors associated with another agent. This variability implies difficulties for any uniform therapeutic manipulation of inflammatory pain. There are similar considerations and perhaps opportunities in differentiating processes associated with acute and chronic inflammatory pain. The undecapeptide substance P acutely produces vasodilation and plasma extravasation, and is said to facilitate the sensitizing and algogenic effects of prostaglandin and bradykinin. By itself, SP does not appear to sensitize nociceptors of normal tissues. On the other hand, SP has been found in higher than normal concentrations in arthritic joints and to enhance disability when injected in the affected joint of an experimental animal model of chronic arthritis. Thus, substances like SP may be important in the process leading to chronic pain while not necessarily being acutely algogenic.

Nociceptors' sensitization by partial peripheral nerve injury

A certain proportion of peripheral or spinal nerve injuries result in syndromes in which "spontaneous" pain, often triggered by nonnoxious stimuli, is the major symptom. These algogenic disorders include syndromes called causalgia (literally burning pain) or reflex sympathetic dystrophies and situations related to neuromas. Classically, causalgia is a sequela in 10–20% of traumatic injuries of large peripheral nerves and typically appears some days or weeks after injury, a time course that roughly parallels that expected for adrenergic supersensitivity occurring in a variety of tissue after sympathetic denervation. Characteristically, pain is triggered by light touch or small temperature changes and often is associated with signs of increased regional sympathetic activity (e.g., vasoconstriction and sweating). Chemical or surgical interruption of the sympathetic supply to the affected region may temporarily or permanently relieve the pain and other signs. The part played by nociceptors in this peculiar form of chronic pain has been uncertain until recently when animal experimentation showed that shortly (days to weeks) after partial nerve injury otherwise intact CPM units of the injured nerve develop a novel excitatory response to norepinephrine and sympathetic stimulation (SS). SS also enhances the sensitization of CPM units produced by repeated noxious heat. Competitive antagonists of the α_2 adrenergic receptors block the adrenergic excitation. Thus, as a consequence of partial nerve injury, the induction or the expression of increased adrenergic receptor action (up-regulation) seems to occur in some remaining intact nociceptors. Such transformations could explain related clinical observations and represent an initiating cause for causalgia and other reflex sympathetic dystrophies. In the case of the tumor formation that follows interruption of a nerve (neuroma), a similar aberrant burning pain sometimes can be elicited with innocuous stimuli, albeit at the point of nerve injury and swelling rather than at the peripheral terminals. Adrenergic responsiveness has been described in the latter situation and may be related to that occurring in partial nerve injury.

The emphasis in this commentary has been on "chronic" (persisting) changes in nociceptor responsiveness. These changes lead to enhanced input messages to the central nervous system from long-term modification in responsiveness of nociceptors and contribute to chronic pain and associated reactions. Another important side of sensitization or novel responsiveness of peripheral nociceptors is the effect of persisting, increased input on the central nervous system and its circuitry. The nervous system of mammals in particular has a remarkable plasticity, and repetition of a normally rare input can be expected to induce central modifications. The potential for changes of this type has been shown at the synapses between primary afferent fibers and neurons of the spinal dorsal horn. For example, there is good evidence that several putative synaptic transmitters of primary afferent nociceptors have the capability to prolong postsynaptic excitatory action, and it appears that these longer-term synaptic effects require repeated or continual presynaptic activation. Such longer-duration, synaptically-generated effects could provoke persisting postsynaptic alterations. Thus, one can expect long-term modifications in the responsiveness of nociceptors to be associated in turn with prolonged modifications in central circuitry and pathways. Such central modifications must contribute to chronic pain even when the initiating factors are clearly peripheral.

Further reading

Sato J, Perl ER (1991): Adrenergic excitation of cutaneous pain receptors induced by peripheral nerve injury. *Science* 251: 1608–1610

Wiesenfeld Hallin Z, Hallin RG (1984): The influence of the sympathetic system on mechanoreception and nociception. A review. *Hum Neurobiol* 3:41–46

Willis W, ed (1991): *Hyperalgesia and Allodyma: The Bristol-Myers Squibb Symposium on Pain Research*. New York: Raven Press

Panic Disorder, Psychobiology

Andrew W. Goddard and Dennis S. Charney

Introduction

Panic disorder (PD) is a chronic illness characterized by the occurrence of spontaneous, unprovoked, and intense episodes of fear. Accompanying these episodes are typical symptoms of autonomic overactivity including sweating, palpitations, flushing, dizziness, and psychological symptoms including depersonalization, fear of dying, and fear of losing control. Hyperventilation and paraesthesia are also common symptoms of a panic attack. These symptoms are persistent and recurrent and may lead to a secondary syndrome of agoraphobia, where patients are afraid to travel far from home because of the fear of having a panic attack away from a safe environment. The syndrome appears to be familial, and twin studies suggest a concordance rate for PD of 30% or more in identical twins.

Historically, panic disorder was recognized by physicians as "neurasthenia" in the nineteenth century and as "anxiety neurosis" in the early and middle part of the twentieth century. However, in the 1960s PD was proposed as a distinct anxiety disorder following the discovery of the efficacy of the antidepressant imipramine in anxiety patients who suffered from recurrent panic attacks. This proposal was officially recognized in 1980 by its inclusion in the Anxiety Disorders section of the *Diagnostic and Statistical Manual* (third edition) of the American Psychiatric Association. The prevalence of the illness in the community is between 2–5%, meaning that at least 5 million Americans are currently affected. The discovery of effective medical treatment has resulted in considerable preclinical and clinical neuroscientific investigation into the pathophysiology of the disorder and the mechanism of action of therapeutic treatments. The brain noradrenergic (NE) system, among others, has been an important object of study in relation to PD.

Noradrenergic neuronal function

Electrical stimulation of the locus coeruleus (LC), the major NE nucleus in the brain, is associated with fear behaviors in animals. This is true of freely moving animals. Also, animals exposed to dangerous situations have a dramatic increase in LC firing rate, while the activity of major serotonergic and dopaminergic nucleii remains unaltered. The specific alpha-2 adrenergic antagonist, yohimbine, which increases LC cell firing, also produces fear responses in animals.

Clinical implications. Studies with yohimbine in clinical populations reveal that PD and posttraumatic stress disorder (PTSD) patients are much more likely to have panic attacks in response to yohimbine administration than are healthy control subjects and patients with other psychiatric disorders like schizophrenia, mood disorders, and obsessive-compulsive disorder. These data suggest that a subgroup of patients with PD and PTSD have a dysregulation of NE function. There are a number of challenge studies with isoprenaline demonstrating a possible role of the beta receptor pathology in panicogenesis.

Medications that affect NE function, like tricyclic antidepressants and monoamine oxidase inhibitors, are effective in blocking spontaneous panic attacks, further suggesting that NE dysfunction is an important part of the pathophysiology of panic for many patients. NE dysfunction may also account for other clinical phenomena related to PD. For instance, psychostimulants like caffeine, cocaine, and amphetamines readily precipitate panic attacks in PD patients compared with healthy controls. Other clinical problems seen in PD patients (but not depressed patients), such as clinical worsening during initiation of tricyclic pharmacotherapy, may be due in part to NE dysfunction, which may underlie active PD.

While the NE dysfunction theory may account for the occurrence of spontaneous panic, it does not fully explain other important clinical manifestations of PD such as anticipatory anxiety, situational panic, and agoraphobic behavior. Also it does not fully account for the average age of onset being in the twenties and early thirties.

Serotonergic neuronal function

Serotonin (5-HT) neurons in the dorsal raphe nuclei appear to have diverse biological roles, including regulation of mood. Pharmacologic depletion of central 5-HT in animal models leads to hyperactive and aggressive behaviors. Also, acute and chronic stress paradigms may produce alterations in the density of 5-HT receptor subtypes. For example, acute immobilization stress induces an elevation of 5-HT2 receptors in rat cortex, while inescapable shock leads to a reduction of ^{3}H-imipramine binding in similar regions.

Clinical implications. Thus far, evaluation of serotonergic challenge agents in PD has led to mixed results. The 5-HT receptor agonist m-chloro-phenyl-piperazine (mCPP), when given orally, may precipitate anxiety in PD but not in healthy controls. In contrast, intravenous mCPP appears to produce similar increases in anxiety in both control subjects and PD patients. However, newer antipanic medications like the serotonin reuptake inhibitors fluvoxamine, fluoxetine, and clomipramine are clinically effective. When administered chronically these compounds appear to generally improve 5–HT neurotransmission. There is data that the reuptake inhibitor fluvoxamine may influence NE dysregulation via a specific serotonin–norepinephrine functional interaction. Preclinical and clinical studies indicate that NE and 5-HT neurons interact in a functionally important manner. Future research in PD will need to evaluate the significance of interactions between these important neuronal systems since neither the NE nor 5-HT hypothesis of panic disorder fully accounts for the clinical phenomena of PD or the response of PD to various pharmacologic and nonpharmacologic treatments.

Benzodiazepine-GABA system function

The benzodiazepine class of drugs has been found to produce profound anxiolysis in animals and humans. This effect is mediated via specific benzodiazepine (BZ) receptors, which produce allosteric modulation of the GABA receptor, which in turn improves neuronal Cl ion conductance. Benzodiazepines like alprazolam and diazepam are full agonists at BZ receptors. Compounds that are inverse agonists, like the beta carboline FG-7142, are proconvulsant and profoundly anxiogenic, while antagonists like flumazenil reverse the effects of the agonists or inverse agonists without altering GABA function substantially. In animal paradigms of uncontrollable stress there appears to an association between behavioral deficits (reduced exploration and reduced escape behaviors) and decreases in GABA-ergic function and brain BZ receptor occupancy.

Clinical implications. Medications like alprazolam and clonazepam are efficacious for PD and exert their therapeutic effect immediately, unlike the antidepressants used for PD. There is evidence that in addition to blocking panic attacks, these agents are effective in reducing anticipatory anxiety, which is a common and disabling symptom of PD. In addition, they are frequently indicated for PD patients experiencing increases in anxiety following initiation of tricyclic compounds. A withdrawal syndrome following discontinuation of chronic administration is well described and is an indication for the cautious use of these medications in PD. Newer medications like the partial BZ agonist abecarnil may produce anxiolysis without much sedation or withdrawal.

An exciting literature has developed on the subject of BZ receptor functioning in PD. The antagonist flumazenil has been found to induce panic attacks in PD patients but not in normal controls. This observation has been interpreted as suggesting a shift in the spectrum of BZ receptor functioning in PD so that the antagonist behaves more as a partial inverse agonist. The availability of numerous ligands for the BZ receptor, coupled with the explosion of new brain imaging techniques like SPECT (single photon emission computed tomography) and PET (positron emission tomography), predict an exciting future for research into the neurobiology of PD. In particular, the SPECT ligand I-123 iomazenil has favorable characteristics for clinical imaging. Measurement of the binding potential of I-123 iomazenil in healthy humans has already been accomplished and evaluation of patient groups is underway.

Neuropeptide systems

Neuropeptides may be involved in the pathophysiology of panic disorder. Preclinical work indicates that corticotrophin releasing factor (CRF) is found in high concentrations in the LC and related regions. In rats, direct application of CRF to the LC dramatically increases unit firing, which in turn is associated with fear behaviors. Also intraventricular administration of CRF in animals induces a panic or terror-like state. Cholecystokinin (CCK), another neuropeptide, is found in high concentrations in the cerebral cortex, amygdala, and hippocampus of the mammalian brain (areas of the brain involved in anxiety responses) and is anxiogenic in rats and some species of nonhuman primate. These behavioral effects can be reversed with benzodiazepines and also specific CCK antagonists.

Clinical implications. CRF may play a significant role in the pathophysiology of panic attacks. CRF administration in animals leads to the typical manifestations of an acute stress response including fearful behaviors, elevated central and peripheral catecholamines, and elevated plasma ACTH levels. Given these diverse effects, CRF may be partly responsible for triggering many of the somatic manifestations of panic attacks such as palpitations, sweating, nausea, blood pressure changes, trembling, and skin flushing. In chronic PD it is possible that CRF levels remain chronically elevated, leading to upregulation of tyrosine hydroxylase and

other catecholamine synthetic enzymes. This mechanism may perpetuate the clinical syndrome. These CRF effects are functionally antagonized by a variety of antidepressants, perhaps accounting for their antipanic efficacy.

CCK-4 produces panic attacks in humans following IV administration. PD patients appear to be more sensitive to this effect than healthy controls. A number of CCK antagonists have been developed in recent years and some, like CI-988, are undergoing evaluation for their efficacy and safety in PD and other anxiety disorders. These compounds could potentially supplant many current antianxiety agents since they appear to carry no abuse potential, and tolerance or withdrawal phenomena have not been described. Whether the anxiogenic effect of CCK-4 is mediated centrally or peripherally remains to be established. CCK may conceivably play a role in PD patients that present with irritable bowel syndrome.

NMDA/glutamate systems

Much data implicates the amygdala as an important structure involved in anxiety and fear behaviors. In rats, lesions of the central nucleus of the amygdala block the natural reactions to fear and stress, while electrical stimulation of this structure elicits fear behaviors. The central nucleus has strong anatomic connections to the hypothalamus and brainstem regions, which are responsible for many of the signs and symptoms of fear. The fear-potentiated startle paradigm provides a specific measure of function of the central nucleus of the amygdala. Preclinical data suggest that an NMDA antagonist (AP5) can block extinction of fear-potentiated startle responses. NMDA antagonists can also block acquisition of potentiated startle to an auditory conditioned stimulus. Thus, NMDA/glutamate systems appear to be involved in learning processes relevant to fear and anxiety, which are coordinated via the amygdala.

Clinical implications. It is conceivable that patients with PD have a deficit in their ability to "switch off" fear responses. Hence they experience situational panic attacks and develop agoraphobia as a compensatory behavior. Clinical studies using the startle paradigm are currently being conducted in healthy controls and PD patients to evaluate this hypothesis. Specific pharmacotherapies or behavior therapies may eventually be designed to correct these putative deficits.

Cerebral blood flow and metabolism

Imaging studies have evaluated some of the components of PD using various experimental approaches. PET studies using the radiotracer ^{15}O-water have compared blood flow in PD patients with controls both in the resting and panic state. At rest, some PD patients appear to have an area of abnormally increased blood flow (and presumably neuronal activity) in the right parahippocampal region. Moreover, it is these same patients that tend to experience lactate-induced panic. Lactate panic attacks were associated with increases in blood flow in both temporal poles. However, this state finding is also found in healthy volunteers anticipating an electric shock, suggesting a final common pathway for the cortical transduction of fear. It is of clinical interest that patients experiencing significant levels of anticipatory anxiety before entering a phobic situation are much more likely to have a panic attack. It is possible that the temporal association areas prime the NE and other neuronal systems in some way, thereby facilitating the onset of panic attacks.

Other studies have evaluated the effect of yohimbine-induced panic on cerebral blood flow using the SPECT ^{99}Tc-HMPAO method. Frontal cortical blood flow was decreased bilaterally in PD patients who experienced yohimbine panic. This finding suggests that transitory impairment of higher cortical function may be part of the panic state. The clinical finding of impairment of reasoning and concentration during panic attacks may be produced by such a mechanism. Careful studies in controls suggest that hyperventilation (an important panic symptom) per se does not produce the regional changes found. These functional studies need to be replicated with further methodologic refinements such as careful MRI coregistration. A preliminary PET study using the CCK-4 chemical panic model failed to replicate the bitemporal increase in blood flow found in previous studies.

Functional imaging studies have the potential to tell us much about the *in vivo* physiology of the brain in anxiety states. Future studies may focus on correlating a variety of brain functions such as glucose metabolism, blood flow, and receptor physiology with specific behaviors and cognitions observed in panic states.

Comment

Panic disorder is the most readily treated major psychiatric syndrome thanks to improvements in pharmacotherapy and behavior therapy over the last two decades. The continuing challenge for science is the careful delineation of the psychobiologic processes occurring in PD. More refined understanding of these processes should result in even more effective treatments. The burgeoning discovery of a number of biologic markers in PD should help geneticists in their quest for detection and study of candidate genes in PD patients. Thus far, evaluation of the alpha 2 receptor gene and the BZ receptor gene in PD probands has not led to the discovery of a specific lesion. It may be that there is pathology in a number of different gene loci, which collectively create the panic diathesis. Study of environmental factors such as life events and their biologic impact may reveal clues to the mechanisms underlying the onset of the disorder. Other potential precipitating and protective factors such as childbirth and pregnancy merit further study. The presence of a neurosteroid receptor site near the GABA complex suggests that a biologic mechanism may be underlying exacerbation and remissions of PD in relation to the female reproductive cycle.

We can expect to enter a time in the coming decades when the clinical ideals of early detection and primary prevention may become an everyday reality with respect to this prevalent and disabling condition. The efforts of preclinical and clinical neuroscientists continue to propel us toward those goals.

Further reading

Ballenger JC, ed (1990): Neurobiology of panic disorder. *Frontiers of Clinical Neuroscience,* vol. 8. New York: Wiley-Liss

Burrows GD, Roth M, Noyes R Jr (1990): *The neurobiology of anxiety.* Vol. 3, *The Handbook of Anxiety.* Amsterdam: Elsevier Science

Charney DS, Woods SW, Nagy LM, Southwick SM, Krystal JH, Heninger GR (1990): Noradrenergic function in panic disorder. *J. Clin Psychiatry* 51, 12 (suppl A):5–11

Nutt D, Lawson C (1992): Panic attacks. A neurochemical overview of models and mechanisms. *Brit J Pyschiatry* 160:165–178

Paraneoplastic Disorders

Jerome B. Posner

Paraneoplastic disorders are symptom complexes occurring in patients with cancer but affecting organs or tissues remote from the site of the cancer and not ascribable to invasion of the involved organ by the cancer. Paraneoplastic disorders affect many structures and cause such diverse symptoms as arthritis, skin rashes, and endocrine dysfunction. Paraneoplastic disorders that affect the nervous system have also been called "remote effects" of cancer. In some paraneoplastic neurological disorders, symptoms are localized to a single neural structure (e.g., cholinergic synapse, retinal photoreceptor, cerebellar Purkinje cell); in others the symptoms are more widespread, affecting simultaneously brain, spinal cord, and peripheral neural structures.

Paraneoplastic disorders of the nervous system are rare, affecting far fewer than 1% of patients with cancer. Despite their rarity, they are clinically important for several reasons: (1) The neurological disorder usually precedes diagnosis of the cancer. Thus, the physician who recognizes the neurological disorder as a paraneoplastic syndrome may localize and identify an occult and potentially curable cancer. (2) Autoantibodies are present in the serum of some patients with paraneoplastic disorders. These antibodies, where present, define the neurological disorder as paraneoplastic and predict the presence of a specific cancer (e.g., small cell lung cancer). (3) Some evidence suggests that the cancers in patients with paraneoplastic syndromes grow more slowly and are more responsive to treatment than are histologically similar cancers in patients without paraneoplastic syndromes. Cancers of patients with paraneoplastic syndromes often show greater lymphocytic infiltration than do other cancers. (4) Early discovery and treatment of the cancer may not only cure the cancer but ameliorate the neurological symptoms. Unfortunately, in most patients, neurological symptoms persist despite successful treatment of the cancer.

Clinical findings

Signs and symptoms. Paraneoplastic disorders can affect virtually any portion of the nervous system, the signs and symptoms depending on the portion of the nervous system most involved by the pathological process. In general, paraneoplastic syndromes begin abruptly and evolve over several weeks or months to cause severe neurological disability. When only one structure in the central or peripheral nervous system is involved, the symptoms tend to be stereotypic. When multiple structures are involved, dysfunction of one structure usually predominates, but the predominant symptoms differ from patient to patient. Examples of single structural involvement include "cancer associated retinopathy" with loss of retinal photoreceptors, resulting in blindness; paraneoplastic cerebellar degeneration with loss of Purkinje cells, causing ataxia, dysarthria, and nystagmus; Lambert-Eaton myasthenic syndrome, in which the cholinergic synapse is the site of pathology associated with weakness and fatigability in the extremities.

When there is both clinical and pathological involvement of more than one neural structure, the disorder is called "paraneoplastic encephalomyelitis." If one clinical symptom predominates but the pathology is more widespread, the disorder is named for the predominant symptom locus. Examples include limbic encephalitis, a disorder clinically characterized by memory loss, abnormal behavior, and sometimes seizures associated with pathological damage to the hippocampus and other structures of the limbic system; brainstem encephalitis, characterized clinically by vertigo, nystagmus, dysphagia, and extrapyramidal dysfunction and associated with pathological involvement of the upper and/or lower brainstem; paraneoplastic myelitis, characterized by both upper and lower motor neuron weakness with the predominant site of pathology in the anterior horn cell; sensory motor peripheral neuropathy, with both weakness and sensory loss in a peripheral nerve distribution, associated with either demyelinization or axonal damage to peripheral nerves.

The pathological changes of paraneoplastic disorders usually include death of neurons, peripheral demyelination, gliosis, and in some patients, inflammatory infiltrates. The inflammatory infiltrates consist of both B and T cells. Recent evidence suggests that some lymphocytes in both the tumor and the neural tissue identify neural proteins that are also recognized by autoantibodies circulating in that patient's serum and cerebrospinal fluid.

In some patients, IgG has been identified in the target neurons of the paraneoplastic syndrome. Much of that IgG appears to be a specific antibody found in the serum and spinal fluid of those patients.

Pathogenesis

The pathogenesis of most neurological paraneoplastic syndromes is unknown. Several hypotheses have been advanced with some evidence to support each. It is likely that no one pathogenetic mechanism plays a role in every paraneoplastic syndrome and that in some syndromes more than one pathogenetic mechanism may be important.

Substances made by the tumor. The first hypothesis formulated in the last century suggested that the tumor makes a substance that directly damages the nervous system. Tumors are known to produce both hormone-like substances and cytokines, some of which have profound effects on the nervous system. For example, the elaboration of parahormone-like substances by a tumor can cause hypercalcemia with weakness and behavioral changes. Ectopic ACTH production by a tumor can cause Cushing's syndrome with behavioral changes. Overproduction of cytokines, such as interleukin-1 and tumor necrosis factor, probably play a role in the cachexia, asthenia, and general fatigability complained of by some patients with cancer.

Competition for essential substrates. Denny-Brown proposed in 1948 that paraneoplastic sensory neuronopathy resulted from competition between the tumor and the dorsal root ganglion cell for a vitamin or essential substrate. This hypothesis has never been proved and current evidence suggests that it is unlikely. However, neurological paraneoplastic syndromes are sometimes associated with carcinoid tumors that appear to compete with the nervous system for tryptophan. Competition between tumor and the nervous system for glucose may also be responsible for some neurological symptoms.

Opportunistic infection. Patients immune-suppressed by cancer may develop opportunistic infections of the nervous system. One such infection, a papovavirus, causes the clinical disorder of *progressive multifocal leukoencephalopathy,* in which large areas of brain white matter are destroyed as a result of the virus infecting oligodendroglial and astrocytic cells. The paraneoplastic syndrome called *subacute motor neuronopathy,* characterized by weakness and loss of anterior horn cells in patients with lymphomas and Hodgkin's disease, also represents an opportunistic viral infection.

Autoimmunity. Many investigators believe that the pathogenesis of most paraneoplastic syndromes is "autoimmune." The hypothesis is that antigens in the tumor elicit an antibody response that cross-reacts with identical or similar antigens in the nervous system. The immune reaction both suppresses tumor growth and damages the nervous system, explaining the small size and indolent growth of the primary cancer as well as the profound neurological dysfunction.

The best evidence for an autoimmune pathogenesis is the Lambert-Eaton myasthenic syndrome. About two thirds of patients with these neurological disorders also have small cell lung cancer. The disorder is apparently caused by an IgG antibody that reacts with calcium channels in both the small cell lung cancer and at the cholinergic synapse. At the cholinergic synapse the reaction of the IgG with calcium channel prevents the ingress of calcium when the action potential reaches the synapse, thus decreasing the release of acetylcholine and resulting in the weakness and fatigability that characterize the syndrome. Plasma exchange, which removes IgG from the serum, effectively treats the disorder; injection of that IgG into experimental animals reproduces the disorder. Treatment of the tumor usually ameliorates the neurological disorder.

An immune pathogenesis for paraneoplastic cerebellar degeneration, subacute sensory neuronopathy, limbic encephalitis, opsoclonus-myoclonus, and a few other syndromes is suggested by the presence of specific antibodies in serum and spinal fluid in these patients. Each antibody is associated with a specific neurological disorder and a specific cancer. For example, the antibody called anti-Yo is associated with paraneoplastic cerebellar degeneration and gynecologic cancers (particularly ovarian and breast), whereas the anti-Hu antibody is found in patients with paraneoplastic sensory neuronopathy or encephalomyelitis and small cell lung cancer. In these paraneoplastic syndromes, the antibody usually has a higher relative specific activity in spinal fluid than in serum. In patients who die with the disorder, the antibody can be identified at higher relative specific activity in brain and tumor than in either spinal fluid or serum. IgG is found within neurons of patients with these specific paraneoplastic disorders. That IgG appears to be largely the specific antibody.

The antigens recognized by these antibodies are usually intracellular and some are intranuclear. Some of the genes coding for these antigens have been cloned from expression libraries using the antibodies as probes. In one series of ex-

periments, the cloned fusion protein reacted with lymphocytes in both the brain and tumor of patients suffering from the paraneoplastic disorder. In some instances, the antigens appear to be homologous to known antigens. For example, the "cancer associated retinopathy" antigen appears to be homologous with the photoreceptor protein recoverin; the myasthenic-associated genes (MysA and B) appear to be homologous with the β subunit of the calcium channel; the sensory neuronopathy/encephalomyelitis antigen (Hu) appears to be homologous to the drosophila developmental antigens, sex lethal and Elav.

Future directions

Many facets of paraneoplastic syndromes remain to be explained. If neurological paraneoplastic syndromes are immune in nature, how does the immune response occur? One possibility is that the gene in the cancer is mutated and seen as foreign by the host. Another is that genes in the tumor and brain are identical, but the nervous system protein sequestered in the "immune-privileged" brain has never encountered the immune system.

Another problem arises from the fact that some antigens (e.g., Hu) appear to be present in all small cell lung cancers but only a small subset of patients generate an immune response. One possible explanation is that there is lesser MHC expression on tumor cells of those tumors that do not elicit an immune response.

If paraneoplastic disorders are immune, how does the immune response damage the nervous system. No evidence has yet indicated that the antibody itself is directly toxic to the central nervous system, as appears to be the case in the Lambert-Eaton syndrome. In many patients, however, few or no T cells are identified in the brain, at least at autopsy. Furthermore, the antigens appear to be intracytoplasmic or intranuclear, raising the question of how the antibody or the T cell encounters the antigen.

Many of these questions will only be answered when an adequate animal model of paraneoplastic syndromes has been developed. With the exception of the Lambert-Eaton syndrome no such models currently exist.

Further reading

Hetzel DJ, Stanhope R, O'Neil BP, Lennon VA (1990): Gynecologic cancer in patients with subacute cerebellar degeneration predicted by anti-Purkinje cell antibodies and limited in metastatic volume. *Mayo Clin Proc* 65:1558–1563

Leys K, Lang B, Johnston I, Newsome-Davis J (1991): Calcium channel autoantibodies in the Lambert-Eaton myasthenic syndrome. *Ann Neurol* 29:307–314

Posner JB, Furneaux HM (1990): Paraneoplastic syndromes. In: *Immunologic Mechanisms in Neurologic and Psychiatric Disease,* BH Waksman, ed. New York: Raven Press

Szabo A, Dalmau J, Manley G, Rosenfeld M, Wong E, Henson J, Posner JB, Furneaux HM (1991): HuD a paraneoplastic encephalomyelitis antigen contains RNA binding domains and is homologous to Elav and Sex-Lethal. *Cell* 67:325–333

Parkinsonism, Effects of Lesions of the Subthalamic Nucleus

Hagai Bergman, Thomas Wichmann, and Mahlon R. DeLong

Parkinson's disease and the MPTP model

Parkinson's disease is the most common movement disorder in humans. It is characterized by akinesia (poverty of movement), rigidity (muscular stiffness), and tremor. Although it is known that Parkinson's disease results from a loss of dopaminergic neurons in the substantia nigra, the resulting alterations in activity in the central nervous system responsible for parkinsonian motor deficits are still poorly characterized.

Recently it was demonstrated that systemic treatment of monkeys with the dopaminergic neurotoxin N-methyl-4-phenyl-1,2,3,6-tetrahydropyridine (MPTP) causes clinical and pathological changes closely resembling human parkinsonism. Since then, the MPTP treated monkey has been used as an outstanding primate model for the human disease. In our experience, akinesia appears as the earliest behavioral effect of this treatment, a few days after the first injection, and increases until the monkeys sit largely motionless in their cages (Fig. 1A). Several days after the appearance of akinesia, muscular rigidity (Fig. 1B), and an intermittent 5 Hz postural tremor involving proximal limb muscles and trunk frequently develops (Fig. 1C, D). Other typical parkinsonian signs such as postural instability and drooling are also observed.

Functional anatomy of the basal-ganglia cortical circuits

The striatum (putamen and caudate nucleus) is the major "input" portion of the basal ganglia circuitry. It receives afferents from most parts of the cortex, including sensory, motor, and association areas, and the activity of its neurons appears to be strongly influenced by the dopaminergic nigrostriatal tract. The striatum in turn projects to the internal segment of the globus pallidus (GPi), which together with the pars reticulata portion of the substantia nigra constitute the "output stage" of the basal ganglia. The basal ganglia–cortical circuit is partially closed by the inhibitory (GABA-ergic) projections from GPi to the motor nuclei of the thalamus, and the excitatory connections between the thalamus and frontal cortical areas (Fig. 2).

Two pathways exist between the input and output stages of the basal ganglia, a *direct* inhibitory (GABA-ergic) pathway from the putamen to GPi, and an *indirect* pathway, which involves the external segment of the globus pallidus (GPe) and the subthalamic nucleus (STN). The net effect of

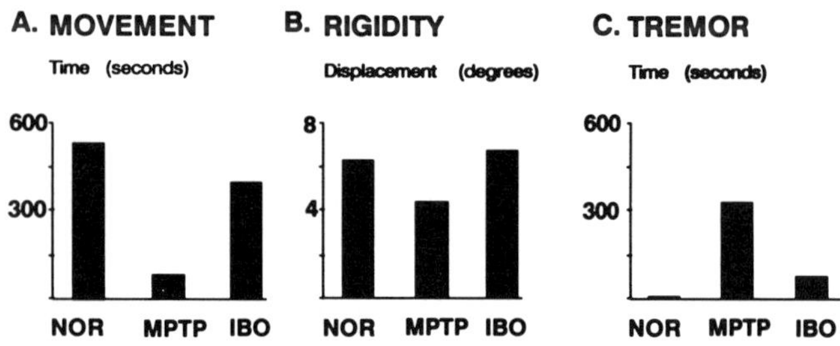

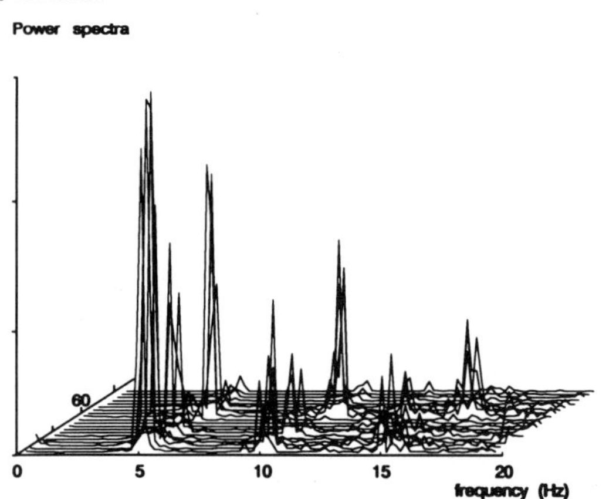

Figure 1. Parkinsonian signs and effects of lesion of the subthalamic nucleus by injections of ibotenic acid (IBO) in parkinsonian monkeys. (A) Akinesia: columns represent the amount of time per 30 minutes that monkey spent in movement before (NOR), after MPTP treatment (MPTP) and after lesion of the STN. (B) Rigidity: columns represent the maximal forearm displacement in monkey, induced by elbow torque pulses. (C) Tremor: columns represent the amount of time per 30 minutes that monkey spent tremoring. (D) Tremor: the figure shows an example of power spectra for successive trains of wrist tremor in green monkey rendered parkinsonian by systemic MPTP treatment.

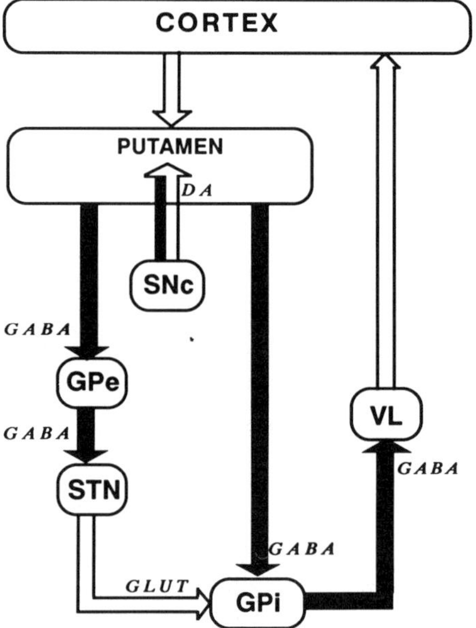

Figure 2. Schematic diagram of the circuitry and the transmitters of the basal ganglia. Open arrows represent excitatory connections, filled arrows represent inhibitory connections. The putamen as "input" stage of the circuit is connected with the internal division of the globus pallidus, which is the "output" stage and regulates the activity in thalamocortical circuits. Abbreviations: GPe, external division of the globus pallidus; GPi, internal division of the globus pallidus; SNc, substantia nigra pars compacta; STN, subthalamic nucleus; VL, ventro lateral nucleus of the thalamus; DA, dopamine; GLUT, glutamate.

the indirect pathway on GPi activity is probably excitatory, since it involves two inhibitory (GABA-ergic) pathways (from putamen to GPe, and from GPe to STN), and one excitatory (glutamatergic) pathway (from STN to GPi).

Electrophysiological and metabolic studies of the basal ganglia of MPTP-treated monkeys have demonstrated that the neural activity in the STN and GPi is increased, and that the activity in GPe is decreased. These and other studies suggest a model where the net action of dopamine is different on two subpopulations of striatal neurons. Under normal conditions, putamen neurons projecting directly to GPi appear to be facilitated by dopamine, whereas neurons projecting to GPe are inhibited by dopamine (Fig. 3A). In Parkinson's disease, loss of striatal dopamine causes, therefore, a decrease in the activity of inhibitory striatal neurons projecting directly to the GPi. In addition, striatal dopamine depletion results in overactivity of striatal projection neurons to GPe, which subsequently releases the STN from tonic inhibition. The increased activity of STN neurons exerts an enhanced (excitatory) drive on neurons in GPi, which gives rise to the major output from the basal ganglia to the thalamus. Dopamine depletion in the putamen therefore leads to both a reduction of activity of the direct inhibitory pathway and an increase of the activity in the indirect excitatory pathway, synergistically leading to an increase in GPi activity. Since the GPi-thalamic projection is inhibitory, increased GPi discharge leads to inhibition of thalamocortical neurons. The resulting reduction of cortical activation would then account for the parkinsonian signs (Fig. 3B).

The effect of subthalamic lesion in parkinsonian monkeys

Green monkeys were rendered parkinsonian by systemic treatment with MPTP. Under electrophysiological guidance, injections of the axon-sparing excitotoxin ibotenic acid were placed in the STN. This treatment effectively lesioned a large part of the STN, resulting in a marked loss of subthalamic neurons, with no evidence of damage to nearby structures. The STN lesion lead to a marked amelioration of all major parkinsonian signs. Within a minute after the injection, the monkeys began to move the extremities contralateral to the injection site (Fig. 1A). The ability of the animals to move their arms purposefully was strongly increased, such that the animals were again able to feed and groom themselves. Tremor was almost completely abolished in the contralateral limbs (Fig. 1C). Neurological examination and assessment of mechanical response of the arm to application of torque pulses to the elbow showed that the muscle tone on the side contralateral to the lesion was markedly reduced (Fig. 1B). Besides the amelioration of parkinsonian signs, the animals also developed transient involuntary movements (dyskinesia) of the contralateral arm and leg. The dyskinesia gradually disappeared, whereas the amelioration of parkinsonian signs remained unchanged.

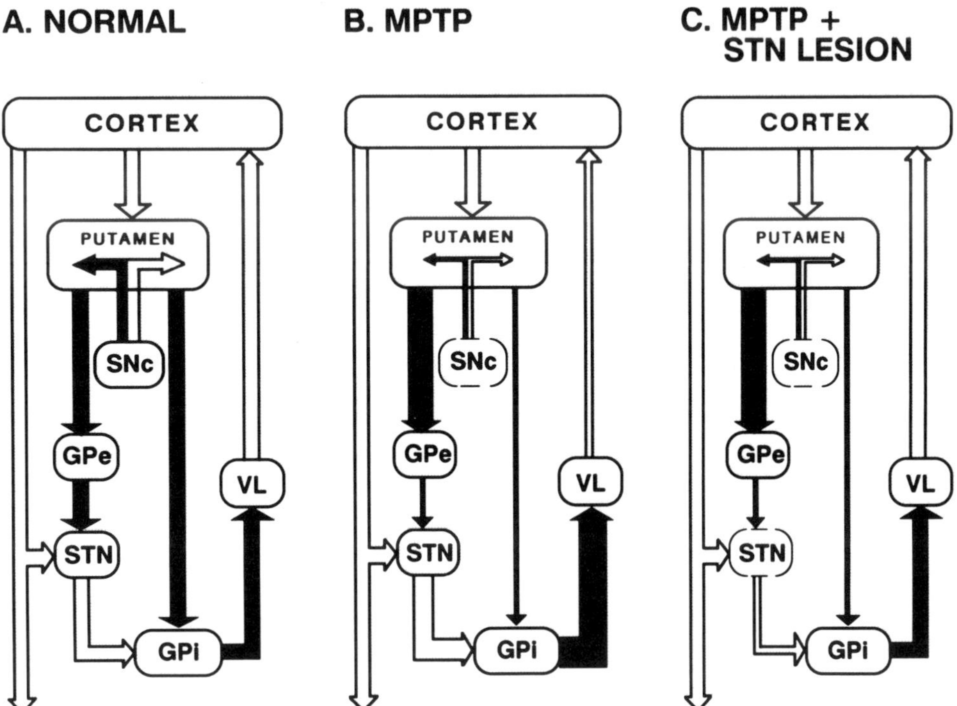

Figure 3. Functional connectivity within the basal ganglia–thalamocortical circuit. (A) Normal: open arrows represent excitatory connections, filled arrows represent inhibitory connections (abbreviations as in Fig. 1). (B) MPTP-induced parkinsonism: After MPTP treatment, the SNc is damaged. Resulting changes in the overall activity in individual projection systems are indicated by changes in the width of arrows. Inactivation of the nigro-putaminal projection increases GPi activity, secondary to an increase in excitatory drive from the STN. The resulting overinhibition of thalamocortical circuits may account for some of the parkinsonian motor signs. (C) Effect of subthalamic nucleus lesions in parkinsonism: inactivation of the STN restores normal GPi output to the thalamus, and as a result, parkinsonian motor signs are reversed. (Reprinted with permission of the AAAS from Bergman et al. (1990): Reversal of experimental parkinsonism by lesions of the subthalamic nucleus. *Science* 249:1436–1438. Copyright 1990 by the AAAS.)

These results strongly support the model whereby in parkinsonism striatal loss of dopamine ultimately results in an increased (inhibitory) output from the basal ganglia to the thalamus (Fig. 3B). Under these circumstances, lesions of the STN may reduce the excessive excitatory drive of the indirect pathway on GPi (Fig. 3C). The role of the decreased inhibitory activity of the direct putamen-GPi pathway in the pathophysiology of parkinsonian signs remains uncertain, but the present results suggest that it is less important than the increased activity in the indirect putamen-GPi pathway.

Revised view of the parkinsonian symptoms and STN lesion "side effects"

Akinesia. Previous surgical attempts to ameliorate akinesia by thalamic lesions have been largely unsuccessful. Akinesia has been therefore viewed as a "negative" sign, that is, a loss of function due to tissue damage that cannot be restored by subsequent lesions. In contrast, rigidity and tremor are viewed as "positive" signs, that is, resulting from the activity of remaining systems. The reversal of akinesia by lesions of the STN strongly suggests instead that akinesia is also due to excessive pallidal inhibition of thalamocortical neurons and therefore should also be viewed as a positive sign.

Tremor. The marked effect of lesions of the STN on tremor is also somewhat surprising, since parkinsonian tremor has generally been postulated to result from oscillatory activity of thalamocortical loops, independent of the basal ganglia. The present findings indicate that the basal ganglia–thalamo pathway is important in the development of parkinsonian tremor. Many neurons in the STN and in the globus pallidus of MPTP-treated monkeys fire in periodic bursts, which are time-locked to the tremor. Conceivably, this may represent pacemaker-like activity in selected populations of basal ganglia neurons, or more probably, oscillations due to increased gain in feedback loops involving the basal ganglia.

Rigidity. The effects of lesions of the STN on rigidity are more readily understood since rigidity is alleviated in parkinsonian patients by lesions of GPi or its projection pathways. Both GPi and STN lesions lower GPi output and thereby normalize muscle tone.

Dyskinesia. The development of transient dyskinesia of the contralateral limbs following lesions of the STN is expected since they are well documented in experimental animals and in humans (hemiballismus). However, it is noteworthy that while dyskinesia gradually decreased, the amelioration of parkinsonian signs remained unchanged.

Future directions

The improvement of all parkinsonian symptoms following lesion of the STN strongly supports the hypothesis that increased activity of the subthalamic nucleus plays a significant role in the motor abnormalities in Parkinson's disease. These results also provide new insights into the pathophysiology of Parkinson's disease, suggesting a novel view of akinesia as a positive rather than a negative sign in this disorder. Certainly the experiments need to be replicated with longer observation periods. They do, however, redirect attention to surgical lesioning approaches, which may emerge again as effective means to control parkinsonian motor signs by direct reduction of basal ganglia output, in cases of advanced parkinsonism.

Further reading

Albin RL, Young AB, Penney JB (1989): The functional anatomy of basal ganglia disorders. *Trends Neurosci* 12:366–379
Bergman H, Wichmann T, DeLong MR (1990): Reversal of experimental parkinsonism by lesions of the subthalamic nucleus. *Science* 249:1436–1438
Miller CW, DeLong MR (1987): Altered tonic activity of neurons in the globus pallidus and subthalamic nucleus in the primate MPTP model of parkinsonism. In: *The Basal Ganglia II. Structure and Function—Current Concepts,* Carpenter MB, Jayaraman A, eds. New York: Plenum

Polyamines in the Nervous System

Michel Baudry and Imad Najm

The polyamines spermidine and spermine, as well as their precursor putrescine, are polycations that have long been recognized as playing critical roles in developing or regenerating tissues. Although the exact mechanisms underlying their physiological functions are not totally understood, it is generally assumed that polyamines act in ways similar to calcium by stimulating the activities of numerous enzymes (protein kinases, phosphatases, etc), by interacting with negatively charged molecules in membranes and other structural elements, and by stabilizing and protecting nucleic acids. Although the presence of polyamines in nervous tissues has been known for a long time, it was assumed that polyamines were playing important roles during neuronal development, and their functions in the adult brain remained largely unknown. Recently, a number of findings have renewed the interest of neuroscientists in studying the role of polyamines in the nervous system: (1) polyamine synthesis is rapidly activated under several pathological conditions, and (2) a polyamine site is present on the NMDA receptor and regulates the function of the receptor. In this review, we will first summarize the evidence linking polyamines to brain development and then discuss the possible roles of polyamines in brain pathology and the applications of these findings in clinical pathology.

Polyamines and brain development

Several reviews have been written concerning the roles of polyamines during brain development, and only a brief summary of the general conclusions will be presented. In good agreement with the general roles of polyamines in growing tissues, the activity of ornithine decarboxylase (ODC), the rate-limiting enzyme in polyamine synthesis, and polyamine levels in brain are high during the early stages of development and decrease gradually to reach lower levels in adult brain. Different brain regions exhibit different patterns of ontogenetic development of the ODC pathway that correspond to their developmental patterns of growth. The availability of an irreversible inhibitor of ODC, difluoromethylornithine (DFMO), has provided additional ways to assess the potential roles of polyamines not only for cell proliferation but also for cell migration and the establishment of mature synaptic connections. Thus, as predicted, prenatal treatment with DFMO produced long-term alterations in cortical structures but not in cerebellum, whereas postnatal treatment mainly affected cerebellum. Furthermore, other cell populations

were selectively vulnerable to postnatal DFMO treatment, such as the granule cells of the dorsal cochlear nucleus. Additional alterations in catecholaminergic neurons were observed with such a treatment. It is important to stress, however, that despite prolonged treatment with DFMO, brain levels of spermine were not significantly decreased. This has been generally attributed to the slow turnover of spermine, as well as to the existence of interconversion pathways from one polyamine into another. It remains therefore possible that polyamines participate in more subtle developmental events than those revealed by DFMO treatment.

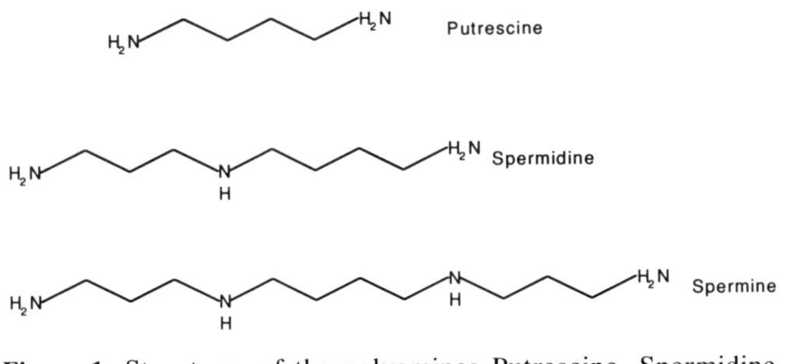

Polyamines and brain pathology

A variety of treatments (e.g., axotomy, seizure activity, ischemia, and hypoglycemia) associated with neuronal pathology produce a rapid and transient increase in ODC transcription and activity and a prolonged increase in putrescine levels. Two main hypotheses have been proposed to relate changes in polyamine metabolism, polyamine functions, and pathological consequences of the triggering treatments: (1) polyamines are part of a "regeneration program" that promotes neuronal survival or recovery from stress-induced damage, or (2) polyamines participate in the development of pathological responses. The first view is supported by evidence indicating that blockade of ODC prevents the regeneration of axotomized peripheral neurons and that peripheral injections of a mixture of polyamines promote the survival of neurons that would have died during the developmental period or following ischemic episodes. Finally, polyamines decreased the set point at which mitochondria buffer extramitochondrial calcium, thereby stimulating the ability of neurons to overcome toxic calcium overloads. On the other hand, a number of arguments suggest that polyamines play an additional function as promoters of neuronal pathology. First, putrescine levels appear to be a good index of neuronal pathology following ischemia and excitotoxin-induced seizure activity. Recently, polyamines have been shown to modulate the function of the NMDA receptor, and drugs acting as antagonists of the polyamine site provide protection against ischemia-induced cell death. Polyamines also potentiate calcium-dependent proteolysis of cytoskeletal proteins, and there is a good correlation between changes in putrescine levels in hippocampus and piriform cortex and *in situ* calpain activity following kainic acid–induced seizure activity. A "killer role" for polyamines has been proposed on the basis of the existence of the interconversion pathway for polyamines. Thus, polyamines can be acetylated by activation of a spermine or spermidine acetyltransferase (SAT), which is rapidly induced by a number of toxic stimuli. In turn, the acetyl derivatives of polyamines regenerate spermidine and putrescine following oxidation by a polyamine oxidase. This process also generates hydrogen peroxide, which has been linked to cell injury under numerous conditions. This mechanism has been proposed to underlie programmed cell death in the embryo and ischemia-induced neuronal death. Finally, putrescine has been shown to modulate calcium fluxes and transmitter release, further contributing to the potentially lethal effect of polyamines. Additional deleterious effects have been suggested as a result of disturbances of the blood-brain barrier produced by

Figure 1. Structures of the polyamines Putrescine, Spermidine, and Spermine.

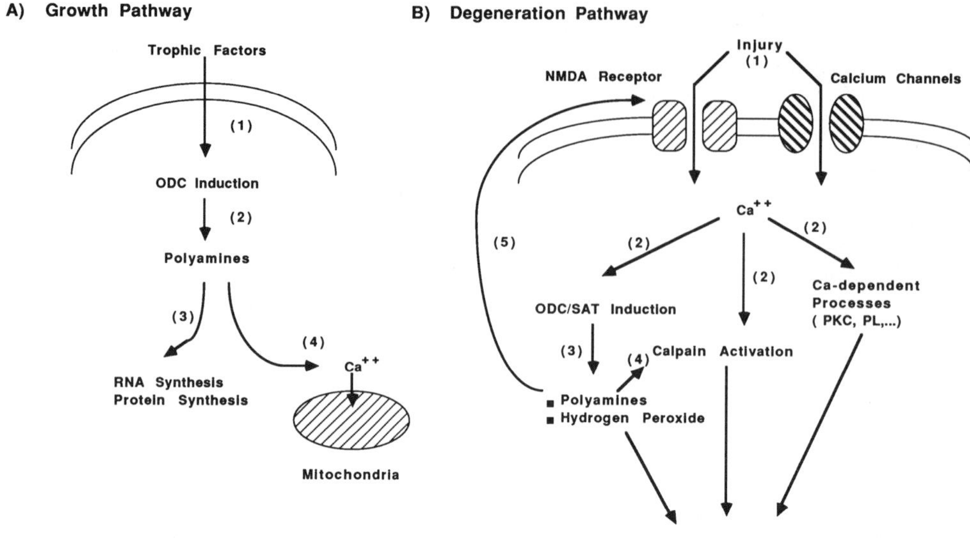

Figure 2. Schematic representation of the mechanisms linking polyamines and neuronal growth and neuronal damage. (A) ODC induction by growth factors or other signals (1); stimulation of polyamine synthesis (2); stimulation of RNA and protein synthesis (3); stimulation of intramitochondrial calcium sequestration and prevention of toxic calcium overload (4). (B) Injury triggers calcium influx via activation of NMDA receptor and other calcium channels (1); calcium activates several calcium-dependent processes (2); induction of ODC/SAT, increased polyamine synthesis, and generation of hydrogen peroxide (3); potentiation of calpain by polyamines (4); modulation of NMDA receptor by polyamines (5).

polyamines. Obviously, more work is needed to establish whether modifications of polyamine metabolism are beneficial or lethal and to identify the cellular functions involved in the effects of polyamines after insults or injuries.

Polyamines in clinical pathology

Clinical interest in polyamines is the result of the known association of polyamines with cell division and accelerated growth. Tumor tissues are characterized by high levels of polyamines, and there is a positive correlation between red blood cell and cerebrospinal fluid polyamine levels and the activity of certain brain tumors (medulloblastoma and other malignant gliomas). As polyamines are necessary for the survival of tumor cells in cultures, these findings have prompted the initiation of clinical trials of ODC inhibitors, such as DFMO, for the treatment of CNS tumors. Their limited success has been proposed to be due to the existence of the interconversion pathway, and the recent findings that a combination of ODC inhibitors, polyamine oxidase inhibitors, and polyamine-free diet inhibits malignant cell proliferation in the gastrointestinal tract have renewed the hope that such an approach can be applied successfully to the treatment of brain tumors. Although the assays of polyamines and metabolites in different biological fluids (blood, urine, cerebrospinal fluid) have been applied to the monitoring of the state of various tumors, a similar approach has not been tested for neurodegenerative diseases.

Conclusion

Although significant progress has been made concerning our understanding of the roles of polyamines in the CNS, it is clear that much more work is needed to answer many important questions: (1) Do polyamines exist in different subcellular compartments in which they play different functions? (2) What are the exact functions of polyamines in pre- and postnatal development? (3) What are the functions of polyamines following brain injury? (4) What are the effects of different inhibitors of polyamine metabolism in the prevention and treatment of different brain pathologies or tumors?

Further reading

Coffino P, Poznanski A (1991): Killer polyamines? *J Cell Biochem* 45:54–58

Fulton DS, Levin VA, Lubich WP, Wilson CB, Marton LJ (1980): Cerebrospinal fluid polyamines in patients with glioblastoma multiforme and anaplastic astrocytoma. *Canc Res* 40:3293–3296

Janne J, Alhoner L, Leinohen P (1991): Polyamines: From molecular biology to clinical applications. *Ann Med* 23:241–259

Najm I, Vanderklish P, Etebari A, Lynch G, Baudry M (1991): Complex interactions between polyamines and calpain-mediated proteolysis in rat brain. *J Neurochem* 57:1151–1158

Pegg AE (1986): Recent advances in the biochemistry of polyamines in eukaryotes. *Biochem J* 234:249–269

Sarhan S, Knodgen B, Seiler N (1989): The gastrointestinal tract as polyamine source for tumor growth. *Anticanc Res* 9:215–224

Seiler N, Bolkenius F (1985): Polyamine reutilization and turnover in brain. *Neurochem Res* 10:529–544

Shaw GG (1979): The polyamines in the central nervous system. *Biochem Pharmacol* 28:106

Slotkin TA, Bartolome J (1986): Role of ornithine decarboxylase and the polyamines in nervous system development: A review. *Brain Res Bull* 17:307–320

Williams K, Romano C, Dichter MA, Molinoff PB (1991): Modulation of the NMDA receptor by polyamines. *Life Sciences* 48:469–498

Post-Traumatic Stress Disorder (PTSD), Psychobiology

Dennis S. Charney, Steven M. Southwick, and John H. Krystal

Post-Traumatic Stress Disorder (PTSD) is a psychiatric disorder of considerable prevalence and morbidity. It may be a consequence of a variety of precipitants including combat, sexual, and physical trauma. However, the clinical syndrome shares common symptomatic elements and, in many patients, may persist for years.

There has been little neurobiological research designed to identify the alterations in brain function that may occur following severe trauma. Preclinical and clinical studies have led to hypotheses that a spectrum of PTSD symptoms may relate to alterations in function of specific brain regions and neurochemical systems. A premise of these hypotheses is that although acute neurobiological responses to massive psychological trauma are adaptive and of survival significance, they may result in long-lasting maladaptive changes in central nervous system function of clinical consequence.

Preclinical investigations of uncontrollable stress are valuable because the behavioral changes induced by uncontrollable stress resemble the symptoms of PTSD (Table 1). Clinical observations indicate that the uncontrollable or inescapable aspects of stress are associated with severe and chronic psychiatric sequellae.

Behavioral effects of uncontrollable stress

There is considerable evidence that the lack of control or predictability of the stressor are critical variables in the development of specific behavioral and somatic responses, including escape deficits, reduced exploration, decreased feeding weight loss, suppressed immunological function, gastric ulcerations, and analgesia. Of particular interest are the learning deficits that cover a wide range of tasks and often are transferred from one task to another. The molecular mechanisms by which lack of stress controllability exerts such profound behavioral effects are not known. However, recent studies of the neuronal mechanisms involved in learning and memory processes suggest long-term potentiation and behavioral sensitization may be important in this regard.

Long-term potentiation. Long-term potentiation (LTP), a long-lasting increase in synaptic responsivity to a constant volley of brief tetanic stimulation of afferent fibers, has been proposed as a putative mechanism involved in learning and memory. Several investigations have found that LTP can be produced in the hippocampus and the amygdala. Hippocampal LTP is impaired in animals exposed to uncontrollable stress, suggesting that the behavioral deficits produced by uncontrollable stress may be mediated, in part, by the development of impaired hippocampal LTP. Projections from

the medial geniculate body to the amygdala may mediate the formation of traumatic memories established by pairing an acoustic stimulus with a foot shock. These findings raise the possibility that dysfunctional amygdala LTP may be related not only to the learning abnormalities associated with uncontrollable stress but also to the encoding, storage, and retrieval of traumatic memories.

The neurochemical systems critical to the interaction between stress and LTP are not established. However, there is evidence implicating the opioid and catecholamine systems in the mechanisms responsible for stress-induced impairment of hippocampal LTP. Consistent with the LTP amygdala findings are also studies identifying a key role for the amygdala in noradrenergic and opioid modulation of memory.

Behavioral sensitization. Acute uncontrollable stress has been shown to result in the development of sensitization to subsequent stressors or related stimuli. The development of sensitization may be strongly related to the memories of the original stress and hormonal and neurochemical (e.g., noradrenergic) brain systems activated by the experience. The intensity of traumatic memory correlates with the degree to which key neurochemical systems have been activated by the trauma.

Neurochemical systems mediating the effects of uncontrollable stress

Uncontrollable stress produces substantial alterations in multiple neurotransmitter and neuropeptide systems. The best studied have been the noradrenergic, benzodiazepine, dopamine, opioid, and hypothalamic-pituitary-adrenal (HPA) systems.

Noradrenergic neuronal function

Uncontrollable stress produces regional increases in norepinephrine turnover in the locus coeruleus (LC), limbic regions (hypothalamus, hippocampus, amygdala), and cortex. However, when animals have mastered a coping task to effectively reduce stress, increased norepinephrine turnover generally does not occur. The LC probably plays a particularly important role since it is activated in association with fear and anxiety states, and the limbic (thalamus, amygdala, hippocampus, septum, hypothalamus) and cortical regions innervated by the LC are those involved in the elaboration of adaptive responses to stress. A series of investigations have shown that uncontrollable, but not controllable stress, produces increased responsibility of LC neurons to excitatory stimulation.

Sensitization and noradrenergic function. It has been suggested that the behavioral sensitization to repeated stress may relate to altered noradrenergic function. There is evidence that sensitization or conditioning of noradrenergic neuronal systems may be induced by environmental stimuli previously paired with uncontrollable shock. Neutral stimuli that have been paired with inescapable shock produce increases in brain norepinephrine metabolism and behavioral deficits similar to that elicited by the shock.

Table 1. Uncontrollable Stress in Laboratory Animals Produces Behaviors Resembling PTSD

Initial alarm state
Disruption of learning and memory
Motoric depression
Social withdrawal
Aggression
Increased sensitivity to stress
Psychosomatic disturbances
Deficits in immune function

Clinical implications. Psychophysiologic studies have documented heightened autonomic or sympathetic nervous system activity in combat veterans with PTSD based upon responses to visual and auditory combat-related stimuli in the laboratory. Further, several psychophysiologic studies have found hyperreactive responses to combat-associated stimuli but not to other stressful non-combat-related stimuli. These data are consistent with the hypothesis that noradrenergic hyperreactivity in patients with PTSD may relate to the conditioned or sensitized responses to specific traumatic stimuli.

Neuroendocrine studies and investigations of peripheral catecholamine receptor systems have also provided evidence of dysregulated peripheral sympathetic nervous system activity in PTSD. Significantly elevated 24-hour urine norepinephrine excretion has been documented in combat veterans with PTSD compared with healthy subjects or patients with schizophrenia or major depression. Consistent with this observation, it has been reported that the density of platelet alpha-2 adrenergic receptors is reduced in PTSD, perhaps reflecting adaptive "downregulation" in response to chronically elevated levels of circulating endogenous catecholamines.

A better probe of brain noradrenergic function is determination of the behavioral, biochemical, and cardiovascular responses to the alpha-2 adrenergic receptor antagonist, yohimbine. As predicted from preclinical studies, combat veterans have been shown to exhibit enhanced behavioral responses to yohimbine. PTSD patients experience a high frequency of yohimbine-induced panic attacks and flashbacks that could not be accounted for by comorbid panic disorder. The incidence of yohimbine-induced panic attacks in PTSD patients approximates that observed in panic disorder patients. In contrast, yohimbine rarely induced panic attacks in healthy subjects or in patients with schizophrenia, major depression, generalized anxiety disorder, or obsessive-compulsive disorder. These findings suggest that PTSD and panic disorder (PD) have similar pathophysiologic dysfunctions in the regulation of noradrenergic function. However, the etiology may differ, with panic disorder due more to genetic factors and PTSD to severe environmental trauma.

Dopaminergic neuronal function

Inescapable acute stress produces increases in dopamine (DA) metabolism in specific brain areas. However, in contrast to the noradrenergic system, it has not been established whether the response of the dopamine system to stress is related to the controllability of the stressor.

The predominant brain DA system involved in the stress response is the prefrontal cortex. Mesocortical neurons are preferentially activated by stress compared with limbic and striatal dopamine areas. Compared with the noradrenergic system, there is less data regarding the role of dopamine neurons in stressor sensitization.

Clinical implications. The increase in mesocortical dopamine neuronal function by uncontrollable stress raise the possibility that behaviors mediated by this region may be manifest in PTSD. Consequently, the hypervigilance and paranoia frequently observed in PTSD patients may be related to DA neuronal hyperactivity in the frontal cortex.

Case reports and small descriptive studies have described the appearance of trauma-related psychosis in Nazi concentration camp survivors, Vietnam veterans and Cambodian refugees. Childhood or adult psychological trauma might ex-

acerbate the course of psychotic illness or perhaps contribute to neuroleptic resistance in schizophrenic patients.

These investigations predict that dopamine agonists would exacerbate PTSD symptomatology. Chronic stress and cocaine administration have similar effects on dopamine function, and daily stress exposure increases the locomotor response to cocaine or amphetamine in animals. PTSD patients may have potentiated behavioral responses to these drugs, to the degree that they may be more vulnerable to the development of paranoia or psychosis after ingestion of cocaine or amphetamine.

Endogenous opiate system function

The primary behavioral effect of uncontrollable stress on endogenous opiate function is stress-induced hypoalgesia. These effects are likely to be mediated, in part, by stress-induced release of endogenous opiates because stress-induced analgesia is blocked by naltrexone and shows cross-tolerance to morphine, and opiate peptides are elevated after acute uncontrollable shock.

Clinical implications. In nontraumatized human populations, naloxone has been shown to reverse stress-induced analgesia after noxious foot shock, after long-distance running, and during an uncontrollable problem-solving task. Similarly, in Vietnam veterans with PTSD, it has been demonstrated that naloxone reverses the analgesia induced by stressful combat films. These findings are consistent with the induction of opioid-mediated stress-induced analgesia in PTSD and with the observation that wounded combatants during WWII required lower doses of narcotics than civilians with less severe injuries.

A role for endogenous opiates in PTSD is also consistent with the observation that opiates represent a preferred substance of abuse among many traumatized veterans. Some investigators have suggested that exogenous opiates decrease hyperarousal symptoms of PTSD by decreasing the firing rate of the locus coeruleus and thus decreasing central noradrenergic output. Opiate withdrawal, on the other hand, is associated with an increase in central noradrenergic activity as well as an increase in PTSD symptoms.

Benzodiazepine-GABA system function

Recent data indicates that brain benzodiazepine-GABA receptor function is linked to the effects of uncontrollable stress. The benzodiazepine drugs lorazepam and chlordiazepoxide prevent the development of the behavioral deficits produced by uncontrollable stress. On the other hand, the inverse benzodiazepine receptor agonist beta carboline, FG-7142 (N-methyl-β-carboline-3-carboxamide), produces similar behavioral deficits as uncontrollable shock.

The behavioral deficits induced by uncontrollable shock have been shown to be associated with a decrease in GABA receptor–mediated chloride ion flux, depolarization-induced hippocampal release of GABA, and brain benzodiazepine receptor occupancy.

There is also a reduction of the density of low-affinity $GABA_A$ receptors and chloride efflux and uptake in the cerebral cortex following uncontrollable stress. The mechanisms responsible for these effects are not known and may be due to changes in endogenous modulators of the benzodiazepine-GABA receptor complex or molecular events at the receptor level.

Clinical implications. Benzodiazepine drugs have had a long history of use in PTSD treatment, particularly for acutely traumatized individuals. They are beneficial when prescribed in the acute posttraumatic period for anxiolysis and restoration of sleep. However, the efficacy of benzodiazepines in chronic PTSD has not been established.

Serotonergic system function

Several studies have suggested that the behavioral deficits induced by uncontrollable stress may involve alterations in the serotonin system activity. There are data to suggest that certain forms of stress stimulate brain serotonin neuronal activity. In addition, stress also alters the density of different 5-HT receptors. ^{3}H-imipramine binding in rat cortex is decreased by inescapable shock. Acute immobilization stress induces an elevation of 5-HT$_2$ binding sites in the same brain region.

Clinical implications. Of relevance to PTSD are data suggesting that serotonin may play an important role in diverse functions relating to the regulation of mood and aggressive, impulsive, and compulsive behaviors. The high prevalence of depression, aggression, and obsessive and compulsive symptoms in PTSD may be due, in part, to dysfunction of serotonin neurons.

Hypothalamic-pituitary-adrenal (HPA) axis function

Acute stress of many types produces increases in ACTH and corticosterone levels in laboratory animals. The mechanism responsible for transient stress-induced hyperadrenocorticalism may involve a downregulation of hippocampal glucocorticoid receptors.

There is preliminary evidence that the learning deficits produced by uncontrollable stress may be related to neurotoxic effects of elevated glucocorticoid levels on hippocampal neurons. The most convincing supportive evidence for this assertion comes from two recent investigations in vervet monkeys. In one study vervet monkeys who died spontaneously after experiencing sustained social stress had marked and preferential hippocampal degeneration. In a subsequent investigation, glucocorticoid administration in vervet monkeys produced very similar damage in the hippocampus.

Clinical implications. There is clinical evidence that acute trauma in humans can produce profound increases in glucocorticoid levels. It is not known if the magnitude of these increases are sufficient to produce hippocampal damage as demonstrated in the nonhuman primate studies. Magnetic resonance imaging studies capable of measuring hippocampal volume in PTSD patients to evaluate this possibility are indicated.

There is comparatively little data on HPA axis function in patients with chronic PTSD. Urinary free cortisol levels in PTSD patients may be reduced compared with healthy subjects and those with other psychiatric disorders. Consistent with a decrease in urinary free cortisol concentrations is the finding that lymphocyte glucocorticoid receptors may be increased in PTSD.

There have been several investigations conducted designed to evaluate HPA system regulatory mechanisms in PTSD. In a small sample of PTSD patients the ACTH response to CRF was reported to be blunted in the presence of normal plasma cortisol levels. Further, there is preliminary evidence that some PTSD patients may be overly sensitive to the cortisol-suppressing effects of dexamethasone. Considered together, these results suggest that central inhibitory mechanisms suppressing CRF and ACTH function may be increased in chronic PTSD. This is consistent with preclinical investigations demonstrating adaptive HPA responses to chronic stress.

Despite the foregoing findings, there is also evidence that substantial cortisol increases can be elicited from PTSD patients in response to intense emotional stimuli and pharmacological agents, such as yohimbine. Thus, more investigation is needed to elucidate HPA axis function in acute and chronic PTSD in relation to basal activity and regulatory mechanisms involving stimulatory and inhibitory processes.

Comment

The preclinical investigations strongly suggest that acute, severe psychological trauma may result in the *simultaneous,*

Table 2. Neurochemical Responses to Severe Stress Related to Primary Symptoms of PTSD. Adapted with permission from Charney et al. (1993).

Neurochemical System	Functional Alteration	Brain Regions Involved	PTSD Symptoms
Noradrenergic	Increased regional norepinephrine turnover Increased responsivity of locus coeruleus neurons	Locus coeruleus Hippocampus Amygdala Hypothalamus Cerebral cortex	Anxiety, fear, autonomic hyperarousal, "fight or flight" readiness, encoding of traumatic memories, facilitation of sensory-motor responses
Dopamine	Increased dopamine release in frontal cortex and nucleus accumbens Activation of mesocortical dopamine neurons	Prefrontal cortex Nucleus accumbens	
Opiate	Increased endogenous opiate release Decreased density of mu opiate receptors	Periaqeductal gray Cerebral cortex Amygdala	Analgesia Emotional blunting Encoding of traumatic memories
Hypothalamic-pituitary-adrenal	Acutely elevated glucocorticoid levels Elevated corticotropin releasing factor	Hippocampus Locus coeruleus Amygdala	Metabolic activation, learned behavioral responses, anxiety and fear responses

Table 3. Hypothesized Persistent Maladaptive Neurobiological Sequellae from Severe Uncontrollable Stress

Sequelae	Neurobiological Mechanisms	Clinical Correlates
Sensitized neurochemical and behavioral responses	Turnover	Anxiety, flashbacks, and hyperarousal symptoms, alcohol, opiates and benzodiazepines used to reduce above symptoms
	Dopamine: enhanced prefrontal dopamine metabolism to repeated stress	Persistent hypervigilance, may show potentiated responses to agents that increase dopamine metabolism, such as cocaine, which may produce intense paranoia in PTSD patients May show impairment in reward-dependent behaviors
	Opiates: hypoalgesia occurs following less intense stress	Persistent stress-induced hypoalgesia and emotional blunting
Neurotoxicity	HPA: stress-induced hippocampal damage related to elevated glucocorticoids	Persistent learning and memory difficulties

interactive functional alteration of noradrenergic, benzodiazepine, opiate, dopamine, and HPA systems producing a compendium of behavioral and physiological responses (Table 2). The initial terror, anxiety, and autonomic arousal elicited by the trauma could be due to perturbations in the noradrenergic and benzodiazepine systems.

The hypervigilance associated with the behavioral response to trauma may relate to the activation of prefrontal dopamine neurons by stress. The analgesia-reducing pain sensitivity of trauma victims and the blunted emotional responses frequently associated with physical and psychological trauma may relate to the increased release of endogenous opioids by stress.

The behavioral responses produced by the simultaneous alteration of numerous brain neurochemical systems and structures by psychological and physical trauma likely represent adaptive responses critical for survival in a dangerous environment. The fear, autonomic hyperarousal, analgesia, and hypervigilence will facilitate appropriate rapid behavioral reactions to threat. The trauma-induced cortisol increases may promote the metabolic activation necessary for sustained physical demands required to avoid further injury and to survive. Activation of the norepinephrine and opiate systems may relate to the encoding of the traumatic memory, thereby facilitating appropriate behavioral responses to danger in the future.

Chronic neurobiological sequelae

Although initially beneficial, there appear to be longer-term negative consequences to these neurobiological responses that may be responsible for many of the chronic symptoms of PTSD (Table 3). The acute activation of noradrenergic neurons may result in a system that is persistently overly responsive to stress both related and unrelated to original trauma resulting in chronic anxiety (e.g., panic attacks) and hyperarousal symptoms. This notion is supported by preclinical studies demonstrating stress-induced alpha-2 autoreceptor dysfunction and stressor sensitization of noradrenergic neurons.

The flashbacks and other dissociative phenomena may also relate to chronic noradrenergic system hyperactivity. Since the original traumatic memory encoding is probably associated with heightened noradrenergic activity, stressors of many types that activate the noradrenergic neurons may evoke these memories and produce flashbacks. Recent clinical studies identifying potentiated behavioral and biochemical responses to yohimbine in PTSD support this view.

More research is needed to identify individual differences in vulnerability to stress. Vulnerability factors may be associated with specific genetic and environmental influences. For example, genetic abnormalities in any of the neurochemical systems or brain regions linked to stress responses may render an individual more sensitive to trauma. Moreover, previous stress exposure via a variety of neurobiological mechanisms may alter subsequent stress responsiveness. Perhaps most important is the need to apply the increased understanding of the psychobiology of traumatic stress toward the development of more effective treatment approaches for PTSD.

Further reading

Charney DS, Deutch A, Krystal JH, Southwick SM, Davis M (1993): Psychobiological mechanisms of Post Traumatic Stress Disorder. *Archives General Psychiatry* (in press)

Krystal JH, Kosten TR, Perry BD, Southwick S, Mason JW, Giller EL (1989): Neurobiological aspects of PTSD: Review of clinical and preclinical studies. *Behav Ther* 20:177–198

LeDoux JE, Romanski L, Xagoraris A: Indelibility of subcortical emotional memories. *J Cog Neurosci* 1:238–243

Maier SF (1986): Stressor controllability and stress induced analgesia. *Ann NY Acad Sci* 467:55–72

Maier SF, Ryan SM, Barksdale CM, Kalin NH (1986): Stressor controllability and the pituitary-adrenal system. *Behav Neurosci* 100:669–674

McGaugh JL (1989): Involvement of hormonal and neuromodulatory systems in the regulation of memory storage. *Ann Rev of Neurosci* 2:255–287

Roth RH, Tam S-Y, Ida Y, Yang J-XX, Deutch AY (1988): Stress and the mesocorticolimbic dopamine systems. *Ann NY Acad Sci* 537:138–147

Sapolsky RM, Uno H, Rebert CS, Finch CE (1990): Hippocampal damage associated with prolonged glucocorticoid exposure in primates. *J Neurosci* 10:2897–2902

Shors TJ, Seib TB, Levine S, Thompson RF (1989): Inescapable versus escapable shock modulates long-term potentiation (LTP) in the rat hippocampus. *Science* 244:224–226

Simpson PE, Weiss JM (1988): Altered activity of the locus coeruleus in an animal model of depression. *Neuropsychopharmacol* 1:287–295

Southwick SM, Krystal JH, Morgan CA, Johnson D, Nagy LM, Nicolaou A, Heninger GR, Charney DS (1993): Abrnomal noradrenergic function in post traumatic stress disorder. *Archives General Psychiatry* (in press)

Tsuda A, Tanaka M (1985): Differential changes in noradrenaline turnover in specific regions of rat brain produced by controllable and uncontrollable shocks. *Behav Neurosci* 99:802–817

Uno H, Tarara R, Else J, Suleman M, Sapolsky R (1989): Hippocampal damage associated with prolonged and fatal stress in primates. *J Neurosci* 9:1705–1711

vanderKolk B, Greenberg M, Boyd H, Krystal J (1985): Inescapable shock, neurotransmitters, and addiction to trauma: Toward a psychobiology of post traumatic stress. *Biol Psychiatry* 20:314–325

Prader-Willi Syndrome

Joan H. M. Knoll, Joseph Wagstaff, and Marc Lalande

Prader-Willi syndrome (PWS), initially described by Prader, Labhart, and Willi in 1956, is characterized by reduced fetal activity, neonatal hypotonia, mental retardation, obesity, short stature, small hands and feet, hypogonadism, a characteristic facies, and an interstitial deletion of chromosome 15q11q13 in about 60% of patients. PWS is generally sporadic, occurring with a prevalence of 1 in 15,000 to 1 in 30,000.

Clinical features

The clinical course of PWS can be divided into two distinct stages. The most important features of the first stage are nonprogressive neonatal hypotonia, hypothermia, hypogonadism and hypogenitalism, weak or absent cry, and impaired suck and swallowing reflexes resulting in failure to thrive. The hypotonia usually improves by 12 months and the feeding problems, often requiring gavage feeding, improve within the first few months. The second stage is characterized by developmental delay and early-onset obesity as a result of an insatiable appetite. PWS children generally sit by 12 months, walk by 30 months, and develop obesity between 3 and 5 years if their hyperphagia is left untreated. Varying degrees of intellectual impairment, food searching, daytime sleepiness, hypopigmentation, increased pain threshold, skin picking, speech and language problems, hyperthermia, skeletal and dental problems, and a characteristic facies (including narrow bifrontal diameter, almond-shaped eyes, and triangular mouth) are also seen during this stage. Behavior problems including depression, stubbornness, and violent temper tantrums sometimes provoked by withholding food occur in about half of PWS patients aged 3 to 5 years and may continue throughout early adulthood. Short stature and small hands and feet are noticed during adolescence when PWS patients fail to undergo a growth spurt. The diagnosis of PWS is difficult and often awaits expression of clinical features in the second stage.

Genetics

Cytogenetics. A cytogenetic deletion of the proximal long arm of chromosome 15, 15q11q13, is seen in about 60% of patients. Five percent have other chromosome abnormalities such as inversions, duplications, translocations, and small bisatellited additional chromosomes (SBACs) involving 15q11q13. The remaining 35% of patients have apparently normal chromosome 15s. Because of the small region involved and the variability in contraction of proximal 15q, it is essential that high-resolution chromosome analyses, including fluorescent in situ hybridization, be performed when PWS is suspected. Chromosomal polymorphisms of the short arm of chromosome 15 have revealed that the deleted chromosome in PWS is always of paternal origin. The cytogenetic deletion observed in PWS involves the same chromosomal region as that involved in Angelman syndrome (AS), a clinically distinct syndrome characterized by severe mental retardation, ataxic movements, seizures, unfounded bouts of smiling and laughter, and a characteristic facies.

Molecular genetics. The ability to construct genomic libraries enriched for chromosome 15q11q13 by flow sorting of SBACs has allowed the isolation of specific cloned DNA markers (3-21, IR39d, IR4-3R, IR10-1, 34, 189-1). Such markers have been used to analyze DNA from PWS patients, quantitatively and qualitatively. Three main molecular classes have been found through the use of quantitative dosage hybridization. Two of these classes represent large molecular deletions that differ only by the presence or absence of one centromere proximal marker (IR39d), and the third class represents a molecular class in which no deletion is detected. There is concordance between the cytogenetic and molecular data: the two deletion classes correspond to patients with cytogenetically visible deletions and the third class corresponds to patients with normal chromosomes. The same molecular classes have been found in Angelman syndrome.

Qualitative studies utilizing restriction fragment length polymorphisms (RFLPs) for the cloned DNA markers have confirmed that the deletion is on the paternally derived chromosome 15, a finding in sharp contrast to that in Angelman syndrome. Interestingly, RFLP analyses in all confirmed nondeletion PWS patients have revealed aberrant inheritance of chromosome 15 with two maternal copies and no paternal copy (maternal uniparental disomy). By contrast, paternal uniparental disomy has been reported in only a small fraction of nondeletion AS patients. It appears that gene(s) in 15q11q13 are imprinted or marked during gametogenesis as being either maternal or paternal in origin and that both maternal and paternal copies of 15q11q13 are required for normal development. Two copies of genes from one parent do not compensate for missing genes from the other parent and deletions from parents of opposite sex manifest clinically distinct syndromes.

A dysfunction in the hypothalamus could cause many of the clinical features present in Prader-Willi syndrome. Postmortem examination of the hypothalamus in PWS patients has not, however, shown any anatomic abnormality. A molecular genetic approach to defining the gene(s) responsible for PWS will require identification of the smallest critical region of deletion that produces PWS. Since the minimum estimated size of 15q11q13 is 5 Mb, this region may contain 100 to 300 genes (10^5 genes/3000 Mb). A number of expressed sequences have been found, but only recently has the first gene of known function been localized to 15q11q13. This gene GABRB3, encodes the β3 subunit of the $GABA_A$ (gamma amino butyric acid) receptor. GABRB3 is deleted on the paternally derived chromosome 15 in PWS patients with cytogenetic deletions but is intact in nondeletion PWS patients and in at least one patient with an unbalanced chromosome 15 translocation, who is deleted for other markers in common to the two deletion classes. This latter finding strongly suggests that GABRB3 does not play a role in PWS. However, these results raise the intriguing possibility that other neurotransmitter receptor genes may be located in nearby regions of 15q11q13 and could be involved in the pathogenesis of PWS.

Further reading

Butler MG (1990): Prader-Willi syndrome: Current understanding of cause and diagnosis. *Am J Med Genet* 35:319–332

Cassidy SB (1984): Prader-Willi syndrome. *Curr Prob Pediat* 14:1–55

Donlon TA, Lalande M, Wyman A, Bruns G, Latt SA (1986): Isolation of molecular probes associated with the chromosome 15 instability in the Prader-Willi syndrome. *Proc Natl Acad Sci (USA)* 83:4408–4412, 6964

Knoll JHM, Nicholls RD, Magenis RE, Graham JM Jr, Lalande M, Latt SA (1989): Angelman and Prader-Willi syndromes share a common chromosome 15 deletion but differ in parental origin of the deletion. *Am J Med Genet* 32:285–290

Knoll JHM, Nicholls RD, Magenis RE, Glatt K, Graham JM Jr,

Kaplan L, Lalande M (1990): Angelman syndrome: Three molecular classes identified with chromosome 15q11q13-specific DNA markers. *Am J Hum Genet* 47:149–155

Nicholls RD, Knoll JHM, Butler MG, Karam S, Lalande M (1989): Genetic imprinting suggested by maternal heterodisomy in nondeletion Prader-Willi syndrome. *Nature* 342:281–285

Wagstaff J, Knoll JHM, Fleming J, Kirkness EF, Martin-Gallardo A, Greenberg F, Graham JM Jr, Menninger J, Ward D, Venter JC, Lalande (1991): Localization of the gene encoding the GABA$_A$ receptor β3 subunit to the Angelman/Prader-Willi region of human chromosome 15. *Am J Hum Genet* 49:330–337

Prosopagnosia

Daniel Tranel and Antonio R. Damasio

Introduction

The inability to recognize familiar faces is known as *prosopagnosia* or *face agnosia*. The phenomenon has been noted since the turn of the century, and although relatively infrequent, it has been the focus of a good deal of scientific inquiry. The fascination with prosopagnosia stems from the oddity of the disorder, along with the opportunity it provides to investigate perception, learning, and recall of complex and unique knowledge. From a clinical standpoint, prosopagnosia is no less intriguing, and given its devastating effect on its victims, accurate diagnosis and management of the condition are essential. Investigation of prosopagnosia has been marked by controversy regarding the relative importance of *perceptual* versus *mnestic* factors in the development of the condition, and a dispute about whether prosopagnosia requires *bilateral* lesions or can be caused by *unilateral* damage.

Neuropsychological and neuroanatomical correlates of prosopagnosia

The face recognition defect in prosopagnosia typically covers *both* the retrograde and anterograde compartments, that is, patients can no longer recognize the faces of previously known individuals and are unable to learn new faces. Patients are unable to recognize the faces of family members, close friends, and, in the most paradigmatic instances, even their own face in a mirror. Upon seeing those faces, the patients experience no sense of familiarity, no inkling that those faces are *known* to them, that is, they fail to conjure up consciously any pertinent information that would trigger recognition. Prosopagnosia must be distinguished from disorders of *naming*, i.e., it is not an inability to name faces of persons who are otherwise recognized as familiar. There are numerous examples of face-naming failure, from both brain-injured populations and from the realm of normal everyday experience, but in such instances, the unnamed face is invariably detected as familiar, and the precise identity of the possessor of the face may also be apprehended accurately. In prosopagnosia, however, the defect sets in at the level of recognition. Needless to say, the patients will also manifest a face-naming impairment. There are several major types of prosopagnosia.

Pure associative prosopagnosia. In this variety, the recognition impairment is relatively pure, in the sense that it is confined to the visual modality, and occurs in the setting of normal or near-normal visual perception. Associative prosopagnosia largely conforms to the classic notion of agnosia, that is, "a normal percept stripped of its meaning." The patients perform normally on standard neuropsychological tests of visuoperceptual discrimination and visuospatial judgment. Recognition via other modalities is unaffected; for example, upon hearing the voices of individuals whose faces go unrecognized, the patients will instantly recognize the identities of those individuals. Even within the visual modality, the defect is highly circumscribed. For instance, patients may be able to recognize individuals on the basis of a distinctive feature (e.g., hairstyle), or gait or posture.

Most patients with pure associative prosopagnosia also have achromatopsia, an acquired impairment of color perception. The combination of the defects is due to the contiguity of the neural processing systems for form and color, and their sweeping damage by one lesion. The color perception defects per se, however, cannot account for the face recognition impairment, since normal individuals can easily recognize faces in black and white. Another common correlate of face agnosia is defective visual appreciation of texture, although the relationship of this defect to the prosopagnosia is unclear. The ability to read may or may not be affected in prosopagnosic patients, depending upon the location of the lesion in the left occipital region. When the lesion encompasses *both* the left occipitotemporal region and the left periventricular region (the white matter beside, beneath, and behind the occipital horn), reading impairment (alexia) coexists with prosopagnosia.

Pure associative prosopagnosia is caused by bilateral damage in inferior occipital and temporal visual association cortices, that is, in the inferior component of cytoarchitectonic areas 18 and 19, and part of the nearby cytoarchitectonic area 37. Most cases are due to cerebral infarctions caused by occlusion in posterior cerebral artery branches. Head injury and cerebral tumors, especially gliomas originating in one occipital lobe and traversing into the opposite hemisphere via the splenium of the corpus callosum, can also produce prosopagnosia. Prosopagnosia in connection with lesions (bilateral or unilateral) located exclusively *above* the calcarine fissure, in superior visual association cortices, has never been reported, although there are cases of face agnosia in which on one or both sides the inferior lesions extend upward above the calcarine fissure.

Apperceptive prosopagnosia. Face agnosia in the setting of significant visuoperceptual disturbance is termed "apperceptive" prosopagnosia. Apperceptive face agnosics have defects in basic visual perception, demonstrable on neuropsychological testing, that compromise abilities such as matching of unfamiliar faces, judgment of line orientations, and mental manipulation of pictures and picture fragments. In making the clinical distinction between apperceptive and associative forms of face agnosia, however, it must be emphasized that perceptual processes, on the one hand, and recognition or recall processes, on the other, cannot be rigidly compartmentalized. Those processes are part of a continuum, and there is no sharp demarcation point at which perceptual processes stop and recognition processes take over. Even in "pure" cases of associative prosopagnosia, there may be fairly subtle, albeit important, disturbances of high-level integrative abilities, which cannot be detected by available probes. Any explanation of face agnosia tied to perceptual factors must reconcile the fact that most patients with visual perceptual defects, even severe ones, do *not* lose their ability to recognize faces. That is, visuoperceptual disturbance as detected by neuropsychological probes does not necessarily cause face agnosia.

Apperceptive face agnosia is most often associated with damage in right visual association cortices within the occipital and parietal regions. It appears that the damage must involve *both* the inferior and superior components of posterior visual association cortices (areas 18 and 19), mesially and laterally, for severe and lasting face agnosia to develop. In most cases, parts of areas 39 and 37 on the right will also be damaged.

Amnesic associative prosopagnosia. In this type of prosopagnosia, patients have normal visual perception. In amnesic prosopagnosia the face recognition defect is part of a broader recognition impairment that is equal across multiple sensory channels and covers multiple categories of stimuli. Thus, neither viewing a face nor listening to the appropriate voice will trigger a sense of familiarity or identity recognition. As in other types of prosopagnosia, the face recognition impairment may cover virtually all faces from the retrograde compartment, including the faces of family members and the self.

Face agnosia of the amnesic associative type is associated with bilateral damage in anterior temporal regions. The lesions compromise the hippocampal system (entorhinal cortex, hippocampal formation, and amygdala) and paralimbic and neocortical fields in cytoarchitectonic areas 38, 20/21, and 22. The posterior occipitotemporal cortices, however, are spared. Such damage is typically caused by herpes simplex encephalitis or Alzheimer's disease.

Developmental prosopagnosia. Some individuals never develop a normal capacity for learning faces. They have a life-long deficiency in learning and recognizing faces that ought to have been readily mastered, and since the problem begins in childhood, it is appropriate to call it "developmental" prosopagnosia. No neural correlates for this condition have been identified yet. Developmental prosopagnosia has not been widely reported or studied, but it may be far more frequent than suspected, its apparent rarity being due to the fact that affected persons tend to conceal their disability. It is probably associated with learning disability for other classes of visual stimuli that, like faces, require individual identification and have many similar exemplars. Developmental prosopagnosics tend to be profoundly embarrassed by their deficiencies and often show adjustment difficulties.

Other aspects of face processing

The nature and extent of the defect. In prosopagnosia, the recognition impairment occurs at the most *subordinate* taxonomic level, that is, at the level of identification of unique faces. Prosopagnosics are fully capable of recognizing faces as faces, that is, performance is normal at the *superordinate* taxonomic level. Also, most prosopagnosics can recognize facial expressions and facial gender, and can make accurate estimations of facial age. These dissociations highlight the fact that recognizing faces at the level of unique identity is a highly demanding task that requires the brain to distinguish between numerous different exemplars that bear a high degree of resemblance to one another.

Although face agnosics will recognize any number of visual entities at the *basic object level* (e.g., cars as cars, buildings as buildings, dogs as dogs), they will often fail to recognize these items at the subordinate level of unique identity. Thus, similar to the problem with faces, they are unable to recognize the specific identity of a particular car or building. These recognition impairments are common in prosopagnosia and underscore the notion that the core defect in face agnosia is the inability to disambiguate fully individual visual stimuli.

Nonconscious discrimination of familiar faces. Recent evidence has revealed that many prosopagnosic patients show accurate *covert* or *nonconscious* discrimination of familiar faces, despite their complete inability to recognize those faces at the overt level, for example, based on self-report or verbal ratings of familiarity. For example, it has been shown that prosopagnosics generate large, discriminatory electrodermal responses to familiar faces that are otherwise unrecognized. Preserved covert face discrimination has been demonstrated in other experimental paradigms, such as reaction-time tasks and forced-choice procedures. In the electrodermal paradigm, covert face discrimination has even been demonstrated for faces from the anterograde compartment, indicating that the brain can continue to learn new visual information even without conscious influence.

Recovery. Prosopagnosic patients often become quite adroit at utilizing nonface information to recognize the persons around them. For example, they rely on voice or gait or a distinctive visual feature. Regarding the latter, it is often helpful to add a distinctive feature in situations where recognition is highly demanded but difficult. At a crowded social gathering, for example, the patient might have his or her spouse wear a special hat or other article of clothing, which would facilitate rapid and accurate identification. The earlier in the course of recovery that compensatory strategies such as these can be taught, the better the chances for healthy adaptation to the disability.

Unanswered questions

The reason why some prosopagnosics demonstrate preserved covert discrimination of familiar faces while others do not remains unclear. A related question concerns the relationship between different indices of covert discrimination; for example, are skin conductance responses, reaction times, and event-related potentials all tapping into the same discrimination mechanism? Another question regards the nature of "visual ambiguity" in faces, that is, what factors account for the high degree of visual similarity among faces, and how are these factors decoded by the brain? Many issues regarding the neural basis of face recognition have yet to be clarified fully. Finally, there has been only very preliminary investigation of the condition of developmental prosopagnosia.

Further reading

Damasio AR (1985): Disorders of complex visual processing: Agnosias, achromatopsia, Balint's syndrome, and related difficulties of orientation and construction. In: *Principles of Behavioral Neurology,* Mesulam M-M, ed. Philadelphia: FA Davis, pp 259–288

Damasio AR, Damasio H, Tranel D (1990): Impairments of visual recognition as clues to the processes of categorization and memory. In: *Signal and Sense: Local and Global Order in Perceptual Maps,* Edelman GM, Gall WE, Cowan WM, eds. New York: Wiley-Liss, pp 451–473

Damasio AR, Damasio H, Van Hoesen GW (1982): Prosopagnosia: Anatomic basis and behavioral mechanisms. *Neurol* 32:331–341

Damasio AR, Tranel D, Damasio H (1990): Face agnosia and the neural substrates of memory. *Ann Rev Neurosci* 13:89–109

Sergent J, Poncet M (1990): From covert to overt recognition of faces in a prosopagnosic patient. *Brain* 113:989–1004

Sergent J, Villemure JG (1989): Prosopagnosia in a right hemispherectomized patient. *Brain* 112:975–995

Tranel D, Damasio AR (1985): Knowledge without awareness: An autonomic index of facial recognition by prosopagnosics. *Science* 228:1453–1454

Tranel D, Damasio AR, Damasio H (1988): Intact recognition of facial expression, gender, and age in patients with impaired recognition of face identity. *Neurol* 38:690–696

Prostaglandins in the CNS: Arachidonic Acid

Leonhard S. Wolfe

The dietary essentiality in mammals of certain polyunsaturated fatty acids of the n-6 family, particularly linoleic acid ((Z,Z)-9, 12-octadecadienoic acid) and arachidonic acid ((Z,Z,Z,Z)-5, 8, 11, 14-eicosatetraenoic acid), has been known since the pioneering studies of George and Mildred Burr in 1929. A few years later von Euler in Sweden and Goldblatt in England reported a smooth muscle–stimulating principle in lipid extracts of human seminal fluid and seminal vesicles of sheep and goats that was called by von Euler "prostaglandin." Thirty years passed before the chemical structures of prostaglandins were determined and the metabolic connection between arachidonic acid and prostaglandins was established by Bergström, Samuelsson, and coworkers at the Karolinska Institutet in Stockholm. In the 30 years since these pivotal discoveries, a multitude of potent biologically important arachidonic acid metabolites have been discovered. Although it was known that dietary essential fatty acids were required for normal brain growth and development, synaptic function, and myelination, it is only in the past 10 years that the important involvement of arachidonic acid metabolites in synaptic receptor signal transduction pathways, ion channel activities, and modulation of neurotransmitter release has been discovered.

The brain is particularly rich in phospholipid species that contain arachidonate esterified to carbon-2 of the glycerol backbone. The concentration of free arachidonic acid in the brain is maintained at a very low level in the basal state owing to the activities of arachidonyl-CoA synthetase and arachidonyl-CoA transferase, which catalyse reacylation of the fatty acid into membrane phospholipids. Release from phospholipids is generally thought to be the rate-limiting step in the formation of prostaglandins and the other oxygenated metabolites. Receptor-mediated generation of free arachidonic acid occurs either through the mediation of a G-protein, or a calcium-dependent activation of phospholipases. In the nervous system two main pathways for arachidonate release have been identified, one controlled by phospholipase A_2, and the other involving the phospholipase C–diacylglycerol lipase pathway that uses phosphoinositides as the source of arachidonic acid; however, the former predominates.

When cells, including neurons, are stimulated chemically, electrically, or immunologically, arachidonic acid is preferentially released from stores in membrane phospholipids by the action of a Ca^{2+}-dependent cytosolic phospholipase A_2 (cPLA$_2$). Physiological stimuli that incrementally raise intracellular Ca^{2+} cause the translocation of cPLA$_2$ to the plasma membrane, with a greatly stimulated activity. Recent molecular cloning and expression studies indicate that cPLA$_2$ has significant similarities to protein kinase C isoforms and lung surfactant proteolipid-associated protein. Following release, arachidonic acid can be metabolized in mammals by three main enzymatic pathways: (1) cyclooxygenase or prostaglandin G/H synthase, which forms prostaglandins and thromboxanes; (2) lipoxygenases, which form leukotrienes, lipoxins, and other specific hydroperoxyeicosatetraenoic acids; and (3) the monooxygenase activity of cytochrome P-450 (epoxygenase), which forms epoxyeicosatrienoic acids. These oxidized metabolites of arachidonic acid are collectively termed eicosanoids, and function as local autocoids when released extracellularly, or as second messengers when produced intracellularly in response to specific receptor stimulation.

The major prostaglandins in the brain of mammals are PGD$_2$, PGE$_2$, and PGF$_{2\alpha}$. Prostaglandin D$_2$ synthase (prostaglandin-H$_2$ D-isomerase) is a glutathione-independent membrane-associated enzyme. Recent reports on the isolation of cDNAs encoding this enzyme show from the deduced amino acid sequences that this enzyme is a member of the lipocalin superfamily, which comprises secretory molecule transporters. A large body of evidence is accumulating that PGD$_2$ in the brain functions as a sleep-inducing agent in the preoptic area of the hypothalamus. It also induces hypothermia and suppresses luteinizing hormone release. Prostaglandin E$_2$ synthase (prostaglandin-H$_2$ E-isomerase) is an anionic form of glutathione S-transferase. In the anterior hypothalamic/preoptic area, PGE$_2$ is an endogenous compound that induces wakefulness and hyperthermia, particularly in response to systemic pyrogens and the cytokine interleukin-1. It is also involved in dopamine-mediated luteinizing hormone release. Thus, in the hypothalamus these two prostaglandins are involved in the regulation of the sleep-wake cycle, temperature, and hypothalamo-pituitary functions. The function of prostaglandin F$_{2\alpha}$ in the brain is still uncertain. It is of interest that cloning studies show that prostaglandin-H$_2$ PGF-synthase is the same enzyme as PGD$_2$ 11-ketoreductase and has an unexpected homology to human liver aldehyde reductase and lens crystallin.

The interaction of prostaglandins in autonomic neurotransmission has been given much attention. There is evidence both for and against the thesis that PGE$_2$ inhibits the exocytotic release of noradrenaline from sympathetic nerve endings. In brain slices PGE$_2$ inhibits the release of noradrenaline, but on the other hand it will facilitate release

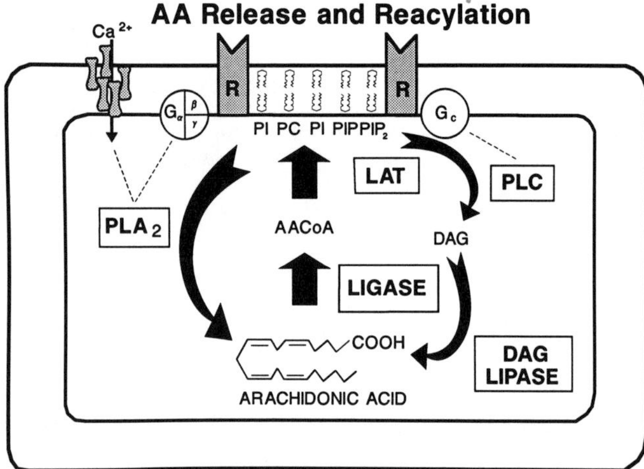

Figure 1. Routes for receptor-mediated release of esterified arachidonic acid from phospholipids and the reacylation mechanisms. The major pathway is controlled by phospholipase A_2 (PLA$_2$) and the second by phospholipase C (PLC) to form diacylglycerol (DAG). The reacylation pathway involves two steps, arachidonyl-CoA synthetase (AACoA) and lysophospholipid acyltransferase (LAT), which combines lysophospholipids and arachidonyl CoA to form phospholipids.

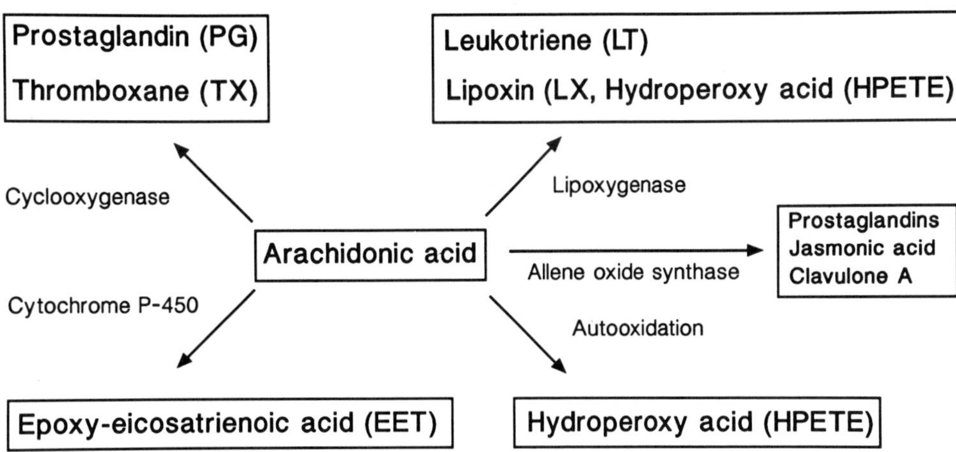

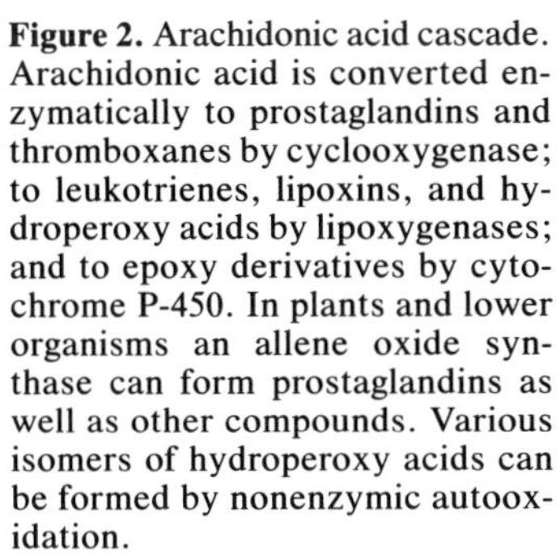

Figure 2. Arachidonic acid cascade. Arachidonic acid is converted enzymatically to prostaglandins and thromboxanes by cyclooxygenase; to leukotrienes, lipoxins, and hydroperoxy acids by lipoxygenases; and to epoxy derivatives by cytochrome P-450. In plants and lower organisms an allene oxide synthase can form prostaglandins as well as other compounds. Various isomers of hydroperoxy acids can be formed by nonenzymic autooxidation.

from brain synaptosomes. Subtypes of PGE_2 receptors have been identified in brain and adrenal medulla and have been found to be coupled to a G-protein.

The peptidoleukotrienes (LTC_4, LTD_4, and LTE_4) and LTB_4 synthesized by the 5-lipoxygenase pathway from arachidonic acid are present in the brain in small amounts, but their cellular source is uncertain. Their action *in vivo* has been linked to the control of neuronal excitability and neuroendocrine function. A pathogenetic role of these compounds in inflammatory processes and reactive changes to injury is more certain. Platelet activating factor is known to increase cerebrospinal fluid leukotrienes.

The invertebrate and mammalian nervous systems have an active 12-(S)-lipoxygenase that catalyzes the conversion of arachidonic acid into a reactive intermediate, 12-S-hydroperoxyeicosatetraenoic acid (12-HPETE). This intermediate can be further metabolized by at least six different pathways. This would have had little interest to the neuroscientist if it were not for the discovery of some remarkable properties of these compounds in the mechanosensory neurons of the marine mollusk *Aplysia californica*. In these neurons 12-lipoxygenase metabolites of arachidonic acid act as second messengers and may also participate in the communication of local groups of cells. In the sensory neurons of *Aplysia* serotonin and the molluscan tetrapeptide, FMRFamide, close and open, respectively, a specialized subclass of potassium ion channels termed K^+S, named because serotonin inactivates the channel. Patch-clamp studies have shown that opening of the K^+S channel can be mimicked by 12-HPETE. Furthermore, this metabolite exerts a dual action, causing first a fast depolarization followed by a slow hyperpolarization. The fast depolarization appears to be mediated by the 12-keto metabolite of 12-HPETE, whereas the slow hyperpolarization is due to the activity of a cytochrome-P-450 metabolism to the 11,12 epoxy metabolite, hepoxilin A_3. All these metabolites are actively produced by neural tissue of *Aplysia*. Of further interest is the finding that 12-HPETE is a potent inhibitor of Ca^{2+}/calmodulin-dependent protein kinase, which catalyzes the phosphorylation of synapsin 1 in mammalian synaptic endings. Thus, 12-lipoxygenase metabolites may also regulate neurotransmitter release and modulate synaptic strength. There are other examples of the modulation of potassium currents by arachidonic acid and its lipoxygenase metabolites, for example, the G-protein-gated muscarinic K^+-channel in atrial cardiac myocytes and similar channels in smooth muscle.

Long-term potentiation of synaptic transmission (LTP) in the hippocampus of the vertebrate nervous system was first described by Bliss and Lømo in 1973. Briefly, LTP can be defined as a persistent enhancement of synaptic efficacy usually triggered by brief high-frequency tetanic activation of glutamate excitatory pathways. This form of synaptic plasticity has been loosely implicated in learning and memory and has been the subject of intense investigation and controversy, particularly on whether its maintenance is determined by presynaptic or postsynaptic events. It seems clear that induction of LTP involves activation of glutamate receptors of the kainate/quisqualate class, which cause sufficient depolarization of the postsynaptic cell membrane to relieve the Mg^{2+} blockade of the N-methyl-D-aspartate (NMDA) type of receptors. Calcium enters the cell via the NMDA-receptor ion channel and initiates the cascade of events that results in persistent enhancement of synaptic transmission.

Recently it has been shown that arachidonic acid and lipoxygenase metabolites were likely involved in the induction and expression of LTP. Glutamate and NMDA release arachidonic acid from hippocampal slices and markedly stimulate the formation of 12-(S)-hydroxyeicosatetraenoic acid (12-HETE). It has also been shown that following the induction of LTP there is a small but rapid release of arachidonic acid and the synthesis of 12-HETE. The noncompetitive NMDA receptor antagonist, MK-801, prevents this release. Lipoxygenase inhibitors and MK-801 also block the induction of LTP. It is uncertain what are the individual roles of arachidonic acid and/or the 12-lipoxygenase metabolites. A current hypothesis is that arachidonic acid, released from postsynaptic membranes during induction of LTP, reaches the presynaptic terminals by free diffusion and facilitates the release of glutamate, that is, it acts as a retrograde messenger. Caution is needed in this interpretation since another candidate molecule is nitric oxide synthesized from L-arginine by the constitutive calmodulin-dependent NO synthase. Analogs such as N^G-Nitro-L-arginine block the induction of LTP in an arginine-reversible manner and so does bathing the hippocampal slices with hemoglobin, which binds NO but does not penetrate cells. The most recent data favor nitric oxide rather than arachidonic acid as a retrograde messenger in LTP, but as yet this is far from proven.

In summary, arachidonic acid and eicosanoids exert both intracellular and intercellular actions in the nervous system. They also contribute to pathophysiological responses, particularly in brain injury, cerebral edema, and ischemia, and in inflammation. In the years to come their roles as second messengers in neurotransmitter signal transduction and their

Long-Term Potentiation

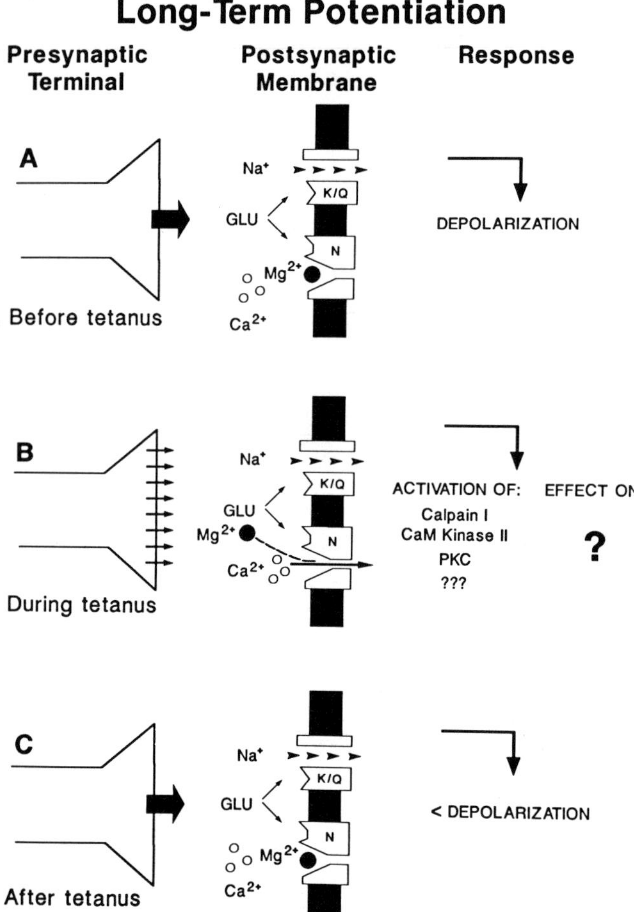

Figure 3. The three steps leading to long-term potentiation (LTP). (A) During normal transmission that precedes induction of LTP, stimulation of afferent fibers leads to depolarization of the postsynaptic membrane, mostly due to the activation of kainate/quisqualate type of receptors by the excitatory neurotransmitter glutamate. (B) A series of repetitive, high-frequency stimulation or tetanus applied to the afferent fibers will cause massive release of glutamate in the synaptic cleft, which will in turn induce a sustained depolarization of the postsynaptic membrane. This allows removal of the Mg^{2+} that normally prevents activation of the NMDA receptor. Binding of glutamate to the receptor now leads to Ca^{2+} influx through the channel associated with the receptor. A cascade of Ca^{2+}-dependent processes is then initiated, a number of these still not characterized. (C) Finally, upon normal afferent fiber stimulation some time after the tetanus, a greater depolarization of the postsynaptic membrane is obtained. A major feature of this phenomenon is that this potentiating effect persists for hours, days, and even weeks *in vivo* after its induction. (Modified from Collingridge and Bliss, *TINS* 7 (1987):288.)

paracrine influences on neighboring neurons and glia will be increasingly clarified.

Further reading

Bliss TVP, Errington ML, Lynch MA, Williams JH (1990): Presynaptic mechanisms in hippocampal long-term potentiation. In: *The Brain,* Cold Spring Harbor Symposia on Quantitative Biology 55. Cold Spring Harbor, NY: Cold Spring Harbor Laboratory, pp 119–129

Hayaishi O (1991): Molecular mechanisms of sleep-wake regulation: Roles of prostaglandins D_2 and E_2. *FASEB J* 5:2575–2581

O'Dell TJ, Hawkins RD, Kandel ER, Arancio O (1991): Tests of the roles of two diffusible substances in long-term potentiation. *Proc Nat Acad Sci* (USA) 88:11285–11289

Piomelli D, Greengard P (1990): Lipoxygenase metabolites of arachidonic acid in neuronal transmembrane signalling. *Trends Pharmacol Sci* 11:367–373

Samuelsson B, Dahlén SE, Lindren JÅ, Rouzer CA, Serhan CN (1987): Leukotrienes and lipoxins: Structures, biosynthesis, and biological effects. *Science* 237:1171–1176

Shimizu T, Wolfe LS (1990): Arachidonic acid cascade and signal transduction. *J Neurochem* 55:1–15

Protein Phosphorylation and Neuronal Modulation

Irwin B. Levitan

Protein phosphorylation catalyzed by a number of different kinds of protein kinases is a ubiquitous mechanism for modulating the biological activities of proteins. Esterification of the hydroxyl moieties of serine, threonine, or tyrosine residues with a high-charge-density phosphate group (Fig. 1) is a rapid and effective way of altering protein structure and function. Among the many proteins whose activities are regulated by phosphorylation are structural proteins (including components of the cytoskeleton), membrane ion channels, and a variety of enzymes including protein kinases themselves. The biological phenomena in which protein phosphorylation plays a prominent role range from control of cell proliferation to regulation of electrical activity in excitable cells. Thus, it is not surprising that phosphorylation is the neuromodulatory mechanism that has been most thoroughly investigated and appears to be most widespread. This brief update will focus on selected recent information concerning the role of phosphorylation in regulating neurotransmitter release, synaptic plasticity, and properties of neurotransmitter receptors and membrane ion channels.

Modulation of ion channel activity by protein phosphorylation

Extensive information exists concerning the actions of protein kinases on the functional properties of ion channels. Measurements of whole-cell voltage-clamp currents and single ion-channel activities have demonstrated that the result of phosphorylation is often a change in channel-open probability or the number of functional channels in the membrane; in contrast, the single-channel conductance does not appear to be regulated by phosphorylation. Changes in channel activity have now been documented for sodium channels, calcium channels, chloride channels, a variety of different classes of potassium channels, and ligand-gated channels including the nicotinic acetylcholine receptor/channel. For example, phosphorylation by protein kinase C causes a slowing of channel inactivation and a reduction of peak current of one cloned subtype of rat-brain sodium channel expressed in a heterologous expression system. On the other hand, phosphorylation of skeletal muscle calcium

channels by cyclic AMP-dependent protein kinase causes an increase in channel-open probability. Different potassium channels may have their activities increased or decreased by phosphorylation, and the same channel may be modulated in different directions by different protein kinases. The only unifying theme to date is that modulation of ion-channel activity by phosphorylation is a widespread phenomenon, and that a cell will use any and all modulatory tricks at its disposal to regulate ion-channel properties.

One important question about modulation of channel properties by phosphorylation concerns the identity of the phosphorylation target. Among those few ion-channel proteins that have been purified several have been shown, using biochemical techniques *in vitro,* to be excellent substrates for protein kinases. These include sodium channels, calcium channels, and the nicotinic acetylcholine receptor/channel. Furthermore, functional studies with purified or cloned channels have made it clear that the phosphorylation sites involved in channel modulation often exist within the sequences of the ion-channel proteins themselves, rather than in some nonchannel regulatory component. For example, in the case of the cloned rat-brain sodium channel referred to previously, removal of one particular serine residue, by site-directed mutagenesis, eliminates channel modulation by protein kinase C in the heterologous expression system. Findings such as this imply that regulatability by phosphorylation is an essential inherent characteristic of many ion channels.

Modulation of sensory and neurotransmitter receptors by protein phosphorylation

Visual and olfactory receptors share with certain neurotransmitter receptors commonalities of structure and transduction mechanism. As detailed elsewhere in this volume and the *Encyclopedia* many of these receptors are characterized by having seven membrane-spanning sequence domains and produce their biological responses via guanyl nucleotide-binding proteins (G-proteins). Furthermore, there is another striking similarity among these receptors: they are down-regulated as a result of phosphorylation by a special class of protein kinases. This has been documented most clearly for the receptor for light, the visual pigment rhodopsin, and for

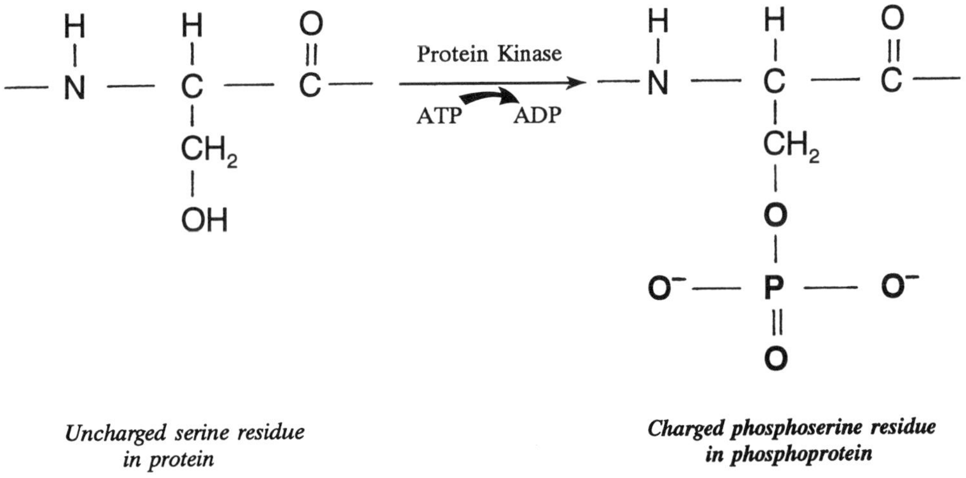

Figure 1. Phosphorylation causes a change in the charge and probably the conformation of target proteins. The figure illustrates the esterification of a protein serine residue with a high charge density and bulky phosphate group. The hydroxyl moieties of threonine or tyrosine residues may also be targets for protein kinases.

Uncharged serine residue in protein

Charged phosphoserine residue in phosphoprotein

the β-adrenergic receptor. In both of these cases the particular protein kinases involved (rhodopsin kinase and β-adrenergic receptor kinase, respectively) have been isolated and characterized in detail. The kinases are specific for the agonist-occupied (activated) forms of the receptors, and phosphorylate multiple serine and threonine residues near their carboxyl termini. This results in uncoupling of the receptor from its G-protein and, at least in the case of the β-adrenergic receptor, sequestration of the receptor inside the cell so it is no longer exposed to agonist. This down-regulation of G-protein–linked receptors can modulate profoundly the responsiveness of cells to sensory stimuli and neurotransmitters.

The nicotinic acetylcholine receptor/channel will also be considered here because its modulation by phosphorylation results in down-regulation that resembles, in some respects, that of the G-protein–linked receptors. The nicotinic receptor/channel exhibits a phenomenon known as desensitization, a decrement in response upon prolonged or repeated presentation of agonist. At the single-channel level desensitization is evident as a decrease in the rate at which the channel opens from the closed state. Phosphorylation of the nicotinic receptor/channel at identified sites in the amino acid sequence, by either cyclic AMP-dependent or tyrosine protein kinases, increases the rate of desensitization and hence downregulates the response of the receptor/channel to agonist. It has also become clear that phosphorylation of the nicotinic receptor/channel can have a longer-term modulatory influence, by regulating the assembly and membrane insertion of receptor/channel subunits at the neuromuscular junction.

Modulation of neurotransmitter release by protein phosphorylation

Molecular mechanisms involved in the regulated secretion of proteins and small molecules from a variety of cell types, including neurons, have been the subject of intense investigation in recent years. Evidence for a modulatory role for protein phosphorylation in the exocytotic secretion of neurotransmitters has come from studies of the squid giant synapse. Because of the large size of both its presynaptic and postsynaptic elements, the squid giant synapse has been a favored preparation for studying mechanisms and modulation of synaptic transmission. Although the chemical events that underlie neurotransmitter release remain only poorly understood, there is evidence to suggest that a particular protein kinase, the type 2 calcium/calmodulin-dependent protein kinase, plays a role in the regulation of release. Injection of this enzyme into the squid presynaptic terminal causes an increase in the amount of transmitter released by a presynaptic action potential. An attractive hypothesis is that this results from phosphorylation of synapsin 1, a major substrate for type 2 calcium/calmodulin-dependent protein kinase in nerve cells. Synapsin 1 is closely associated with neurotransmitter-containing synaptic vesicles and may hinder their fusion with the presynaptic plasma membrane; following its phosphorylation by the kinase, synapsin is released from the vesicles, and they can undergo exocytotic fusion more readily. It remains to be determined whether phosphorylation of synapsin 1 or a synapsin-like protein is involved in regulation of secretion in other kinds of cells.

Modulation of synaptic strength by protein phosphorylation

An enduring dogma in neuroscience is that the synapse must be an important site of neuronal plasticity, and that plastic changes in synaptic strength contribute to behavioral plasticity. Evidence from several model systems points to an important role for protein phosphorylation in modulation of synaptic strength. For example, it is well established that phosphorylation alters ion-channel properties in several different kinds of nerve cells from the marine mollusc *Aplysia,* leading to changes in excitability that, in turn, influence the way the cells respond to synaptic and hormonal stimuli. In addition, a series of studies have demonstrated modulation of neurotransmitter release, again as a result of ion-channel phosphorylation, in *Aplysia* sensory neurons that participate in a defined behavior.

An increasingly popular model for long-term synaptic plasticity in mammals is long-term potentiation (LTP), a very long-lasting increase in synaptic strength that follows an intense presynaptic stimulus at certain synapses in the hippocampus and other parts of the brain. Although the question of a presynaptic versus a postsynaptic locus for the long-term maintenance phase of hippocampal LTP remains controversial, it is clear that the initial induction of LTP requires calcium entry into the postsynaptic cell via the NMDA-type of glutamate receptor/channel. One way the resulting rise in postsynaptic calcium contributes to the induction of LTP is via the activation of calcium-dependent protein kinases. Experiments with peptide inhibitors specific for the type 2 calcium/calmodulin-dependent protein kinase, or for protein kinase C, implicate the requirement for both of these enzymes in LTP induction. More recent evidence with tyrosine kinase inhibitors suggests that tyrosine phosphorylation may also be required for LTP. The target substrates for the protein kinases involved in LTP have not yet been identified.

Summary

This review is by no means comprehensive. For example, space constraints preclude a summary of emerging data concerning the role of protein phosphorylation in neuronal development. Many growth factor receptors are protein tyrosine kinases, and some protooncogenes that are important for neuronal differentiation exert their effects via phosphorylation. This story is at an early stage, and we will be hearing much more about it in the future. Other aspects of neuromodulation that involve phosphorylation similarly have had to be covered in a cursory fashion or not at all. The necessity of being selective serves to emphasize a major theme of this update: protein phosphorylation plays a central role in a large and diverse range of phenomena essential for neuronal function.

Further reading

Hemmings HC, Nairn AC, McGuinness TL, Huganir RL, Greengard P (1989): Role of protein phosphorylation in neuronal signal transduction. *Faseb J* 3:1583–1592

Huganir RL, Greengard P (1990): Regulation of neurotransmitter receptor desensitization by protein phosphorylation. *Neuron* 5:555–567

Levitan IB (1988): Modulation of ion channels in neurons and other cells. *Ann Rev Neurosci* 11:119–136

Madison DV, Malenka RC, Nicoll RA (1991): Mechanisms underlying long-term potentiation of synaptic transmission. *Ann Rev Neurosci* 14:379–397

Pseudotumor Cerebri

Thomas H. Milhorat

Pseudotumor cerebri (benign intracranial hypertension) is a clinical syndrome of obscure origin that is characterized by (1) raised intracranial pressure; (2) small to slitlike cerebral ventricles; (3) normal cerebrospinal fluid (CSF) chemistry; and (4) the absence of clear-cut neuropathological findings. The syndrome has been reported in all age groups but is most frequently encountered in obese females between the ages of 10 and 40 years. In typical cases, the syndrome runs a short course, responds to simple medical management, and exhibits little tendency to recur. However, because pseudotumor cerebri can lead to severe visual loss if not recognized and treated promptly, the reassuring appellation ''benign intracranial hypertension'' is somewhat misleading.

Etiology

Pseudotumor cerebri has been reported in association with a wide variety of etiologic factors. These include menstrual disturbances, pregnancy, the use of oral contraceptives, Addison's disease, prolonged corticosteroid therapy, withdrawal of corticosteroid therapy, otitis media, mastoiditis, basilar skull fractures, severe closed head injury, jugular vein ligation, superior vena cava syndrome, cor pulmonale, and right heart failure. In 5–30% of patients, cerebral angiography will reveal evidence of dural sinus thrombosis. Additional etiologic factors include presumed drug reactions following the administration of tetracycline, sulfamethoxazole, and nitrofurantoin, vitamin A intoxication, vitamin A deficiency, allergic reactions (e.g., bee sting), fever, severe anemias, and polycythemia vera. Despite this multitude of causes, a precise etiology is established in less than 50% of the cases and the diagnosis of pseudotumor cerebri remains one of exclusion.

Pathogenesis

It is widely assumed that pseudotumor cerebri results from generalized edema of the cerebral parenchyma. The hypothesis is supported by the often noted "smallness" of the cerebral ventricles, the frequent occurrence of the syndrome in patients with endocrine or electrolyte disturbances, and the occasional finding of "edema" in brain biopsy material. A distinctive feature of the syndrome is that patients appear remarkably well and alert despite extraordinary elevations of intracranial pressure. In recent years, considerable attention has been focused on abnormalities in CSF absorption at the level of the arachnoid villi, resulting either from increased pressure within the dural sinuses or from increased resistance to drainage across the arachnoid membrane. Supporting this view are reports of delayed clearance of radiopharmaceuticals from the CSF, abnormalities on CSF infusion testing, and evidence in some cases of increased cerebral blood volume. Table 1 classifies the many and diverse etiologic factors associated with pseudotumor cerebri according to two pathogenetic mechanisms.

Cerebral edema. The most common cause of pseudotumor cerebri is probably extracellular (interstitial) edema. This conclusion is consistent with a disorder characterized by raised intracranial pressure, normal neurologic findings, and slitlike cerebral ventricles. Osmotically induced edema, oc-

curring as a consequence of plasma hypoosmolarity, is thought to be responsible for the syndrome in certain patients with menstrual disturbances, inappropriate antidiuretic hormone (vasopressin) secretion, and following the use of oral contraceptives. A significant elevation of CSF vasopressin levels has been reported in pregnant and pubescent patients with this condition. Vasogenic edema, occurring as a consequence of a disturbance of brain capillary permeability, is a possible factor in patients with drug reactions or intoxications, allergic reactions, withdrawal of steroid therapy, and Addison's disease.

Cerebral hyperemia. Expansion of the cerebrovascular bed occasioned either by venous obstruction or arterial dilatation, is a presumed cause of pseudotumor cerebri in approximately 25% of the cases. In patients in whom no etiologic factors can be identified, cerebral angiography will occasionally reveal an unsuspected sinus thrombosis, suggesting that occult occlusions may explain the development of the syndrome in association with otitis media, mastoiditis, basilar skull fractures, and other types of closed head injury. Venous hypertension with expansion of the cerebrovascular bed also appears to be a common feature of extracranial venous occlusions, cor pulmonale, and right heart failure. Expansion of the cerebrovascular bed secondary to arterial dilatation may explain the occasional development of the syndrome in patients with severe anemias and febrile illnesses.

Clinical aspects

The symptoms and signs of pseudotumor cerebri are usually uniform and consist of headaches, vomiting, papilledema, and, occasionally, abducens palsy. The patient is typically bright and alert, and exhibits an appearance of well-being

Table 1. Causes of Pseudotumor Cerebri

1. Cerebral Edema
    Osmotic Edema
        Menarche and obesity
        Menstrual disturbances
        Pregnancy
        Oral contraceptives
        Inappropriate ADH secretion
    Vasogenic Edema
        Intoxications
        Allergic reactions
        Drug reactions
        Withdrawal of glucocorticoid therapy
        Addison's disease
2. Cerebral Hyperemia
    Venous Hypertension
        Dural sinus thrombosis (trauma, mastoiditis)
        Jugular vein ligation
        Superior vena cava syndrome
        Cor pulmonale
        Right heart failure
        Polycythemia vera
    Arterial Dilatation
        Severe anemias
        Febrile illnesses

that is out of keeping with the magnitude of intracranial hypertension. Young children are frequently listless or irritable, possibly as a consequence of headache. In chronic cases, visual loss and optic atrophy supervene. Seizures are not part of this syndrome and focal neurologic deficits do not occur.

The diagnostic procedure of choice for evaluating patients with presumed pseudotumor cerebri is computed tomography (CT) or magnetic resonance (MR) scanning. This should be carried out before performing a lumbar puncture, even when the diagnosis appears fairly certain. With few exceptions, the demonstration of a midline ventricular system of normal or smaller-than-normal size is sufficient to exclude other differential diagnoses including hydrocephalus, brain tumors, and encephalitis. Roentgenograms of the skull, CSF infusion tests, and EEG are of limited value in differential diagnosis. If the syndrome develops without obvious cause, or following head injury or otitis media, MR angiography may be indicated to rule out dural sinus thrombosis.

Once the presence of a structural mass lesion or hydrocephalus has been excluded, a lumbar puncture is performed to measure CSF pressure and to collect fluid for laboratory analysis. Despite extremely high opening pressures, occasionally in the range of 400–600 mm of water, a closing pressure in the normal range (100–150 mm) is often achieved by draining a small volume of fluid (10–15 cc). The CSF cell count is normal and the CSF protein content is normal or low.

Treatment

The treatment of pseudotumor cerebri is aimed at reducing intracranial pressure and preventing loss of vision consequent to optic atrophy. In most cases, conservative measures will be effective. Diagnostic lumbar puncture often provides immediate relief of symptoms and this should be regarded as a first step in treatment. If an obvious etiologic cause can be established, specific therapy should be provided. Weight reduction and salt restriction have been recommended as part of the therapeutic regimen in obese children. Severe or refractory elevations of intracranial pressure can sometimes be effectively controlled by Diamox or short-term, high-dose glucocorticoid therapy. Steroid therapy is obviously contraindicated when the syndrome develops following the administration of glucocorticoids, in which case the dose of the drug should be tapered gradually.

If symptoms persist following an appropriate trial of medical management, serial lumbar punctures can be carried out. These are performed as often as necessary and are spaced at longer and longer intervals as symptoms improve. With each tap, all available fluid is drained. The amount of CSF recovered will vary from 15 to 75 cc depending on the age of the patient.

If symptoms persist despite medical therapy and serial lumbar punctures, or if visual acuity declines, surgical intervention is indicated. The preferred treatment is a lumbar subarachnoid-peritoneal shunt. This is a simple procedure to perform and usually provides good results with little or no morbidity. When all else fails, subtemporal decompression, once the procedure of choice for this condition, can be considered.

Follow-up

Patients undergoing treatment for pseudotumor cerebri should be followed carefully with neurologic, ophthalmoscopic, and visual acuity examinations. Persistent symptoms should raise the possibility of some other diagnosis. The prognosis for most patients is excellent if the condition is recognized early and managed appropriately.

Further reading

Hammer M, Sorensen PS, Gjerris F, Larsen K (1982): Vasopressin in the cerebrospinal fluid of patients with normal pressure hydrocephalus and benign intracranial hypertension. *Acta Endocrin* 100:211–224

Milhorat TH (1978): *Pediatric Neurosurgery*. Philadelphia: FA Davis, pp 369–371

Milhorat TH (1987): *Cerebrospinal Fluid and the Brain Edemas*. New York: Neuroscience Society of New York, pp 89–95

Ray BS, Dunbar HS (1951): Thrombosis of the dural venous sinuses as a cause of "pseudotumor cerebri." *Ann Surg* 134:376–385

Sklar FH (1985): Pseudotumor cerebri. In: *Neurosurgery*. New York: McGraw-Hill, pp 350–357

Reading Disorders, Nonaphasic, and Their Treatment

Josef Zihl

The ability to read can be impaired after brain damage either because the comprehension of written language is disturbed or because perceptual capacities are affected that represent essential prerequisites for reading. Among these perceptual capacities the field of vision, foveal visual functions, and eye movements are the most important. All of them can be affected after posterior brain damage and, as a consequence, can impair the ability to read.

Visual field disorders

Damage to the postchiasmatic visual pathways affects vision in corresponding (homonymous) parts of both contralateral hemifields. In the majority of patients vision is completely lost in one hemifield (hemianopia) or quadrant (quadranopia), or in a small portion of the visual field near the fovea (paracentral scotoma). Visual field disorders can impair reading because the perception and therefore apprehension of words is severely affected or even lost. Obviously the degree of the resulting disability depends on the extent of field sparing. In about 70% of patients with homonymous visual field loss, sparing is less than 4 degrees of visual angle. As a rule, these patients show more or less severe reading difficulties. Based on this observation it can be concluded that sparing of vision in the perifoveal region up to about 4 degrees of visual angle is a crucial prerequisite for normal reading.

Patients with left-sided visual field disorders typically have difficulties in finding the beginning of the line or word and very often omit smaller words or the prefix of polysyllabic words. Patients with right-sided visual field disorders, in contrast, have great difficulties in finding the ending of the line or word and omit syllables in the second half of a word. If patients with this reading impairment are told to move their gaze either to the beginning or the end of the word, they can perceive the whole word by means of several eye fixations and then read it correctly. In searching for an explanation for the "macular-hemianopic reading disorder" it was found that the reading eye movements have lost their organization, that is, the typical sequence of fixation and saccadic jump of the eyes is disturbed. Thus, the paracentral (macular) visual field is crucial not only for the receptive part but also for the guidance of eye movements in reading.

Visual neglect

A similar but much more severe reading impairment can be observed in patients with visual hemineglect. Briefly, visual neglect is a condition where patients are no longer aware of the hemispace contralateral to the side of brain damage and therefore do not attend to stimuli in this hemifield. Probably because of the different hemispheric organization, left-sided visual neglect occurs much more frequently, is more severe, and lasts much longer than right-sided neglect. Concerning reading, patients with visual neglect usually omit the left half of a page and begin their reading in the middle or even the right half of a page. Even after repeated instruction these patients are unable to shift their attention to the affected side and to start their reading at the beginning of the line or word.

Foveal visual dysfunctions

Visual acuity is usually spared after brain damage but can be reduced when foveal fibers are involved, especially in patients with bilateral postchiasmatic damage. Reading can also be severely impaired as a consequence of reduced spatial contrast sensitivity or temporal instability of vision resulting in visual blurring. Both symptoms can occur after uni- and bilateral brain damage. Patients complaining of visual blurring usually report that they are no longer able to read because letters, words, and lines merge one into another, appear to be indistinct or obscured by shadows. These patients show a significant reduction in spatial contrast sensitivity whereby near vision usually is more severely affected. Patients who experience fluctuation and even extinction of words report that at the beginning of reading, vision is normal; but it degrades shortly after. They can, therefore, not continue to read because letters become "grayish" and the whole page appears to be covered with a "dense fog." It is important to note that neither "blurred vision" nor degradation of vision are always accompanied by a reduction in visual acuity or an impairment in accommodation or convergence. This is probably because the short presentation of a single item under high-contrast conditions, as happens in routine testing of visual acuity, is not sufficient to provoke "visual blurring" or temporal instability of vision.

Treatment

Most frequently, nonaphasic reading disabilities in brain-damaged patients result from visual field disorders. Follow-up observations have shown that patients are unable to adapt their eye movements to the visual field disorder so as to apprehend the whole word and to read normally. Despite the frequency of patients with reading impairments due to field loss no standard training procedure has hitherto existed that is widely accepted in treating these patients. Poppelreuter was the first to develop a method of treatment in a neurologic rehabilitation setting. The method of treatment he used was based on his observations on the disorganization of the sequence of fixations and saccadic jumps of the eyes. The rationale for this type of treatment is the acquisition of an oculomotor strategy that allows the patient to compensate for the lost perifoveal field region. More recently, electronic reading aids have proven to be very helpful in treating patients with reading disability due to visual field loss. The method of treatment is basically guided by the principles described by Poppelreuter. Briefly, patients with left-sided field

Table 1. Mean Improvement of Reading Performance in a Group of 95 Patients with Homonymous Visual Field Loss

	L[a] ($n = 53$)	R[b] ($n = 42$)
Field sparing (deg)	2.4 (1–3.5)[c]	2.6 (1–3.5)
Reading time (min)		
Before treatment	5.6 (2.4–9.5)	6.6 (2.4–4.3)
After treatment	2.2 (1.0–4.4)	2.9 (1.3–5.9)
Reading errors		
Before treatment	16.4 (6–33)	18.9 (8–35)
After treatment	3.4 (1–14)	4.5 (0–17)
Training sessions	25 (8–56)	29 (14–56)

[a]Patients with left-sided field loss.
[b]Patients with right-sided field loss.
[c]Range in parentheses.
Time and errors were measured for reading aloud a text consisting of 190 words. The corresponding values for a group of 35 age-matched normal control subjects are 1.6 min (0.6–2.1) and 0.4 errors (0–1). Duration of sessions was about 30 min each.

loss are forced to search first for the beginning of the line and the first letter of every single word in the line by means of saccadic gaze shifts, and then to read what they have seen. In contrast, patients with right-sided field loss must not read a word before they have shifted their gaze to the end of the word and thereby perceived the whole word. Thus, the patients have to learn how to perceive intentionally the whole word irrespective of the loss of vision, left or right, to the fovea. The adaptation of reading eye movements to the field defect is a good example of the improvement of a complex perceptual ability by means of compensation. The substitution of one component (visual field) by another one (eye fixations), both of which are essential for normal reading, allows the patient to regain the ability to read normally. The electronic reading aid facilitates the acquisition of the compensatory oculomotor strategy in that words are presented in horizontal motion; this facilitates the shifts of eye fixations to the beginning or to the end of a word. Table 1 shows the results of treatment in a group of 95 patients using an electronic reading aid. The observations can be taken as evidence that reading disability caused by visual field loss can be treated very effectively. Furthermore, as follow-up observations have shown, the improvement remains stable even after training is finished.

The treatment of patients with left-sided visual neglect is similar to the procedure used in patients with left-sided homonymous visual field loss except that these patients first have to regain the ability to shift their gaze toward the left hemispace in order to find the beginning of the line. Depending on the severity of neglect this may involve great effort, and treatment may thus last much longer than in patients with left-sided visual field loss.

Concerning the treatment of reading impairments caused by visual blurring or temporal degradation of vision, no attempt has been made hitherto. It may well be that contrast sensitivity and temporal stability of vision can be improved by systematic practice. Undoubtedly this, and the development of standardized methods of treatment to rehabilitate brain-damaged patients with nonaphasic reading disorders, would be an important issue for future research.

Reading and vision: Some theoretical considerations

What is the significance of these clinical observations for understanding the functional organization of reading? The observation that perifoveal visual field disorders can impair severely the acquisition of words can be taken as evidence that reading is based on the acquisition of visual word forms and not of successive letters, which at a later stage of processing are eventually synthesized to a whole. For this Gestalt-type visual word-form processing, a "window" of visual field of at least 4–5 degrees visual angle on both sides of the fovea is a crucial condition. This window is also essential for the spatial guidance of eye movements during the apprehension of words, which allows a parafoveal preview and is characterized by the typical pattern of fixation periods and subsequent saccadic eye movements. The adaptation of this eye movement pattern to a written text reflects the activity of a highly flexible top-down mechanism, which is cognitive in nature and guides the peripheral visual-processing part of reading.

The impairment of either the foveal visual functions or the loss of the perifoveal visual field (in the case of visual neglect: the field of visual attention) shows that the top-down mechanism alone is not sufficient for reading, since spontaneous adaptation to the altered visual sensory condition(s) does, as a rule, not take place. The main effect of treatment is, therefore, to adopt a strategy that enables the patient again to make use of his or her eye movement pattern in the absence of the visual field region that normally allows the shifting of attention and gaze along a line of words. This compensatory strategy is the result of an interplay between bottom-up and top-down mechanisms, whereby it is assumed that the top-down mechanism plays the dominant role in this process of adaptation. It is still not clear how these mechanisms interact, but it seems very likely that a "visual attention control subsystem" (Shallice, 1988, p. 313) determines the size and locus of the "window" over which words or parts of words are apprehended. This idea is supported by observations in patients with visual hemineglect; as long as they cannot shift their attention, their reading remains highly impaired. The ability to shift attention to the affected side is, therefore, an essential prerequisite for this type of visual-motor adaptation, which also allows the spatial guidance of eye movements in the absence of a parafoveal preview because of visual field loss or visual neglect. It is interesting to note that only after patients have learned to adapt successfully their reading eye movement pattern to their visual field loss (also with respect to the speed of acquisition of words) can they immediately fully understand what they read, and when reading aloud they show normal prosody. The locus of this cognitive top-down mechanism can therefore be assumed to be between the "peripheral" visual sensory processing stage and the semantic and phonological mechanisms, and it is probably identical with the "word form system" (see Shallice, 1988, p. 68).

Further reading

Poppelreuter W (1917, 1990): *Disturbances of Lower and Higher Visual Capacities Caused by Occipital Damage,* J Zihl, L Weiskrantz, trans. Oxford: Clarendon Press

Shallice T (1988): *From Neuropsychology to Mental Structure.* Cambridge, New York, New Rochelle, Melbourne, Sydney: Cambridge University Press

Zihl J (1989): Cerebral disturbances of elementary visual functions. In: *Neuropsychology of Vision,* JW Brown, ed. Hillsdale, NJ, Hove, London: Lawrence Erlbaum

Zihl J (1990): Treatment of patients with homonymous visual field disorders. *Zeitschrift für Neuropsychologie* 2:95–101

Serotonin (5-Hydroxytryptamine) Receptor Subtypes: Clinical Relevance

Stephen J. Peroutka

Significant advances in the analysis of 5-hydroxytryptamine (5-HT) receptor subtypes occurred in the 1980s. This progress was widespread and included multiple areas of both basic and clinical research. This review focuses on the available pharmacological agents that act at the major subtypes of 5-HT receptors and summarizes their utility in the treatment of human neuropsychiatric diseases.

5-HT$_{1A}$ receptors

The 5-HT$_{1A}$ receptor is the most characterized of all 5-HT receptor subtypes in terms of its molecular biology, pharmacology, biochemistry, and clinical significance. A number of potent and selective pharmacological agents exist for this receptor (see Table 1). The possible role of this receptor in the treatment of anxiety and other neuropsychiatric disorders has stimulated a great deal of interest in 5-HT$_{1A}$ receptors.

Anxiety. In clinical trials, the 5-HT$_{1A}$ partial agonist buspirone has been found to be useful in the treatment of anxiety. Buspirone lacks sedative side effects, a fact that is considered advantageous for patients required to work during anxiolytic therapy. Consequently, buspirone and other 5-HT$_{1A}$ agonists may be a major breakthrough in the treatment of anxiety. Interestingly, the onset of anxiolytic action occurs over a period of weeks for buspirone, suggesting that chronic modulation of 5-HT$_{1A}$ receptors may be necessary for anxiolysis. A number of other partial 5-HT$_{1A}$ receptor agonists, such as gepirone, ipsapirone, and tandospirone, are undergoing either preclinical or clinical trials (Table 1).

Depression. The prototypical 5-HT$_{1A}$ agonist 8-hydroxy-2-(di-*n*-propyl-amino) tetralin (8-OH-DPAT), has been shown to possess antidepressant properties in certain animal models of depression. Indeed, buspirone, gepirone, ipsapirone, and tandospirone are, or will be in the near future, undergoing clinical trials for depression. The preliminary clinical results indicate that 5-HT$_{1A}$ drugs may prove to be useful in the treatment of depression.

5-HT$_{1B}$ receptors

The 5-HT$_{1B}$ receptor has been identified in rat and mouse brain but not in guinea pig, cow, chicken, turtle, frog, or human brain. The fact that this 5-HT receptor binding site does not appear to exist in the human appears to have greatly diminished the interest of the pharmaceutical industry and of academic medicinal chemists in further pharmacological analyses of this site. Therefore, no selective 5-HT$_{1B}$ receptor agents are being developed for the treatment of human diseases.

5-HT$_{1C}$ receptors

The localization of 5-HT$_{1C}$ receptors in the choroid plexus suggests that they may modulate the production and/or absorption of cerebrospinal fluid. However, at the present time, pharmacologically selective agents for the 5-HT$_{1C}$ receptor do not exist and, therefore, it is not yet possible to determine whether the 5-HT$_{1C}$ receptor plays a specific role in any human neuropsychiatric disorders. The fact that many antipsychotic agents, as well as migraine prophylactic medications, display high affinity for this receptor has led to speculation that it may play a role in these two disorders. Moreover, a number of psychotomimetics (e.g., LSD, DOI) display agonist activity at this receptor. However, selective pharmacological agents are needed to confirm these speculations.

5-HT$_{1D}$ receptors

The 5-HT$_{1D}$ receptor was first characterized in 1987 and since then has been shown to be widely distributed in the human

Table 1. Summary of 5-HT Receptor Active Drugs

Receptor	Drugs
5-HT$_{1A}$	8-OH-DPAT
	Buspirone
	Gepirone
	Ipsapirone
	Tandospirone
5-HT$_{1D}$	Sumatriptan
5-HT$_2$	Amitriptyline
	Ketanserin
	Methysergide
	Mianserin
	Nefazodone
	Pizotifen
	Ritanserin
5-HT$_3$	Granisetron
	ICS 205-930
	MDL 72222
	Ondansetron
	Zacopride
5-HT Uptake Blockers	Amitriptyline
	Chlorimipramine
	Fluoxetine
	Imipramine
	Nortriptyline
	Paroxetine
	Sertraline
	Zimelidine

brain. A novel serotonergic agent, sumatriptan (formerly called GR 43175), is a potent and extremely selective agonist at the 5-HT_{1D} receptor. Therefore, the potential clinical relevance of the 5-HT_{1D} receptor rests largely on the clinical effects of sumatriptan.

Migraine (acute therapy). Sumatriptan has recently been reported to be extremely effective in the acute treatment of migraine. Furthermore, only mild side effects have been reported. Although the mechanism of action of sumatriptan in migraine remains unclear, it has been suggested that the drug may selectively interact with the 5-HT_{1D} receptor. The drug is essentially inactive at other neurotransmitter receptor binding sites. These observations are significant since sumatriptan appears to display unique pharmacological properties in both *in vitro* vascular studies and in the acute treatment of migraine. At present, two major theories have been proposed to explain the clinical efficacy of sumatriptan: selective intracranial vasoconstriction and presynaptic inhibition of neurotransmitter release in the vascular wall. Sumatriptan is now in phase III clinical trials and it continues to show promise as an acute migraine therapy with minimal side effects.

5-HT_2 receptors

The 5-HT_2 class of receptors has been extensively analyzed owing to the availability of a large number of potent and selective antagonists. The 5-HT_2 receptor has been implicated in a wide range of diseases including migraine, anxiety, and depression. The extensive characterization of a 5-HT_2 receptor, including its molecular biological features, includes the development of a number of potent 5-HT_2 antagonists. However, although the 5-HT_2 receptor appears to be involved in a variety of clinical disorders, dysfunction of the 5-HT_2 receptor alone has not been related to a specific disease state. While numerous agents have been developed that display nanomolar affinity for the 5-HT_2 receptor, their clinical usefulness is still not clearly defined. A few specific examples are provided as follows.

Migraine (prophylactic therapy). The ergot compound, methysergide, is a potent 5-HT_2 antagonist currently marketed for migraine prophylaxis. The ability of methysergide to reduce the frequency of migraine headaches was tested extensively in the 1960s. Pizotifen is another potent 5-HT_2 antagonist prescribed for the prophylactic treatment of migraine. Pizotifen is only marketed outside the United States. Amitriptyline, which is a potent 5-HT_2 antagonist in addition to being a 5-HT uptake blocker, is also an extremely effective antimigraine agent. This antimigraine activity appears to be independent of the antidepressant effect of the drug. The antipsychotic agent chlorpromazine also has proven to be of benefit in the prophylactic treatment of migraine. Other 5-HT_2 antagonists such as cyproheptadine, metergoline, and lisuride have also been shown, in limited clinical trials, to benefit migraine patients. Consequently, at least some 5-HT_2 antagonists appear to be effective in the prophylactic, but not the acute, treatment of migraine.

Depression. The exact role played by the 5-HT_2 receptor in depressive illness has yet to be determined. Chronic administration of classical antidepressant drugs causes a significant decrease in the number of 5-HT_2 receptors in the brain. Mianserin, a potent 5-HT_2 antagonist, is marketed outside the United States for treatment of depression and is in phase III clinical trials within the United States. Nefazodone, another 5-HT_2 antagonist, is in phase II clinical trials in the United States for treatment of depression.

Other potential uses. 5-HT_2 antagonists have been suggested as a possible treatment for other disorders as well. For example, the 5-HT_2 antagonist ketanserin has been shown to block 5-HT–induced vasoconstriction, bronchoconstriction, and platelet aggregation and continues to be investigated for use in various cardiovascular diseases such as hypertension, peripheral vascular disease, thrombotic or embolic episodes, and cardiopulmonary emergencies. Ketanserin is currently marketed for hypertension outside the United States.

5-HT_3 receptor agents

5-HT_3 receptors were extensively characterized in the 1980s. Within a relatively short period of time, a number of potent and selective antagonists were developed such as granisetron, ondansetron, quipazine, zacopride, and ICS 205-930 (Table 1).

Antiemetics. 5-HT_3 receptor antagonists appear to possess unique and potent antiemetic effects in both animals and man. In human clinical trials, 5-HT_3 antagonists such as ondansetron, ICS 205-930, and granisetron provide excellent relief from chemotherapy-induced nausea and vomiting. Furthermore, 5-HT_3 antagonists appear to possess fewer side effects than traditionally used D_2 receptor antagonists. At the present time, a number of selective 5-HT_3 antagonists are undergoing clinical trials for antiemetic activity (Table 1).

Other potential clinical applications. Other therapeutic indications for 5-HT_3 antagonists are also being clinically evaluated. ICS 205-930 and renzapride have been shown to be effective gastrokinetic agents in man. These agents are being evaluated in clinical conditions involving GI disturbances or motility disorders such as the carcinoid syndrome, a condition in which 5-HT overproduction can cause flushing and persistent diarrhea. Possible psychotherapeutic uses of 5-HT_3 receptor antagonists have been suggested on the basis of animal studies. In many behavioral tests predictive of anxiolytic activity, for example, ondansetron has been found to be as active as diazepam.

5-HT uptake blockers

Following release from the presynaptic terminal, 5-HT is removed from the synapse by two major mechanisms. 5-HT can be degraded by monoamine oxidase (MAO). Alternatively, the neurotransmitter can be transported back into the presynaptic terminals via a 5-HT specific uptake system. Drugs that block these mechanisms potentiate the action of 5-HT and have been reported to alleviate a variety of clinical disorders such as depression, panic attacks, obsessive-compulsive disorders, and obesity.

Depression. A correlation between depression and lowered 5-HT activity was first hypothesized in the late 1960s. The ability of tricyclic antidepressants to potently block the reuptake of 5-HT and other biogenic amine neurotransmitters provided strong support for this hypothesis. Importantly, imipramine, the first of these agents to be discovered, significantly improved patients with depression without the pressor-amine complications as with the previously used MAO inhibitors. Other tricyclics, such as chlorimipramine and amitriptyline, soon followed, yielding a similar clinical profile to that of imipramine. These uptake blockers were effective in treating depression in up to 70% of the cases. Unfortunately, undesirable complications also occur with tricyclic antidepressants, including cardiovascular complaints, sedation, and sexual dysfunction.

The search for more selective 5-HT uptake blockers has led to the development of several clinically important compounds such as fluoxetine, paroxetine, citalopram, indalpine, and sertraline. These compounds are at varying stages of clinical evaluation. Fluoxetine is a bicyclic compound that is most commonly used to treat major depression. Fluoxetine displays few of the serious cardiovascular and anticholinergic side effects of the tricyclic antidepressants. However, there have been reports of mild nausea, anxiety, insomnia, anorexia, diarrhea, nervousness, akathasias, and headaches. Fluoxetine was first marketed in the United States in 1988.

Panic disorder. Although depression has been the major indication for 5-HT uptake blockers, clinical studies also indicate that the traditional 5-HT uptake blockers (i.e., amitriptyline, imipramine) appear to be useful in the treatment of panic disorders. Imipramine was one of the first uptake blockers to demonstrate clinical effectiveness in treating panic disorder. Later, most tricyclics appeared to be more or less equivalent in relieving panic disorder and a variety of phobias. Not surprisingly, the reported adverse reactions are similar to those when taking tricyclics for the treatment of depression.

Obsessive-compulsive disease. Chlorimipramine, imipramine, and amitriptyline were the first uptake blockers reported to alleviate obsessive-compulsive symptoms. Of the more selective 5-HT uptake blockers, only fluoxetine and fluvoxamine have been investigated to any significant extent, and both drugs appear to alleviate obsessive-compulsive symptoms. Clearly, the putative efficacy of 5-HT uptake blockers in the treatment of obsessive-compulsive disorders suggests that 5-HT plays a significant role in the pathophysiology of this disorder. The possibility that the selective 5-HT uptake blockers may be more efficacious in the treatment of this disorder than the less selective agents (such as

Table 2. Summary of Clinical Indications for 5-HT Receptor Agents

Disease	Possible 5-HT Receptor(s) Involved
Anxiety	$5\text{-}HT_{1A}$, $5\text{-}HT_2$, $5\text{-}HT_3$
Depression	$5\text{-}HT_{1A}$, $5\text{-}HT_2$, uptake system
Migraine (acute)	$5\text{-}HT_{1D}$
Migraine (prophylactic)	$5\text{-}HT_2$
Obsessive-compulsive disease	Uptake system

imipramine and amitriptyline) is an area of active interest and research.

Conclusions

The past decade has seen significant progress in the clinical development and application of selective serotonergic agents. This progress has resulted from the preclinical development of potent and selective 5-HT receptor drugs, each of which allows for the stimulation or inhibition of very specific components of serotonergic neurotransmission (Table 2). The development of new, even more selective drugs may further clarify the clinical consequence of activation and inhibition of each 5-HT receptor subtype in the central nervous system. Based on the observations to date, it is likely that the clinical indications (e.g., aggression, impulsivity, alcoholism) for selective 5-HT receptor subtype agents will increase significantly in the years ahead.

Acknowledgments. This work was supported by NIH grant NS 23560-04 and the Stanley Foundation.

Further reading

Brown SL, van Praag HM (1991): *The Role of Serotonin in Psychiatric Disorders.* New York: Brunner Mazel

Fozard JR, Saxena P (1991): *Proceedings of the Third IUPHAR Satellite on Serotonin.* Basel: Birkhäuser

Peroutka SJ (1991): *Serotonin Receptor Subtypes: Basic and Clinical Aspects.* New York: John Wiley & Sons

Whitaker-Azmitia PM, Peroutka SJ (1990): *The Neuropharmacology of Serotonin.* New York: The New York Academy of Sciences Press

Signal Transduction in Neurons by Phospholipid Breakdown Products

Jan Krzysztof Blusztajn and Barbara E. Slack

Membranes of neurons, like those of other cells, are formed from arrays of phospholipid molecules organized in bilayers and held together by noncovalent bonds. Based on their chemical structure (Fig. 1), phospholipids are classified as glycerophospholipids, which contain an L-α-glycerol backbone, and sphingophospholipids, which contain sphingosine.

Glycerophospholipids carry a fatty acid (FA) at each of the *sn*-1 and *sn*-2 positions, and a phosphate at the *sn*-3 position. This core structure is called phosphatidic acid (PA). The phosphate group can be esterified to a number of different headgroups to form different phospholipids. The phospholipids carrying the headgroups choline, ethanolamine, serine, and inositol are called phosphatidylcholine (PC), phosphatidylethanolamine (PE), phosphatidylserine (PS), and phosphatidylinositol (PI), respectively. When in the *sn*-1 position of the phospholipid molecule an aldehyde or alcohol is present instead of a FA, the resulting compounds are called alkenyl (plasmalogens), or alkyl phospholipids, respectively. In brain, ethanolamine plasmalogens are as abundant as PE. The basic structural unit of the sphingolipids is called ceramide. It is formed by the addition of a FA to the amino group of sphingosine, a long-chain aliphatic amine. The addition of a phosphocholine group to the terminal hydroxyl group of ceramide forms the sphingophospholipid sphingomyelin, which resembles PC in that it contains two long carbon chains and a phosphocholine group (Fig. 1).

PC and PE are the most abundant phospholipids in brain, and are important structural components of neuronal membranes. The other phospholipids and the sphingolipids represent a smaller proportion of the total membrane phospholipid content. When plasma membrane receptors are activated by neurotransmitters, hormones, or growth factors (Table 1), specific phospholipases (phospholipid hydrolyzing enzymes) are activated, resulting in the formation of breakdown products that either are signaling molecules by themselves (i.e., they stimulate or inhibit the activity of target macromolecules) or are converted to signaling molecules by specific enzymes. Traditionally, the signaling role has been assigned to minor phospholipid components, particularly PI derivatives. However, it has recently become clear that PC and PE and SM can also be metabolized to form biologically active molecules, while PS is important as an activator for a number of amphipathic enzymes (those that require the presence of cell membranes for their activity). Most current knowledge about signaling by phospholipid breakdown products is derived from studies on nonneuronal tissues, and there is insufficient data to develop the general concepts in this field based on studies of the nervous system alone. The data reviewed here derive from investigations of many types of cells; still the experience to date indicates that similar phenomena obtain in the nervous system, and whenever possible the information applicable to neurons is specifically mentioned.

The breakdown products of phospholipids can be divided into two groups: water-soluble compounds which diffuse into the cytoplasm, and lipid-soluble ones which remain associated with the membrane. The salient features of these compounds are reviewed below.

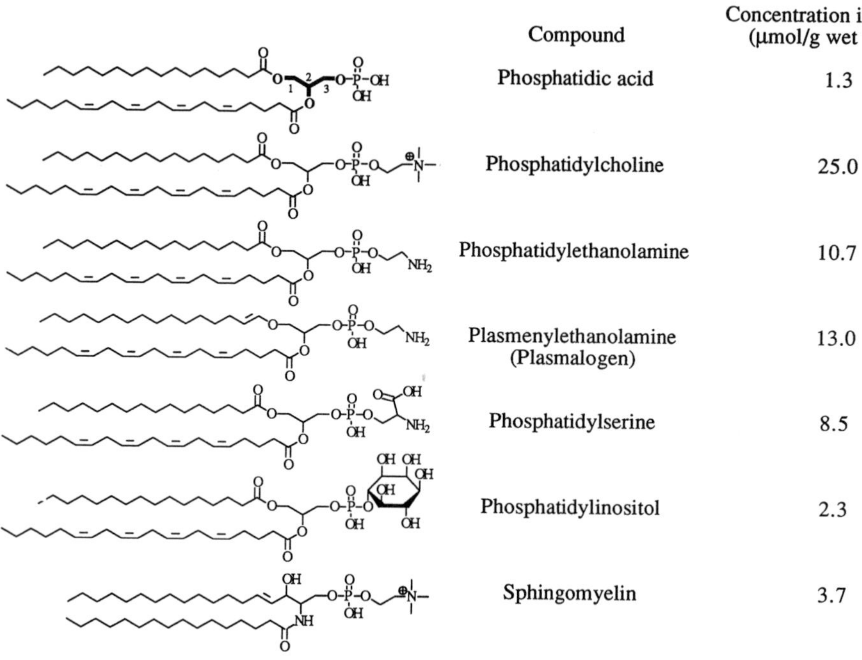

Compound	Concentration in rat brain (μmol/g wet weight)
Phosphatidic acid	1.3
Phosphatidylcholine	25.0
Phosphatidylethanolamine	10.7
Plasmenylethanolamine (Plasmalogen)	13.0
Phosphatidylserine	8.5
Phosphatidylinositol	2.3
Sphingomyelin	3.7

Figure 1. Structures of major brain phospholipids and their content in adult rat brain. 1-palmitoyl-2-arachidonyl-phosphatidic acid is used as an example and its glycerol backbone is shown in bold. (Data from Wells MA, Dittmer JC (1967): A comprehensive study of the postnatal changes in the concentrations of the lipids of developing rat brain. *Biochemistry* 6:3169–3175.)

Inositol-1,4,5-trisphosphate (IP₃)

Stimulation of receptors linked to the activation of phosphatidylinositol-specific phospholipase C (PLC) causes the hydrolysis of phosphatidylinositols, including phosphatidylinositol-4,5-bisphosphate (PIP$_2$), and the generation of the signaling molecule inositol-1,4,5-trisphosphate (IP$_3$) and 1,2-*sn*-diacylglycerol (DAG) (Fig. 2). Three members of the PLC enzyme family have been purified from brain and are designated PLC-β_1, PLC-γ_1, and PLC-δ_1. These enzymes are the products of three distinct genes, and all require calcium for optimal activity. Most neurotransmitter and hormone receptors stimulate PLC activity by first activating a heterotrimeric GTP-binding protein (G-protein), which transmits the activation signal from the receptors to the enzyme. The PLC-β_1 isozyme is activated by the α subunit of a G$_q$-type G-protein. Activation of PLC-γ_1 occurs without a G-protein intermediary but is enhanced by direct phosphorylation of the enzyme on a tyrosine residue (e.g., by the PDGF receptor that has tyrosine kinase activity). Although IP$_3$ has been

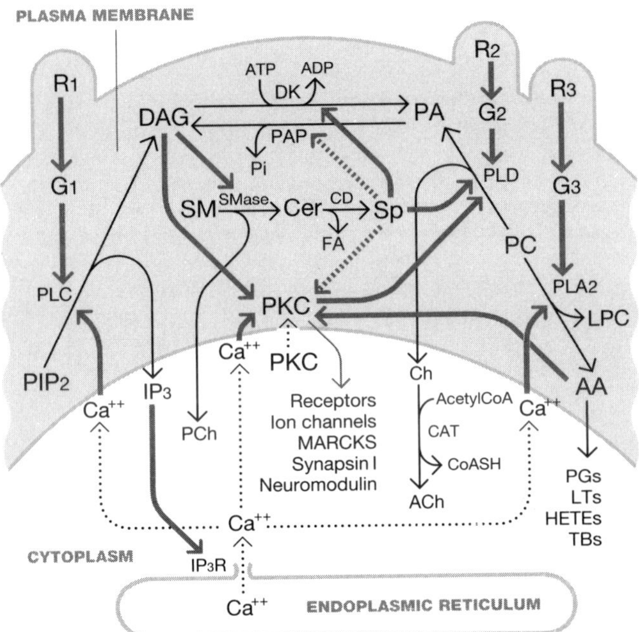

Figure 2. Phospholipids and transmembrane signaling. Phospholipid breakdown products as signaling molecules are described in detail in the text. The diagram shows only the key features of the system. Solid black lines represent enzymatic reactions; thick solid lines represent activation, while thick dashed lines represent inhibition (e.g., DAG activates and Sp inhibits PKC). Dotted lines represent translocation. PKC substrates are indicated by a thin gray arrow. Abbreviations for compounds are AA, arachidonic acid, ACh, acetylcholine; Cer, ceramide; Ch, choline; DAG, diacylglycerol; FA, fatty acid; HETEs, hydroxy eicosatetraenoic acids; IP$_3$, inositol-1,4,5-triphosphate; LPC, lysophosphatidylcholine; LTs, leukotrienes; PA, phosphatidic acid; PC, phosphatidylcholine; PCh, phosphocholine; PGs, prostaglandins; PIP$_2$, phosphatidylinositol-4,5-bisphosphate; SM, sphingomyelin; Sp, sphingosine; TBs, thromboxanes. Abbreviations for macromolecules are CAT, choline acetyltransferase; CD, ceramidase; G, G-proteins; PAP, PA phosphohydrolase; PKC, protein kinase C; PLA$_2$, phospholipase A$_2$; PLC, phospholipase C; PLD, phospholipase D; DK, DAG kinase; R, receptors; SMase, sphingomyelinase.

known for some time to elevate intracellular calcium, the mechanism by which this occurs has only recently been elucidated. We now know that IP$_3$, generated by PLC, diffuses into the cytoplasm and binds to receptors located on the endoplasmic reticulum (ER). The IP$_3$ receptor, originally purified from the Purkinje cells of the cerebellum, contains an intrinsic calcium channel that opens upon IP$_3$ binding, liberating Ca^{2+} ions stored in the ER. It is a membrane protein containing eight putative transmembrane domains. There are several classes of IP$_3$ receptors generated in different tissues by alternative splicing of a common gene. The receptor is subject to phosphorylation by several kinases, including protein kinase C (PKC). Phosphorylation by a cAMP-dependent protein kinase renders the receptor less responsive to IP$_3$. The elevation of cytosolic Ca^{2+} concentration by IP$_3$ leads to the activation of Ca^{2+}-dependent processes (e.g., enzyme activation or neurotransmitter release). Temporal and spatial alterations in intracellular Ca^{2+} levels characterized by oscillations or wavelike propagations throughout the cytoplasm may also be mediated by IP$_3$. Variations in frequency of calcium fluctuations (and therefore of calcium-dependent processes) may provide an additional mechanism by which information is encoded and transmitted from one cell to another.

The IP$_3$ signal is terminated by the hydrolyses of the phosphomonoester bonds catalyzed by specific phosphatases. The pathway is as follows: inositol-1,4,5-trisphosphate $\rightarrow$ inositol-1,4-bisphosphate $\rightarrow$ inositol-4-phosphate$\rightarrow$ inositol. The latter two steps are inhibited by lithium (Li$^+$) at concentrations used clinically to treat manic-depressive illness. These observations led to the hypothesis that the therapeutic effects of Li$^+$ are due to its ability to interfere with IP$_3$-mediated signaling. An alternative pathway for IP$_3$ inactivation involves a phosphorylation reaction catalyzed by 1,4,5-IP$_3$-3 kinase, which produces 1,3,4,5-IP$_4$ (IP$_4$). A purified preparation of this enzyme obtained from rat brain is activated by Ca^{2+}/calmodulin. Thus, it is possible that the Ca^{2+} released by IP$_3$ from intracellular stores could activate 1,4,5-IP$_3$-3 kinase, leading to the removal of IP$_3$ and attenuation of further release of Ca^{2+}. Once formed IP$_4$ may also modulate functions of ion channels.

The generation of IP$_3$ via the hydrolysis of membrane phospholipids, then, constitutes a means by which the cell can translate incoming stimuli into a variety of responses that vary both temporally and spatially, thus greatly increasing the complexity of the information that can be conveyed.

1,2-*sn*-diacylglycerol

DAG, the other product of PLC, remains in the plasma membrane (Fig. 2). Unlike IP$_3$, whose sole function is that of a second messenger, DAG is both a messenger molecule and an intermediate in the metabolism of lipids. Its signaling action probably depends on its subcellular localization. The prime target of DAG is a regulatory enzyme, PKC, which phosphorylates protein substrates on specific serine or threonine residues. Several isozymes of PKC were purified from brain and/or identified by cloning. These PKCs are designated α, β-I, β-II, and γ (all of which contain Ca^{2+}-binding domains and are Ca^{2+}-dependent), and δ, ε, ζ, and η(L) (which are Ca^{2+} independent). Inactive PKC is present in the cytosol and is converted to an active form by associating with membrane lipids. PS, an acidic phospholipid present in the plasma membrane, is a necessary but not sufficient cofactor. The transient generation of DAG by PLC results in the formation of a complex composed of PKC, DAG, and PS on the cytoplasmic surface of plasma membrane (Fig. 2). Under these conditions, the PKC isozymes that are Ca^{2+}-dependent show increased affinity for this ion and assume an active conformation that permits phosphorylation of specific substrate proteins. The biochemical properties of these substrates are altered by phosphorylation, resulting in a range of cellular responses. PKC has been shown to be involved in modulating several neuronal functions including regulation of ion channel activity, neurotransmitter release, and neuronal plasticity. In the hippocampus PKC may be involved in the regulation of long-term potentiation (LTP). This phenomenon refers to long-term changes observed in neurons that have been previously stimulated and is thought to be a model for the process underlying memory. Hence, it is of great interest to understand the mechanism of LTP. Several PKC isozymes are present in the hippocampus, and PKC is translocated from the cytosol to the membrane upon a high frequency stimulation of the perforant pathway. Some features of LTP can be induced in hippocampal neurons by pharmacological activation of PKC with phorbol esters (which mimic DAG), and LTP can be diminished by PKC inhibitors. Identification of the brain PKC protein substrates that might be ultimately responsible for the physiological actions of PKC is continuing. Phosphorylation by PKC of neuromodulin (a protein abundant in growth cones and nerve endings, also designated as GAP-43, B50, F1, and p57) and of neurogranin, both of which bind to calmodulin, may be important in modulating synaptic events and plasticity, including neurotransmitter release and LTP. Another group of proteins abundant in both neurons and glia that are phosphorylated by PKC are known as MARCKS (myristoylated, alanine-rich C-kinase substrates). The physiological role of this class of proteins remains unknown, but some may be involved in neurotransmitter release. PKC also phosphorylates and modulates the conductance of sodium channels in brain. Synapsin-1 is an abundant protein associated with synaptic vesicles and a substrate for phosphorylation by several protein kinases, including PKC. Its functions in regulating neurotransmitter release are controlled by these phosphorylations. The multitude of PKC target molecules involved in synaptic transmission provides support for the hypothesis that PKC plays a significant role in long-term modulation of synapse activity. Modulation of neurotransmission must, however, be a reversible process, and a number of mechanisms exist that curtail the action of PKC. The recent observation that the cytosolic form of synapsin-1 has DAG kinase activity (i.e., converts DAG to PA) provides evidence that this protein may be involved in terminating the DAG signal. However, since PA has a number of biological effects of its own, it is possible that it is the synthesis of PA and not the removal of DAG that is physiologically relevant. There are two additional ways for removing DAG: (1) hydrolysis to monoacylglycerol (MAG) and a fatty acid (FA) by a DAG lipase, and (2) conversion to phospholipids. Among the metabolites of DAG, neither PA nor MAG activate PKC; however, the γ isozyme of cerebral PKC can be activated by unsaturated FAs, and the α and β isozymes are activated by FAs and DAG in a synergistic fashion (Fig. 2).

Fatty acids

Most neuronal FAs exist in esterified form as components of membrane phospholipids. Free FAs are normally present in neurons in low concentrations but can be released from all

Table 1. Receptors or Ligands Linked to Phospholipase Activation

Receptor/Ligand	PLC	PLD	PLA$_2$
Adrenergic (α_1)	+	+	+
Angiotensin	+	+	
ATP (purinergic P$_{2Y}$)	+	+	
Bombesin	+	+	+
Bradykinin	+	+	
Cholecystokinin	+		
Endothelin	+		
EGF	+	+	
Excitatory amino acids (metabotropic)	+		+
fMLP	+	+	+
Histamine (H$_1$)	+		
Interleukin-1			+
Leukotriene (LTB4, LTD4)	+		
Light (rhodopsin in rods)	+		+
Muscarinic (m$_1$, m$_3$, m$_5$)	+	+	+
Neuropeptide Y	+		
Neurotensin	+		
PAF	+		+
PDGF	+	+	
Prostaglandin (EP$_1$, EP$_3$, FP)	+		
Serotonin (5-HT$_{1C}$, 5-HT$_2$)	+		+
Thrombin	+	+	+
Thromboxane	+		
TRH	+		
Tachykinin (substance P, neurokinins)	+		
Vasopressin	+	+	

EGF, epidermal growth factor; fMLP, formylMetLeuPhe chemotactic peptide; PAF, platelet activating factor; PDGF, platelet derived growth factor; TRH, thyrotropin releasing hormone. Designations of receptor subtypes are given in parentheses.

phospholipid classes by the sequential action of phospholipase A$_2$ (PLA$_2$) and lysophospholipase. PLA$_2$ activity is stimulated by certain receptors (Table 1), via a mechanism that involves G-protein coupling. In photoreceptor cells of the retina, transducin has been identified as the G-protein mediating the activation of PLA$_2$ by light. Receptor-mediated release of fatty acids is often Ca^{2+}-dependent, indicating that PLA$_2$ catalyzing this process requires Ca^{2+}. The most important FA released by PLA$_2$ is arachidonic acid (AA; Fig. 2). AA may act as a signaling molecule by modulating ion channel conductance, and by activating isozymes of PKC in the absence of DAG. The latter property of AA is shared by other unsaturated FAs. AA may also be oxidized by cyclooxygenase to prostaglandins and thromboxanes and by lipoxygenase to hydroxyeicosatetraenoic acids (HETEs) and leukotrienes. Both of these enzymes are present in the brain. Prostaglandins, thromboxanes, and leukotrienes, generated in response to activation of specific receptors, act in an autocrine or paracrine fashion on their own receptors and, in some cases, initiate signals that involve phospholipid breakdown (Table 1). The predominant brain prostaglandins, PGD$_2$ and PGE$_2$, may be involved in the regulation of sleep. HETEs have been found to regulate ion channel conductance in neurons of *Aplysia* and in the myocardium.

Under certain pathological conditions, such as hypoxia or during seizures, the free FA concentration in the brain rises rapidly, possibly due to the activation of PLA$_2$.

Phosphatidic acid

Small amounts of phosphatidic acid (PA) are constitutively present in membranes. In addition receptor-mediated PA formation occurs by phospholipase D (PLD)–catalyzed hydrolysis of PC (and to a lesser extent, PE; Table 1). The other product of this reaction, choline, may be converted to the neurotransmitter, acetylcholine (ACh). Stimulation of muscarinic receptors located at cholinergic synapses leads to the release of choline from PC by activation of PLD and may therefore provide choline for ACh synthesis. Receptor-linked activation of PLD is mediated by G-proteins and/or by activation of PKC. Thus, in many cell types, signal transduction includes the activation of PLC, PKC, and PLD in concert (Fig. 2). Although receptor-mediated activation of PLD has been documented in many cell types, including neuronal cell lines and synaptosomes, little is known about its biological function. Exogenous PA (and lysoPA) stimulate Ca^{2+} influx, raise intracellular pH, and are mitogenic in some cells. *In vitro,* PA has been shown to stimulate PLC activity, and to inhibit the GTPase activity of the *ras* protein (the *ras* protein is a product of the *ras* protooncogene and is involved in a variety of cellular regulatory processes). However, it is not clear whether the PA formed by the action of PLD *in situ* has similar actions. PA can be rapidly hydrolyzed by a PA phosphohydrolase to DAG (Fig. 2). In most cells DAG accumulation follows a biphasic time course. The initial rapid phase is due to the hydrolysis of phosphatidylinositides by PLC, and the slow sustained phase is due to the concerted action of PLD and PA phosphohydrolase generating DAG from PC. The majority of DAG that accumulates following receptor stimulation derives from this process (or from PLC-mediated degradation of PC) rather than from the hydrolysis of PI by PLC. Thus, DAG may be the ultimate second messenger generated by receptor-mediated PLD activation.

Sphingolipids

SM is a sphingophospholipid present almost exclusively in plasma membrane. Although few examples of receptor-mediated hydrolysis of SM have been documented, the pharmacological actions of SM breakdown products are striking. In particular, studies have focused on the properties of ceramide, which can be generated by the action of sphingomyelinase, and on sphingosine, which can be formed from ceramide by ceramidase. Formation of sphingosine is enhanced by DAG, which stimulates sphingomyelinase. Exogenous sphingosine is an inhibitor of PKC, and of PA phosphohydrolase, and an activator of DAG kinase. Thus, generation of sphingosine may contribute to the termination of PKC-mediated signaling. This is accomplished by inhibiting PKC directly, by preventing the generation of DAG from PA, and by stimulating the formation of PA from DAG (Fig. 2). Other actions of sphingosine include inhibition of Ca^{2+}/calmodulin-dependent enzymes, and of brain Na^+, K^+-ATPase, activation of PLD (and thus generation of PA), and stimulation of the EGF receptor tyrosine kinase activity. Ceramide shares the latter property with sphingosine. Recent studies indicate that sphingosine and ceramide activate a novel protein kinase that phosphorylates specific threonine residues of target proteins. Thus, ceramide may be a second messenger similar to DAG, that is, it is formed by the removal of a headgroup from a phospholipid and it activates a protein kinase.

Many of the actions of sphingosine can be mimicked by a class of pharmacological agents commonly termed cationic amphiphilic drugs. This group of compounds includes drugs active in the brain, such as tricyclic antidepressants, neuroleptics (phenothiazines), and propranolol. It has been hypothesized that some of the actions of these drugs may be due to interference with sphingosine-mediated signaling.

There are several genetic diseases, commonly classified as sphingolipidoses, in which the affected individuals show a deficit in the activity of a sphingolipid-degrading enzyme. In Niemann-Pick disease a mutation in lysososmal sphingomyelinase leads to accumulation of SM. In Farber lipogranulomatosis there is a ceramidase deficiency; in G_{M2} gangliosidoses (Tay-Sachs disease) there are mutations in the α subunit of β-hexosaminidase, whereas deficiencies in glucocerebrosidase and in galactosylceramidase characterize Gaucher disease and globoid cell leukodystrophy (Krabbe disease), respectively. Although the pathophysiological processes underlying cellular degeneration in these diseases are poorly understood, a hypothesis has been advanced that postulates that they might involve abnormal sphingolipid-mediated signaling or a direct toxic action of lysosphingolipids. The latter notion is supported by data showing that sphingosine as well as lysosphingolipids are cytotoxic to a variety of cells in culture.

Platelet activating factor (PAF)

PAF (1-*O*-alkyl-2-acetyl-*sn*-glycero-3-phosphocholine) is a phospholipid characterized by a fatty alcohol (usually hexadecanol) at the *sn*-1 position of the glycerol backbone, and an acetate residue at the *sn*-2 position. Both the ether linkage at the *sn*-1 position and the acetate group are essential to maintain the potent physiological activity of this compound. When cells are activated by ligand binding, PAF is synthesized via a two-step pathway that involves the initial conversion of alkyl choline phospholipids to the corresponding lyso-intermediate (lyso-PAF) by PLA_2, followed by the addition of an acetate group to the *sn*-2 position. Receptor-mediated synthesis of PAF requires calcium, and is mediated by a G-protein. Synthesis of PAF via this pathway can be stimulated by exposure to complement-coated zymosan particles, calcium ionophores, or phorbol esters in human polymorphonuclear leukocytes, and by thrombin, bradykinin, and ATP in endothelial cells. PAF may also be synthesized *de novo* by a pathway that utilizes cytidinediphosphocholine.

The physiological actions of PAF are exerted via specific receptors that are located on many different cell types. The effects of PAF include the activation of PI-specific PLC, and the release of FAs from phospholipids, possibly via activation of PLA_2. These actions result, in turn, in the generation of the active metabolites DAG, IP_3, and AA, and in the subsequent mobilization of calcium and activation of PKC. The metabolism of AA to eicosanoids (by cycloxygenase and lipoxygenase) represents yet another consequence of PAF activity. Although relatively little is known about the actions of PAF in brain, high-affinity binding sites have been demonstrated both extra- and intracellularly, and there is evidence for PAF involvement in brain responses to ischemia and electroconvulsive shock.

Glycosyl-phosphatidylinositol (GPI)

GPI has been recently implicated as a precursor of a novel second messenger molecule, inositol phosphoglycan (IPG), generated in response to receptor binding of nerve growth factor (NGF), insulin, and interleukin-2. The PI portion of this molecule contains two saturated FAs, and there is a glycosidic linkage of inositol to a glycan via glucosamine. The precise structure of the glycan is unknown. Stimulation of an appropriate receptor (e.g., the low-affinity NGF receptor) leads to the activation of a specific PLC and the generation of IPG and DAG. Because some of the actions of NGF can be mimicked by IPG, and immunoprecipitation of this compound can inhibit the action of NGF, it has been postulated that IPG may be the second messenger in NGF actions mediated by its low-affinity receptor.

An overview

Phospholipids reside in the immediate vicinity of receptor molecules in the membrane. Evolution has taken advantage of this intimacy by placing phospholipases close to the receptors and G-proteins, and by using the phospholipid breakdown products as messenger molecules. Stimulation of receptors activates neighboring phospholipases, generating second messenger molecules that amplify the initial signal, that is, a single ligand-bound receptor can activate more than one phospholipase molecule, which, in turn, can hydrolyze multiple molecules of phospholipids (Fig. 2). The phospholipid breakdown products must activate many effector systems in a regulated fashion in order for the signal to be biologically meaningful. Significantly, the formation of one messenger regulates the levels of another. For example, activation of PLC generates DAG and IP$_3$, resulting in elevated cytoplasmic Ca^{2+} concentrations and activation of PKC. PKC in turn can activate PLD, generating PA, and PA may then be degraded to DAG, further amplifying PKC activation. However, DAG also activates sphingomyelinase, which hydrolyzes SM, ultimately generating sphingosine. Sphingosine helps to bring the system back to the basal state by inhibiting PKC and PA phosphohydrolase, and by activating DAG kinase. The latter two effects would lower DAG levels and therefore reduce the activity of PKC (Fig. 2). These generalizations are derived from studies on the individual constituents of neuronal signaling by phospholipid breakdown products. However, a single receptor type may activate several classes of phospholipases, any one of which may also be activated by several additional categories of receptors. How the neuron responds in a specific and meaningful manner to a particular input given this apparent cross-talk between signaling pathways remains to be determined. In addition it is not clear how neurons distinguish between metabolic pools of phospholipids and their breakdown products, that are involved in the maintenance and remodeling of cellular membranes, from those molecules that are generated as signals.

Further reading

Axelrod J (1990): Receptor-mediated activation of phospholipase A$_2$ and arachidonic acid release in signal transduction. *Biochem Soc Trans* 18:503–508

Berridge MJ (1989): Inositol trisphosphate, calcium, lithium, and cell signaling. *JAMA* 262:1834–1841

Billah MM, Anthes JC (1990): The regulation and cellular functions of phosphatidylcholine hydrolysis. *Biochem J* 269 (suppl): 173–183

Dennis EA, Rhee SG, Billah MM, Hannun YA (1991): Role of phospholipases in generating lipid second messengers in signal transduction. *FASEB J* 5:2068–2077

Hannun YA, Bell RM (1989): Functions of sphingolipids and sphingolipid breakdown products in cellular regulation. *Science* 243:500–507

Nishizuka Y (1989): The family of protein kinase C for signal transduction. *JAMA* 262:1826–1833

Piomelli D, Greengard P (1990): Lipoxygenase metabolites of arachidonic acid in neuronal transmembrane signalling. *TIPS* 11:367–373

Prescott SM, Zimmermann GA, McIntyre TM (1990): Platelet activating factor. *J Biol Chem* 265:17381–17384

Sudden Infant Death Syndrome (SIDS): A Neural Perspective

Laurence Edward Becker

Sudden infant death syndrome (SIDS) is a major cause of death in infants between 1 month and 1 year of age with an incidence that varies in different parts of the world, but ranges from 1 to 2 per 1000 live births in North America and Europe. SIDS is defined as the sudden death of an infant under 1 year of age that remains unexplained after the performance of a complete postmortem investigation including an autopsy, an examination of the scene of death, and a review of the case history. A careful and complete autopsy will exclude SIDS by identifying a specific cause of death in 10–20% of sudden unexpected deaths under 1 year of age: congenital heart disease, myocarditis, cardiomyopathy, encephalitis, meningitis, metabolic encephalopathy (medium-chain acyl coenzyme A dehydrogenase deficiency), and other conditions.

SIDS profile

Although there is a great deal of variability, a characteristic profile of a typical SIDS victim does emerge: 2–4 months of age, male, apparently healthy and growing normally, possible mild diarrhea or upper respiratory tract infection, death during sleep with no apparent struggle or noise. Reported risk factors are many although prone sleeping position, bottle feeding, overheating, and maternal smoking are cited as the most important because they are preventable. Other suggested risk conditions include multiple births (risk is probably related to lower birth weights of such infants), race (SIDS is not a genetic disease but is more common in some races), maternal drug use (morphine, heroin), lower birth weight, prematurity, and lack of prenatal care.

At autopsy, the infant appears well nourished and hydrated. In 85–100% of infants, petechial hemorrhages are present on intrathoracic organs. The presence of petechial hemorrhages over the thymus, lungs, or heart has suggested to many investigators that the petechiae are produced by upper airway obstruction leading to vigorous inspiratory efforts that generate sufficient transmural pressure to rupture intrathoracic capillaries. Petechiae are usually not found with generalized hypoxic-ischemic insults. Normally babies obstruct their upper airway during sleep and open up through normal neural arousal mechanisms. In SIDS the arousal system may not function or some other aspect of the neural control of respiration may be impaired during sleep. That is, the postulated dysfunction of respiratory control may be nonobstructive owing to failure of respiratory rhythmicity such as central apnea or failure of arousal, or obstructive, with respiratory dyskinesia producing a central type of upper respiratory obstruction. Clinical respiratory physiologic studies of apnea and periodic breathing in siblings of children with SIDS and infants with apparent life-threatening events have further supported a role for dysfunction of respiratory regulation. Cardiac arrhythmias may also cause sudden unexpected death and reveal no anatomic cause of death at autopsy, but the evidence for such a mechanism in SIDS has not been established. Therefore, research has focused on the neuroanatomic substrates of respiratory control as they apply to breathing, sleep state, and arousal response.

Neural respiratory regulation

The neural control of cardiorespiratory function is complex and incompletely understood. Our knowledge of relevant neuroanatomy and neurophysiology is based largely on animal studies owing to the restricted techniques that can be applied to the human nervous system at autopsy.

The function of respiratory regulation is to maintain gas exchange appropriate to the metabolic needs of the body, such as exercise and sleep. The respiratory control mechanism is a feedback system with information on the efficiency of gas exchange and the mechanics of ventilation flowing to medullary control centers and from there to the efferent neuromuscular system, thereby maintaining an efficient exchange of oxygen and carbon dioxide. Afferents from the carotid body and neuroendocrine bodies in the lungs, monitoring blood pH, pO_2, and pCo_2, and afferents from mechanoreceptors and irritant receptors, monitoring airway passages and lung expansion, pass through the ninth and tenth cranial nerves, to the nucleus tractus solitarius in the medulla. Connections from there pass to a central rhythm generator located in the medulla, consisting of inspiratory and expiratory neurons acting in a complex oscillating network, but the exact mechanism, location, and nature of neurotransmitters have not been determined. The central rhythm generator also receives modulating influences from more rostral sites in the pontine, limbic, hypothalamic, subthalamic, and thalamic regions. Integrated impulses from the rhythm generator pass to the efferent system, which consists of a dorsal respiratory group (nucleus tractus solitarius) and a ventral respiratory group (nucleus retroambiguus, nucleus paraambiguus, nucleus retrofacialis). Axons from these neuronal groups pass contralaterally to the spinal cord, then to the phrenic and intercostal motor neurons, innervating diaphragm and intercostal muscles. In addition, axons from the dorsal and ventral respiratory groups pass to the nucleus ambiguus, which innervates the skeletal muscles of the pharynx and larynx. The net effect is to achieve respiratory rhythmicity, with fine coordination of respiratory muscles in the thorax, larynx, and pharynx so that relaxation of upper airway muscles occurs in concert with active contraction of diaphragmatic and intercostal muscles.

Brainstem astrogliosis

Because the neurons responsible for respiratory control are located in the medulla, attention has focused on this region. In the central nervous system, scarring is characterized by astrogliosis, which is a nonspecific reactive process to many insults including hypoxia and ischemia. Assessment of astrogliosis in immature brains must take into account developmental changes that are occurring in myelin-associated astrocytes (myelination glia) and the normal background of astrogliosis that exists in the brainstem in such areas as the subpial and subependymal regions.

In order to ascertain the degree of astrogliosis, all recent studies have emphasized the need for quantitation. Considering the difficulties of quantitating astrogliosis in the brain and the probable heterogeneity within SIDS, variability of astroglial counts among different studies would be expected.

Table 1. Brainstem Astrogliosis

Site	Control					SIDS (GFAP)				
	−	+	+ +	+ + +	(N)	−	+	+ +	+ + +	(N)
Medulla										
Midline	0	4	5	10	(19)	0	3	3	38	(44)
Paramedian r.n.	3	14	2	0	(19)	2	4	16	24	(46)
Magnocellular r.n.	3	15	1	0	(19)	2	6	21	17	(46)
Lateral r.n.	7	10	2	0	(19)	3	5	18	20	(46)
NTS	8	11	0	0	(19)	2	10	20	11	(43)
DVN	7	10	2	0	(19)	0	3	9	30	(43)
ION	3	6	5	5	(19)	0	3	5	38	(46)
Pons										
Midline	1	2	3	13	(19)	0	2	3	38	(43)
Midline r.n.	6	9	4	0	(19)	2	19	15	10	(46)
Lateral r.n.	12	7	0	0	(19)	2	28	12	4	(46)
Midbrain										
Midline	0	2	6	11	(19)	0	4	7	25	(36)
Mesen. r.n.	7	12	0	0	(19)	3	19	16	1	(39)
S.C.	12	6	1	0	(19)	11	21	4	1	(37)

Modified from Becker 1990.

r = reticular
DVN = dorsal vagal nucleus
\+ = <5 astrocytes/600 μm^2
mesen. = mesencephalic

n = nucleus
ION = inferior olivary nucleus
\+ + = 6–10 astrocytes/600 μm^2
S.C. = superior colliculus

NTS = nucleus tractus solitarius
− = no astrocytes/600 μm^2
\+ + + = >11 astrocytes/600 μm^2
GFAP = glial fibrillary acidic protein

What is remarkable is the concordance among many reports that have evaluated astrogliosis. The location of the astrogliosis is primarily in the tegmentum of the brainstem in those areas of the medulla that contain the neurons believed to be responsible for respiratory control (Table 1). These geographic regions are also susceptible to pathology in infants who have been exposed to severe hypoxic-ischemic insults in early gestation or in the early neonatal period. In these infants, frank infarction may be present, but the damaged areas are in the same neuroanatomical distribution as the subtle astrogliosis in SIDS, leading to the suggestion that this region of the brainstem is developmentally vulnerable to hypoxic-ischemic insults. Because astroglial counts of SIDS overlap with controls, it is not possible to use these counts on an individual basis as a diagnostic test of SIDS.

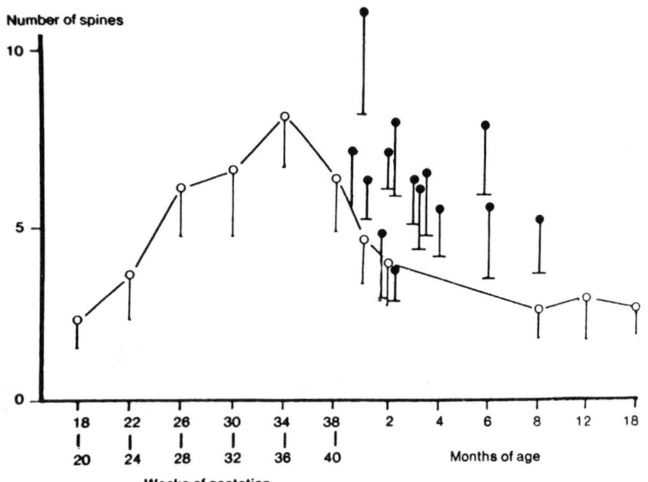

Figure 1. Developmental changes of dendritic spine density (number of spines per 25 μm) in the medullary reticular formation of SIDS victims (●) and controls (O). From Becker 1990.

Density of dendritic spines in brainstem neurons

Two separate research groups have reported a persistence of dendritic spines in SIDS victims and have interpreted this observation as evidence of altered brainstem neural maturation. The neurons that were examined were located in the tegmentum of the medulla. In normal development, the number of spines in this region reaches a peak at 34–36 weeks gestation and after birth decreases rapidly. In SIDS, the number of spines is maintained beyond this period of development and does not decrease until early infancy (Fig. 1). Because these neurons are located in the region of the brainstem involved in the neural control of cardiorespiratory function, there may be a functional impairment or immaturity of the higher levels of respiratory control in SIDS victims. In SIDS, other components of the respiratory control system, such as the motor neurons of the hypoglossal nucleus and the cervical anterior horn cells of the phrenic nerve, show no maturational delays.

Neurotransmitters in brainstem

Some preliminary studies, although not entirely in agreement, have been reported examining levels of selected enzymes in the brains of infants dying of SIDS and other causes. The activity of dopamine beta hydroxylase, the noradrenaline-synthesizing enzyme, and phenolethanolamine methyltransferase, the adrenaline-synthesizing enzyme, have been low in the brainstem in SIDS, compared with control infants, using small numbers of cases and limited anatomic sampling. Delayed dendritic development of catecholaminergic neurons in the ventrolateral medulla has also been reported in SIDS.

Because of the known effects of opiates on respiratory control, several groups have attempted to evaluate the endogenous opiate system. Postmortem analysis has shown high brainstem concentrations of endorphins in a child with Leigh encephalopathy who had frequent periods of apnea clinically reversed by Naloxone. One unconfirmed study has shown increased beta endorphin activity in the CSF of infants with apnea and siblings of victims of SIDS. This has led to the suggestion that SIDS may be associated with an increased activity of the endogenous opiate system. In a single study, Met-enkephalin levels were reported to be twice as high in SIDS victims as in controls; however, in another report, no differences were found between SIDS and controls. Therefore, there is some equivocal evidence of altered endogenous opiate levels in SIDS.

Promising studies are underway by the group at Boston Children's Hospital, which is embarking on a correlative neurochemical and neuroanatomical approach using receptor autoradiography, which may resolve some of the discrepancies that are now apparent in the literature. This method will permit quantitation of neurotransmitter receptor densities in specific areas of the brain.

Chemoreceptors in carotid body

The carotid bodies, located at the bifurcation of the common carotid artery, respond reflexly to changes in arterial pO_2, pCO_2, and pH with alterations of respiration through the respiratory control feedback mechanism. Because of this homeostatic role, carotid bodies have been hypothesized to be of possible relevance to the pathogenesis of SIDS. Although some reports have described hyperplasia and hypoplasia of carotid bodies, others have found no such abnormalities.

We have examined the catacholamine content of the carotid bodies of SIDS victims and age-matched controls and found that the mean levels of noradrenaline, dopamine, and adrenaline were elevated compared with the control group. The most striking difference was an eight-fold increase in the concentration of dopamine (Table 2). In both animals and man, dopamine administration inhibits respiration by acting directly on the carotid bodies. Dopamine diminishes the neural discharge from the carotid bodies, which decreases ventilation, particularly the frequency of respiration and tidal volume. Dopamine also appears to inhibit the response of the carotid bodies to hypoxia. Therefore, the finding of high levels of dopamine and raised levels of noradrenaline in the carotid bodies of SIDS victims suggests that their chemosensors may be inhibited, making them vulnerable to hypoxia and less able to be aroused. If the increased dopamine levels were secondary to chronic hypoxia, the chief cells should be hyperplastic. Morphologic and morphometric studies of carotid bodies in SIDS have shown no increase in size or volume of the chemoreceptor cells, therefore suggesting that the increased catacholamine content in SIDS may be related to an abnormal control of the release of neurotransmitter in the carotid bodies or due to short episodes of acute hypoxia insufficient to produce chief cell hyperplasia.

Table 2. Catecholamine Content of Carotid Bodies in SIDS Victims and Control Victims

Catecholamine (ng/g Tissue)	SIDS Victims	Controls	Significance $p = <$
Dopamine	1622	202	0.02
Noradrenaline	502	276	0.02
Adrenaline	125	47	0.10

From Becker 1990.

Other neuropathologic observations

A number of interesting neuropathologic observations have been made, although their significance in some instances is not apparent.

Brain growth. The brains of SIDS victims are significantly heavier than reference values matched for both age and body length. Some data suggest that the brains did not grow disproportionately after birth, but rather that the children were born with larger brains. The cause for the increased brain size in SIDS is unknown although the possibility of preterminal mild cerebral edema has not been excluded.

Pineal gland. The function of the pineal gland in normal physiology is obscure. Since SIDS occurs during sleep and the pineal gland is thought to influence diurnal rhythm, there may be a relationship of pineal gland function to SIDS. One group has observed that the pineal gland was reduced in size as determined morphometrically compared with age-matched controls. The significance of reduction in size of the pineal gland is unknown.

Periventricular and subcortical leukomalacia. Periventricular and subcortical leukomalacia are found in 20–25% of SIDS victims. Both lesions are due to hypoperfusion and ischemic injury. Subcortical leukomalacia has been described as being associated with discrete boundaries, absence of coagulative necrosis, distinct location, minimal astrogliosis, rare swollen axons, and no proliferation of microglial cells.

Myelination. Certain regions of the brain in SIDS appear to have delayed myelination. These include the temporal lobe at the level of lateral geniculate body, frontal lobe, and medial crus pedunculi. It is suggested that this delayed myelination preferentially affects the late and slowly myelinating tracts, but the areas that are delayed do not correspond to a particular anatomical or functional region. Specifically, they do not appear to be delayed in those parts of the brainstem that are thought to play a role in respiratory control.

In the peripheral nervous system, the myelination of efferent and afferent nerves involved in respiratory control have been studied. The vagus nerve is well myelinated at birth. However, in SIDS, an altered pattern with a decreased proportion of small myelinated axons has been recorded. The significance of this is unclear, but it suggests delayed development of myelinated fibers.

Hypothalamus. A recent single report examining the hypothalamus of a small number of SIDS and control infants suggests that changes in serotonergic chemistry could disrupt the rhythms of sleep and cholinergic alterations could disturb arousal from sleep. Further exploration is important.

Hypoxanthine. Important results from Oslo show that the hypoxanthine levels in vitreous humor are significantly higher in SIDS victims than in children who die acutely and violently, indicating that sudden death in SIDS is preceded by a relatively long course of hypoxic episodes. Further support of hypoxic exposure comes from the observation of high lactate and low pH values from multiple brain sites in one third to one half of SIDS victims.

Neuropathology and possible pathogenetic mechanisms

The preceding observations taken together suggest a multifactorial and multistage process leading to SIDS (Fig. 2).

Prenatal factors as indicated by epidemiologic data play an important role in SIDS. These include maternal factors, such as low weight gain, anemia, urinary tract infection, higher rate of smoking, lower incidence of prenatal care, and higher rate of illicit drug use. SIDS infants also have characteristics that suggest the occurrence of adverse factors in early gestation: retarded prenatal growth, lower birth weight, slower postnatal growth, lower average body weight, length, and head size, and higher incidence of minor anomalies (hernias, club feet, hemangiomas, minor cardiac problems). These observations suggest that infants who die of SIDS have one or more vulnerabilities that arise from early gestation, and these impair normal responses to internal and external demands.

The second link is the normally immature respiratory sleep system, which undergoes significant alterations during the early months of infancy and creates susceptibility to environmental insults. During the first month, the infant sleeps 18 hours a day, and 70% of this time is in REM sleep. Breathing and temperature control vary according to sleep stage. During REM sleep, there is better temperature control. By three months, major developmental changes are occurring with less time spent sleeping and much less in REM sleep. In the neonate, breathing is slow and stable, but relatively insensitive to metabolic needs. Over the first 3 months, breathing becomes more sensitive and rapid, but less stable. At 3 months, breathing has an oscillatory pattern with a few breaths followed by a pause occurring in an irregular repetitive pattern. Over 4 months of age, breathing becomes highly sensitive, responding rapidly to metabolic demands and becoming more stable. In fact, many important maturational alterations and adjustments are occurring in the infant in the first 6 months of life, and some of these changes are increasing the susceptibility of these infants to SIDS.

A third link in the SIDS chain is postulated subclinical respiratory instability, which may manifest as upper airway obstruction. Infants are both anatomically and physiologically vulnerable to obstructive events at the pharyngeal level. At

birth, the pharyngeal area is crowded and remains so for the first year of life. During this time, there is a growth of the mandible and a remodeling of the oropharyngeal area, perhaps accounting for the fact that SIDS does not occur after 1 year of age. In addition, the mandible in the infant is short and the temporal mandibular joint is shallow, allowing posterior displacement of the mandible to occlude the pharyngeal airway. A recent report documents that the tongue in SIDS victims is heavier as well as longer, wider, and thicker than in control infants, but this finding has not yet been confirmed by other studies.

Airway patency is a balance between airway-dilating and airway-constricting forces. The major constricting forces are suction during inspiration and airway compression caused by neck flexion. The dilating forces are contraction of the upper airway muscles, which act to open the airway. The problem is that these muscles are inhibited during REM sleep, and infants spend a large part of their time in REM sleep, which puts them at increased risk for upper airway collapse. Respiratory tract infections and associated secretions could further compromise the patency of the airway.

The fourth, and possibly critical, link is hypoxic injury to the brain. Strong evidence of hypoxic-ischemic injury to the brain includes the increased incidence of brainstem astrogliosis, periventricular and subcortical leukomalacia, elevated hypoxanthine levels, and elevated neurotransmitters in the carotid body. Although not well documented, hypoxic-ischemic insults to the brain could also affect the maturation process and produce neural developmental delays. Damage to the medulla indicated by astrogliosis may further impair arousal and ventilatory responses.

A fifth link is the neural maturation delay manifest by delayed dendritic spine development and delayed myelination caused by unknown prenatal factors or possibly by hypoxic-ischemic insults. Such immaturity could selectively affect the infant already made vulnerable by prenatal factors, destabilizing the cardiorespiratory-sleep-arousal system.

Complicating this scenario is the sixth possible link, triggering or stressor factors, which could tip the balance of this finely tuned system and produce SIDS. Triggering factors such as altered sleep patterns, hyperthermia, prone sleeping position, and infections would be insufficient to cause death, but by themselves in the setting of a vulnerable respiratory system would be sufficient to alter the balance.

This pathogenetic scheme brings together the major observations in SIDS and emphasizes that the neuropathologic evidence indicates that developmental delay and hypoxic-ischemic injury are important links in the chain of events ultimately leading to SIDS.

Links in the Chain of Events Leading to SIDS

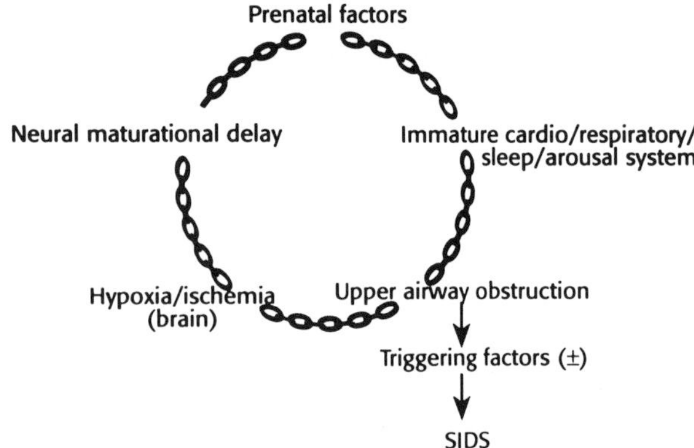

Figure 2. Pathogenesis of SIDS hypothesized as a multistage process.

Further reading

Becker LE (1990): Neural maturational delay as a link in the chain of events leading to SIDS. *Can J Neurol Sci* 17:361–371

Quattrochi JJ, McBride PT, Yates AJ (1985): Brainstem immaturity in sudden infant death syndrome: A quantitative rapid Golgi study of dendritic spines in 95 infants. *Brain Res* 325:39–48

Rognum TO, Saugstad OD, Oyasaeter S, et al (1988): Elevated levels of hypoxanthine in vitreous humor indicate prolonged cerebral hypoxia in victims of sudden infant death syndrome. *Pediatrics* 82:615–618

Takashima S, Armstrong D, Becker L, et al (1978): Cerebral hypoperfusion in the sudden infant death syndrome? Brain gliosis and vasculature. *Ann Neurol* 4:257–262

Taste Transduction

Stephen D. Roper

Taste transduction can be defined as the conversion of the chemical potential inherent in a solution of sapid molecules into a change in the plasma membrane voltage of a taste receptor cell. (For some taste stimuli, researchers have postulated that transduction occurs by activating intracellular second messengers, without a change in the plasma membrane voltage). Taste transduction takes place in the peripheral sensory organs of taste, namely taste buds. The membrane events that occur during transduction begin with the interaction of ions or molecules with the apical tips of sensory receptor cells in taste buds. This usually results in a flow of electrical current from the apical tip to the basolateral membrane within those cells.

Membrane events in taste transduction

Sensory receptor cells in taste buds possess properties found in epithelial and nerve tissues. As in many other epithelial structures, taste buds represent a renewing population of cells. Stem cells provide a constant source of precursor taste receptors that differentiate into mature receptor cells and then die. Furthermore, as in other epithelial cells, taste receptor cells possess an apical membrane that is separated from the basolateral region by a junctional complex, including a band of tight junctions (*zonula occludens*) that seals off the mucosal (apical) from serosal (basolateral) compartments. In epithelial cells, the properties of the apical membrane differ from those of the basolateral membrane, and taste receptor cells are no exception. For instance, there are membrane specializations, such as microvilli where the initial events in taste transduction take place, that are unique to the apical region. There are ion channels in the apical membrane, specifically voltage-gated K^+ channels, that are localized to that region and are not found in the basolateral membrane of receptor cells. Yet taste receptor cells also resemble neurons. For example, synapses are situated on the basolateral membranes of receptor cells. Furthermore, taste cells are electrically excitable and generate action potentials in response to chemical stimulation.

Membrane properties. In cases where it has been possible to make reliable measurements, taste receptor cells have resting membrane potentials between -50 and -80 mV. Taste cells possess voltage-gated K^+, Na^+, and Ca^{2+} channels, and these channels are responsible for the cells' electrical excitability. Taste cells also have Ca^{2+}-mediated K^+ channels, Ca^{2+}-mediated Cl^- channels, and an inwardly rectifying K conductance. The membrane location of most of these conductances includes apical and basolateral regions. However, where it has been explored in detail, namely in the large taste receptor cells of *Necturus maculosus*, voltage-gated K^+ channels have been found only on the *apical*, chemosensitive tips of taste receptor cells. This voltage-dependent K^+ conductance is non-inactivating and is partially active even at normal resting potentials. Consequently,

agents such as tetraethylammonium chloride (TEA) that are known to block voltage-dependent K^+ channels have a profound effect on the resting membrane potential when the agents are applied to the surface of the tongue. Namely, they depolarize taste receptor cells. The depolarization caused by blocking *apical* K-channels is not shunted or counteracted by any consequent voltage-activated increase in K^+ conductance in the basolateral membrane, which is bathed in tissue fluid (i.e., low $[K]_o$). Voltage-dependent K^+ conductance is only present on the apical membrane. As will be described presently, this is one mode of chemosensory transduction. That is, some taste stimuli mimic TEA and block the resting apical K^+ conductance, thereby producing a depolarizing receptor potential.

When a chemical stimulus interacts with the apical membrane in such a way as to produce an influx of cations (or to block the efflux of K^+), the resulting inward current is transmitted to the basolateral membrane along the length of the elongate receptor cells. Current returns to the mucosal surface to close the loop by way of the paracellular pathway (Fig. 1). Electrotonic decrement of the signal along the length of the receptor cell would result in some fraction of the apical depolarization reaching the basal end of the receptor cells. This electrotonic loss is overcome by two strategies: (1) synapses are distributed all along the receptor cells, not just at the basal end and (2), taste cells generate action potentials that distribute the excitatory response undiminished in amplitude throughout the cell.

Salt sensitivity. Taste cells respond to table salt (NaCl) by the influx of Na^+ through amiloride-sensitive apical Na^+ channels. Amiloride-sensitive channels are passive channels, unlike the voltage-gated, tetrodotoxin-sensitive Na^+ channels that underlie the action potential in taste cells. This means that the amiloride-sensitive Na^+ channels in the apical membrane of taste receptor cells are open at rest. The driving force for the Na^+ influx is the electrochemical gradient that the applied NaCl creates, and the influx of cations produces the depolarizing generator current, referred to previously. These facts help explain why the threshold for sensing NaCl is in the millimole range and above: the electrochemical gradient for Na^+ must overcome the contribution from internal Na^+ concentration (estimated to be between 5 and 15 mM).

Potassium salts, such as KCl, act in the same manner, via apical channels through which K^+ ions permeate. One such channel is the non-inactivating, TEA-sensitive delayed rectifier K conductance, described previously, that is partially active at normal resting potentials. Thus, a passive influx of K^+ ions results from the increased electrochemical gradient due to the presence of K^+ salts at the apical membrane.

It is believed that calcium salts act in a profoundly different manner. Apical Ca^{2+} channels are voltage dependent and are not open at the normal resting potential. Apical Ca^{2+} channels are not activated until the membrane potential is

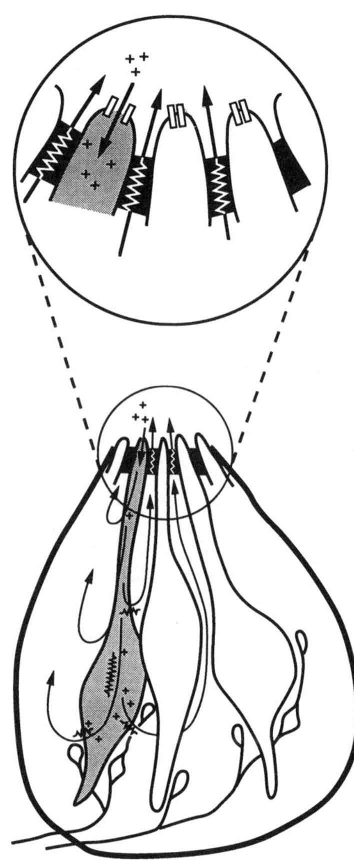

Figure 1. Schematic drawing of a vertebrate taste bud with three receptor cells shown. Synapses are illustrated in the bottom half of the taste bud. One taste receptor cell has been stimulated at the apical pore. Typically, chemostimulation elicits an inward receptor current (arrows) that is carried by cations ($+$) through ion-selective channels embedded in the apical membrane of taste cells, as shown here (cf. Fig. 2). The receptor current passes throughout the receptor cell along the cell's longitudinal internal resistance and out across the basolateral membrane, representing a distributed membrane resistance. The receptor current depolarizes the membrane and presumably opens voltage-activated Ca^{2+} channels situated near the synapses. This evokes transmitter release. The return pathway for the receptor current includes the paracellular route illustrated here, namely the junctional complexes between receptor cells (shaded region in insert). The junctional complexes are situated just below the apical tips of the receptor cells.

quite depolarized ($V_m >$ ca. -20 mV), and at rest there is little direct permeation of Ca^{2+} into the receptor cell through those channels. Instead, Ca^{2+} at the taste pore has been shown to act *extracellularly* by blocking the apical K^+ conductance, much like TEA. The consequence is that the receptor cell is depolarized. If the depolarization is of sufficient magnitude, voltage-dependent Ca^{2+} channels subsequently may open and an influx of Ca^{2+} can presumably occur secondarily.

The anion in salts is known to affect taste. However, the detailed membrane mechanisms for the anion involvement are not well understood. Putative mechanisms include the selective permeation of anions through Ca-dependent Cl^- channels that are present in taste cells or the selective permeability of paracellular epithelial pathways that bypass cell membranes.

Sour. Acids taste sour to humans and evoke strong responses in many animals. The sourness is due primarily to protons (H^+). In some species, such as *Necturus*, protons block the resting apical K^+ conductance, and this produces a depolarizing receptor potential. In mammals, protons appear to permeate amiloride-sensitive apical Na^+ channels and thereby directly depolarize some taste receptor cells. In addition, acids exert other actions on receptor cells, and the combination of multiple actions may explain how mammals distinguish acids from Na^+ salts. For example, as well as blocking K^+ channels, protons may also open Ca^{2+} channels, thereby allowing an influx of cations, or may open Cl^- channels, leading to an efflux of anions. Findings from a number of laboratories, though not yet conclusive, support all of these possibilities in taste cells from a number of species.

Bitter. A diverse collection of substances elicits bitterness, including simple salts (e.g., $BaCl_2$), peptides (e.g., Arg-Pro-Gly), sugar derivatives (e.g., sucrose octaacetate), and more complex molecules such as denatonium salts, caffeine, and strychnine. Some bitter compounds, such as sucrose octaacetate, may initially be adsorbed onto the plasma membrane of taste cells. Others, such as quinine, are lipophilic and may enter the cell via solubilization in the plasma membrane. Quinine and bitter salts may also act extracellularly by inhibiting ion channels in the apical membrane.

Bitter compounds that are adsorbed onto the taste cells presumably interact with specific receptors on the apical membrane. These receptors are believed to be coupled, by way of a G-protein, to phospholipase C. One consequence is the generation of inositol 1,4,5-triphosphate (IP_3), which, in turn, stimulates the release of Ca^{2+} from intracellular stores. Other bitter compounds have been shown to block K^+ conductance, thereby depolarizing the cell. Indeed, most known K^+ channel blockers have a bitter taste. These two mechanisms—G-protein–coupled mechanism and K^+ channel blockage—are not mutually exclusive and may occur concurrently in response to the same bitter stimulus.

Sweet. Recent data indicate that some compounds that taste sweet, such as sucrose or saccharin, act by way of a G-protein–mediated increase in intracellular cAMP. Presently, the working hypothesis is that intracellular cAMP either acts directly on K^+ channels or stimulates the phosphorylation of K^+ channels in the surface membrane such that K^+ conductance in the taste receptor cell is reduced. This leads to a depolarizing receptor potential.

Other. Many animals, particularly some aquatic species such as catfish, are exquisitely sensitive to amino acids such as alanine and arginine. Amino acids elicit a variety of taste qualities in humans, from sweet (e.g., glycine and alanine) to bitter (e.g., leucine and tyrosine) to a complex meaty taste termed "umami" (e.g., glutamate). It is unlikely that there is a common membrane mechanism. Indeed, studies have indicated that some amino acids, such as L-arginine, bind to and directly activate ion channels (i.e., ligand-activated channels), whereas other amino acids, such as L-alanine, stimulate the production of cAMP by a G-protein–coupled mechanism.

Conclusions

It is evident that there is no one membrane event that describes the response of taste receptor cells to chemical

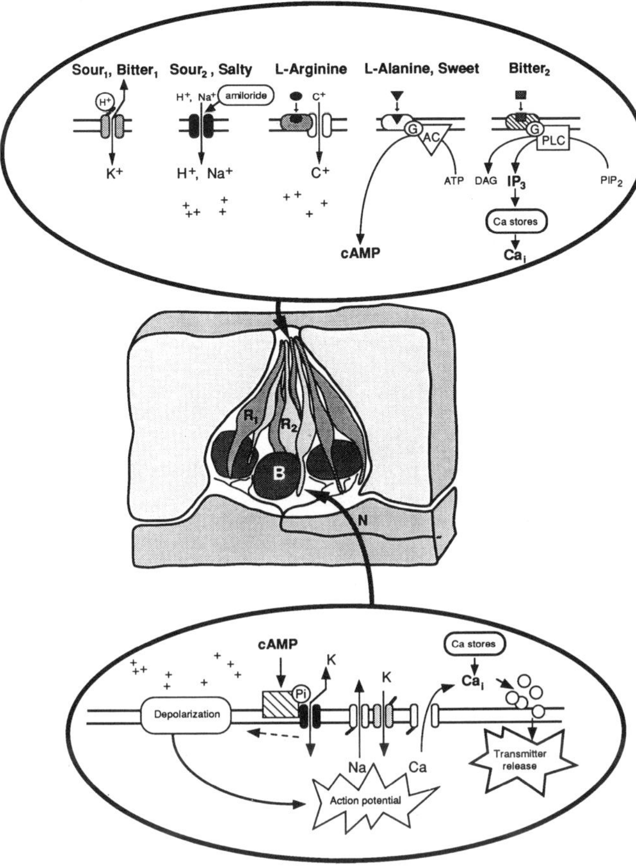

Figure 2. Summary of chemosensory transduction mechanisms occurring in taste buds. Events illustrated at the apical membrane of taste receptor cells include protons (sour) and bitter compounds blocking K channels; protons and Na⁺ passing through apical, amiloride-sensitive sodium channels; L-arginine binding to a ligand-gated, nonselective cation channel; L-alanine and sweet compounds activating a G-protein by binding to appropriate membrane receptors and stimulating adenylyl cyclase; bitter compounds activating a G-protein that is coupled to phospholipase C and resulting in an increase in inositol triphosphate. Events illustrated on the basolateral membranes of taste receptor cells include depolarization of the membrane consequent to an influx of cations at the apical membrane; reduction of potassium conductance due to cyclic AMP buildup, which also depolarizes the membrane; generation of an action potential when the depolarization reaches threshold; influx of Ca^{2+} ions; release of neurotransmitter stores consequent to increased $[Ca^{2+}]_i$. Abbreviations: C^+, cations; G, G-protein; AC, adenylyl cyclase; PLC, phospholipase C; DAG, diacyl glycerol; IP_3, inositol 1,4,5-triphosphate; PIP_2, inositol 4,5-bisphosphate; R, taste receptor cell; B, basal taste cell; N, nerve bundle. (Reprinted with permission of Oxford University Press from Roper SD (1992): The microphysiology of peripheral taste organs. *J Neurosci* 12:1127–1134.)

stimulation. A diversity of mechanisms exists. Figure 2 summarizes what is known to date about many of the better studied membrane events in transduction.

To date, specific membrane receptors for sweet or bitter taste stimuli have not been isolated. No genes that encode taste receptors have been cloned and sequenced, though presumed taste receptors for amino acids have been expressed in *Xenopus* oocytes. Researchers have constructed models of binding sites in receptors for sweet compounds, based upon molecular features that a large number of chemicals that elicit sweet taste have in common. Similarly, models have been constructed for the bitter receptor site. However, it is probable that there are multiple receptor sites for sweet and for bitter compounds. The ion channels that are involved in taste transduction for some taste qualities, such as amiloride-sensitive Na⁺ channels and apical K⁺ channels, have properties similar to those channels in other tissues. It is likely that their molecular structure will resemble that for these channels in other excitable tissues, for which the amino acid sequences are known.

Researchers do not know whether a single taste receptor cell is "tuned" to only one class of stimuli (e.g., sweet), although this might be a reasonable expectation. Older findings from intracellular microelectrode impalements indicate that single cells *in situ* respond to several different chemical stimuli. However, the experiments are likely to have been confounded by microelectrode damage to the small cells and possible artifacts in the receptor potentials that would interfere with the interpretation. Thus, it remains to be seen whether a single taste receptor cell undergoes only one or several of the membrane events previously described during chemostimulation. Further, clearly there is more than one mechanism for some qualities, such as sour and bitter substances. These mechanisms may differ among species, and even within a given species the transduction events may not be mutually exclusive.

Further reading

Avenet PA, Kinnamon SC (1991): Cellular basis of taste reception. *Curr Opin Neurobiol* 1:198–203

Corey D, Roper SD (1992): *Sensory Transduction*. New York: Rockefeller University Press

Kinnamon SC, Cummings TA (1992): Chemosensory transduction mechanisms in taste. *Ann Rev Physiol* 54:715–731

Roper SD (1992): The microphysiology of peripheral taste organs. *J Neurosci* 12:1127–1134

Simon SA, Roper SD (1993): *Mechanisms of Taste Transduction*. Boca Raton: CRC Press

Transcranial Sonography

N. Venketasubramanian and J. P. Mohr

Introduction

Transcranial sonography is the imaging of intracranial structures using sound waves that pass through the skull and reflect back from moving objects, mostly arterial blood. Up to a decade ago, intracranial structures were difficult to investigate by ultrasound owing to severe attenuation of sound waves by the bony skull. The landmark paper by Aaslid et al. (1982) described a technique for studying the large intracranial vessels at the base of the skull using low-frequency sound waves, which resulted in an explosion of knowledge of intracranial hemodynamics.

Principles

The technique exploits the Doppler effect of a shift in frequency of sound waves that occurs when they are reflected off moving objects. A probe is placed on the skull, and low-frequency (1 to 2 MHz) sound waves produced from electrical impulses converted by piezoelectric material are transmitted through the skull and focused onto the deep structures. These signals are reflected off moving objects, the most important of which are red blood cells within arteries, which reflect the sound waves back to the probe where they are converted back into electrical impulses. A microprocessor analyzes these signals by real-time fast-Fourier transformation and produces a special display of Doppler velocities, seen on an oscilloscope as a waveform. By varying the time after emission when the signal is processed, the depth of the source of the reflected signal can be determined, a process known as range-gating, which makes it possible to insonate a specific vessel at various points along its length. From these studies, an assessment of cerebral blood velocity may be obtained, which has proved to be a useful estimate of cerebral blood flow.

At the present time, transcranial sonography is restricted to the use of Doppler techniques. However, real-time imaging is already on the horizon.

Windows

Although the bone of the skull provides a natural barrier to sound waves, certain sites (windows) of thinner bone and foramina allow the passage of sound waves into the interior of the skull.

The temporal window lies above the zygomatic arch. It is divided into thirds: the anterior, middle, and posterior temporal windows. Through it, part of the intracranial internal carotid artery and the first or proximal (M1) segment of the middle cerebral artery may be insonated. At greater depths, slight anterior angulation of the probe insonates the first or proximal (A1) segment of the anterior cerebral artery, while slight posterior angulation insonates the first (P1) and even the second (P2) segments of the posterior cerebral artery. Sometimes, the posterior communicating artery may also be insonated.

Placing the probe over the eyeball and directing the beam slightly medially and superiorly allows insonation through the transorbital window of the ophthalmic artery and supra- and infraclinoid portions of the carotid siphon.

A suboccipital window is accessed by flexing the neck to open up the foramen magnum, and placing the probe below the occipital protruberance. Directing the beam intracranially allows the insonation of the distal vertebral, and lower and mid-basilar, arteries.

Placement of the probe over the inion allows the insonation of the calcarine artery and straight sinus.

In 5 to 15% of patients, satisfactory windows may not be obtained. This is especially so in the elderly, females, blacks, and Orientals. Anatomic variations may result in the absence or unusual configuration of Doppler signals. The development of injectable materials that generate their own acoustic signals could assist in the detection of flow in otherwise silent vessels.

Factors affecting flow velocities

A number of factors alter the velocities of blood flow in the basal cerebral arteries. These include the patient's age, blood pressure and heart rate (i.e., cardiac output), temperature, carbon dioxide tension, blood viscosity, diameter and curvature of the blood vessels, volume of blood flow, and even brain activation. Different angles of insonation may produce different readings of flow velocities, but do not cause major alteration in values. As the test is performed using hand-held probes, obtaining accurate results would depend on the knowledge, skill, diligence, and experience of the sonographer.

Indices

In addition to detecting the direction of flow, various velocity indices are measured: peak systolic (V_s), end diastolic (V_d), and mean (V_m) velocities, Gosling's pulsatility index (PI), [$(V_s - V_d)/V_m$], pulsatility transmission index [100 × ipsilateral PI/contralateral PI] and resistivity index [$(V_s - V_d)/V_s$] (Figs. 1 and 2). Values obtained are compared with

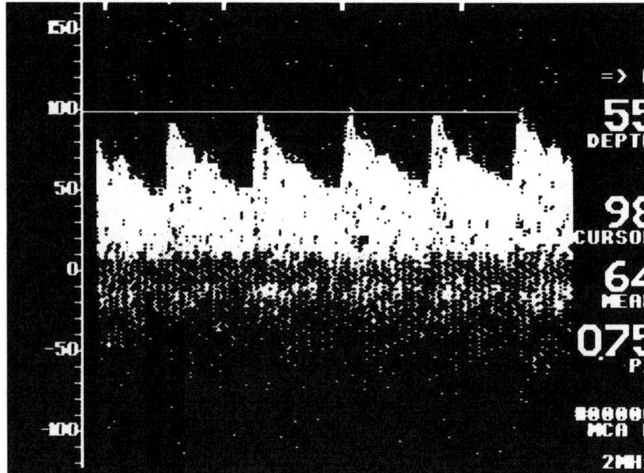

Figure 1. Normal right middle cerebral artery at 55 mm, with flow toward the probe, V_s = 98 cm/s, V_m = 64 cm/s, and PI = 0.75.

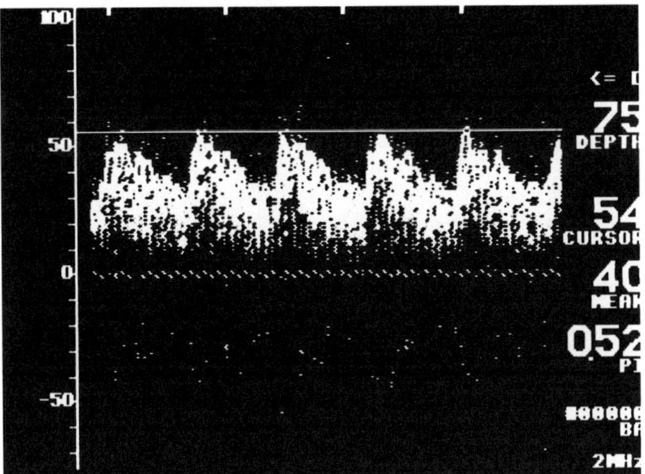

Figure 2. Normal basilar artery at 75 mm, with flow away from the probe, $V_s = 54$ cm/s, $V_m = 40$ cm/s, and PI = 0.52.

the contralateral side, or against expected normal values. To the trained ear, the pitch of the sound and turbulence provide further information about abnormalities in blood flow.

Increased velocities are seen at areas of stenosis or high volume of blood flow; low velocities are seen distal to a stenosis or proximal to an occlusion. Pulsatility is increased proximal to an occlusion or in raised intracranial pressure, and decreased poststenosis or in vessels feeding low-resistance channels such as arteriovenous malformations.

Certain maneuvers are performed to further assess cerebral hemodynamics, such as postural changes, arm exercises, common carotid and vertebral artery occlusions, and hypo- and hypercapnic stresses.

Applications

Transcranial sonography requires inexpensive portable equipment. It may be repeated safely as often as necessary. The use of the monitoring probe even allows continuous and instantaneous information on changes in cerebral hemodynamics.

The current status of transcranial Doppler has recently been clarified (see Report of the American Academy of Neurology, 1990). It is of established value in (a) detecting severe stenosis (>65%) in major basal intracranial arteries; (b) assessing patterns and extent of collateral circulation in patients with known regions of severe stenosis or occlusion; (c) evaluating and following patients with vasoconstriction of any cause, especially after subarachnoid hemorrhage; (d) detecting arteriovenous malformations and studying their supply arteries and flow patterns; and (e) assessing patients with suspected brain death.

Under investigation are uses in (a) evaluating children with various vasculopathies such as sickle-cell disease, moyamoya, and neurofibromatosis; (b) monitoring during cerebral endarterectomy, cardiopulmonary bypass, and other cerebrovascular and cardiovascular interventions, and surgical procedures; (c) evaluating patients with dilated vasculopathies such as fusiform aneurysms; (d) assessing autoregulation, physiologic, and pharmacologic responses of cerebral arteries; (e) assessing patients with migraine.

Assessment of intracranial stenosis

Significant stenosis causes increased velocities maximal at the site of obstruction; the amount of increase correlates with the degree of stenosis. Marked acceleration is seen at stenosis exceeding 80% (Fig. 3). Greater stenosis leads to turbulent signals, aliasing, and finally decreasing velocities as occlusion occurs. Pulsatility is increased proximal to the stenosis and reduced distal to the stenosis, especially in the presence of poststenotic dilatation.

Occluded vessels show a loss of Doppler signals at the site of occlusion, and reduced velocities with increased pulsatility proximally. Recanalization may be heralded by the return of a high-velocity signal at the site of previous occlusion. A problem posed by the missing signal of an occluded vessel is the possibility that a healthy artery has not been properly insonated.

Adequacy of collateral flow

The adequacy of collateral flow due to severe stenosis or occlusion may be assessed. In internal carotid artery occlu-

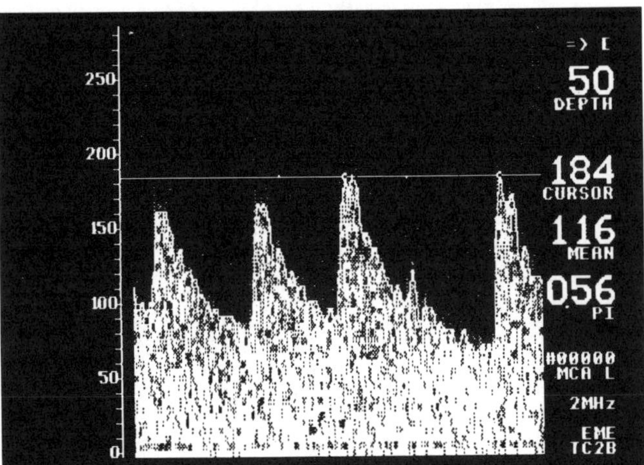

Figure 3. Stenosis of the left middle cerebral artery at 50 mm, with $V_s = 184$ cm/s.

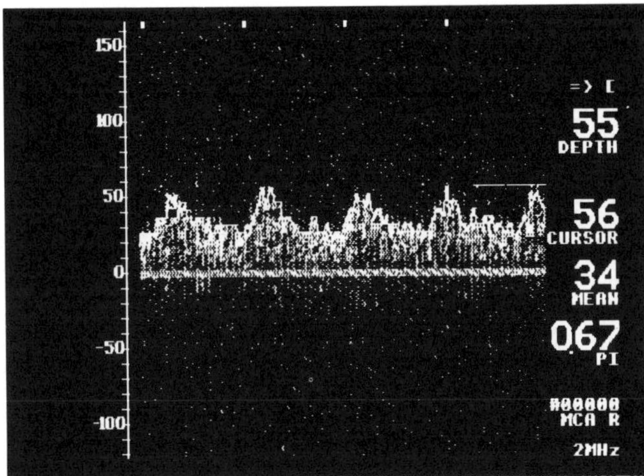

Figure 4. Low V_s and PI in the middle cerebral artery ipsilateral to an occluded right internal carotid artery.

sion, reduced velocities and pulsatilities occur in the ipsilateral middle cerebral artery (Fig. 4). Reversed and markedly accelerated flow in the ipsilateral anterior cerebral artery suggests the presence of collateral flow across the anterior communicating artery from the contralateral circulation; collateral flow from the posterior circulation via the posterior communicating artery may also be demonstrated. Reversal of flow in the ipsilateral ophthalmic artery may also be seen as collateral circulation enters from the ipsilateral external carotid artery. Recanalization of the occluded vessel may lead to reversal of these changes.

Vasospasm

Vasospasm is a dreaded complication of subarachnoid hemorrhage. It results in accelerated velocities along the length of the vasospastic vessel. These changes may be detected before symptoms begin, allowing early therapeutic intervention. Posttraumatic vasospasm may be similarly detected. Progression, and the response of vasospasm to treatment, may be readily monitored as worsening or normalization, respectively, of flow indices.

Arteriovenous malformations

Arteriovenous malformations (AVMs) are high-flow, low-resistance systems. The main arteries from which feeders arise show increased velocities and reduced pulsatilities (Fig. 5). When multiple vessels are involved, the one with the most marked abnormality is likely to be responsible for most of the flow to the AVM. Feeding vessels also suffer a failure in autoregulation: occluding the ipsilateral common carotid artery does not elicit compensatory changes seen in normal vessels; carbon dioxide reactivity is also lost. Successful removal or obliteration of the AVM leads to a reversal of these findings. Recurrence of the AVM may result in the return of the abnormalities.

Intracranial pressure and brain death

Raised intracranial pressure (ICP) impedes intracranial blood flow. This is reflected in falling diastolic velocities, falling systolic velocities, and rising pulsatilities. The waveform appears more spiked and diastolic flow ceases completely. Further ICP increases lead to reversal of diastolic flow. Such reverberating patterns of flow correlate well with the flat electroencephalogram of clinically brain-dead patients. Successful reduction of ICP results in normalization of indices.

Detection of right-to-left shunts

Right-to-left shunts (e.g., through a patent foramen ovale, pulmonary arteriovenous fistulae) allow the passage of peripheral venous blood clots into the systemic arterial circulation. Such clots may subsequently occlude cerebral arteries, resulting in a stroke. Agitated saline injected into a peripheral vein may cross such a shunt, assisted by Valsalva maneuvers, and be detected as blips on sonography (Fig. 6). Subsequent cardiac and other evaluation may demonstrate the shunt more definitively. Shunt closure would eliminate the abnormality.

Childhood diseases

Premature and hypoxic neonates have been found to have abnormalities on sonography. Occlusive diseases of the basal cerebral arteries such as in moya-moya and sickle-cell disease may be studied in a manner similar to occlusive disease in the adult. Such studies may save the child the risk of angiography.

Intraoperative monitoring

Intraoperative monitoring of cerebral blood flow (e.g., during coronary bypass surgery, clamping of the carotid artery for carotid endarterectomy) would provide early warning of reduced cerebral perfusion and collateral failure. Emboli into cerebral vessels during surgery and angiography may be detected. Monitoring would allow graded occlusion of vessels feeding inoperable giant aneurysms. Postendarterectomy hyperperfusion syndromes may also be diagnosed.

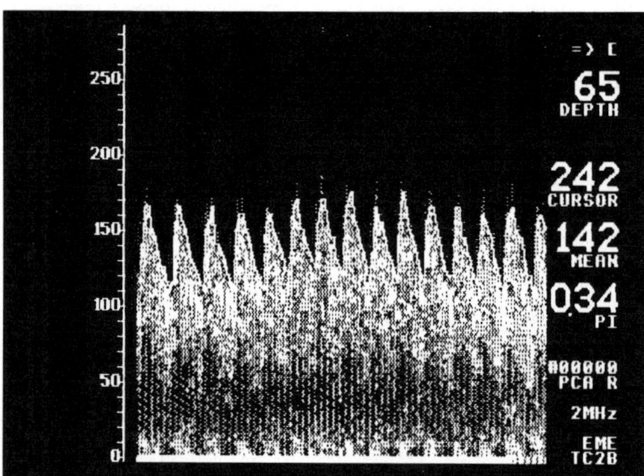

Figure 5. Very high V_s and low PI in the right posterior cerebral artery with feeders to a large arteriovenous malformation.

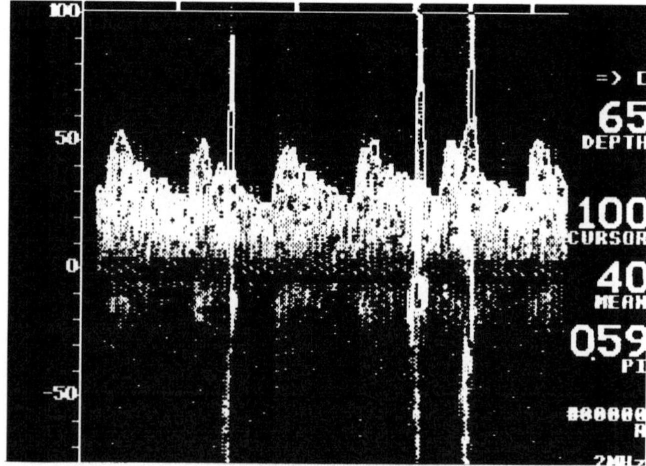

Figure 6. Vertical lines (artifacts) caused by air bubbles entering the MCA after intravenous agitated saline injection in patient with a patent foramen ovale.

Cerebral hemodynamics

There is a fairly good correlation between velocity profiles and cerebral blood flow as demonstrated by single photon emission computed tomography (SPECT); this extends to carbon dioxide reactivity as well. Changes in indices during postural changes, correction of severe anemia, and visual stimuli may also be studied. Studies in migraine sufferers are also being conducted.

Three-dimensional color Doppler flow monitoring

With this relatively new technique, velocities are sampled using two scanning arms, and a composite picture of the basal arteries in three dimensions (viz. coronal, horizontal, and lateral) are presented simultaneously. The various colors reflect the speed and direction of blood flow. Thus, accurate vessel identification and the velocity and direction of flow in these vessels may be demonstrated.

Transcranial duplex imaging

Very early but promising studies of intracranial structures combining real-time imaging with Doppler insonation are only just emerging. It is hoped that refinement of the technology over the next few years would result in its availability for clinical use in the near future.

Conclusion

Transcranial sonography has come a long way in the decade since its discovery. Its many advantages and widespread applications render it an extremely useful investigative tool. Further developments including new applications and duplex capabilities are eagerly awaited.

Further reading

Aaslid R, ed (1986): *Transcranial Doppler Sonography*. Wein, New York: Springer-Verlag

Caplan LR et al (1990): Transcranial Doppler ultrasound: Present status. *Neurol* 40:696–700

Petty GW et al (1990): Transcranial Doppler ultrasonography: Clinical applications in cerebrovascular disease. *Mayo Clin Proc* 65:1350–1364

Report of the American Academy of Neurology, Therapeutics and Technology Assessment Subcommittee (1990): Assessment: Transcranial Doppler. *Neurol* 40:680–681

Tremor

Rodger J. Elble

Tremor is an approximately rhythmic, roughly sinusoidal involuntary movement. Tremor is classified according to the circumstances in which it occurs. *Rest tremor* occurs when the bodily part is in repose. The best known example is the pill-rolling hand tremor of Parkinson's disease. *Postural tremor* occurs during the steady maintenance of posture, such as extending the upper extremities horizontally against the force of gravity. *Kinetic tremor* occurs during voluntary movement. As the limb approaches its target, kinetic tremor frequently increases, and this terminal accentuation of kinetic tremor is referred to as *intention tremor*. *Task-specific* or *occupational tremor* occurs during a specific behavioral activity such as writing (primary writing tremor) and standing (orthostatic tremor). The terms *static tremor* and *action tremor* are ambiguous. Static tremor refers to tremor that is present when there is no voluntary movement and therefore includes both rest tremor and postural tremor. Action tremor is any tremor that occurs during voluntary muscle contraction and encompasses all forms of postural tremor and kinetic tremor.

Physiologic tremor is exhibited by everyone but is barely visible to the unaided eye. It is symptomatic only during activities that require extreme precision. Physiologic tremor consists of two distinct oscillations, mechanical-reflex and 8- to 12-Hz. These two oscillations are superimposed upon a background of irregular fluctuations in muscle force and limb displacement. These background irregularities have a frequency of 0–15 Hz and are produced by motor units that fire near their threshold. The low-pass filtering property of skeletal muscle attenuates the amplitude of these irregularities at frequencies above 3–5 Hz.

The mechanical-reflex component is the larger of the two distinct oscillations of physiologic tremor and is due to the inertial, viscous, and elastic properties of the limb or bodily part in question. The mechanical attributes of most bodily parts are such that damped oscillations occur in response to pulsatile perturbations. The frequency of these mechanical-reflex oscillations is determined largely by the inertia and stiffness of the bodily part. Consequently, normal elbow tremor has a frequency of 3–5 Hz, wrist tremor 8–12 Hz, metacarpophalangeal joint tremor 17–30 Hz, and ocular tremor 35–40 Hz. Irregularities in subtetanic motor unit firing are the major source of continuous forcing to a limb during volitional movement and during the steady maintenance of posture. The ejection of blood at cardiac systole provides an additional forcing to the body, which accounts for physiologic tremor at rest but only a fraction of physiologic postural and kinetic tremor. While somatosensory receptors (e.g., muscle spindles) respond to these passive mechanical oscillations, this response is usually too weak to entrain motoneurons at the frequency of tremor. The stretch-reflex response to oscillation increases during fatigue and anxiety and in response to some medications, producing a modulation of motor unit activity and so-called enhanced physiologic tremor. This involvement of the stretch-reflex can increase tremor or suppress it, depending upon the dynamics of the reflex loop and limb mechanics.

In contrast to the mechanical-reflex oscillation, the 8- to 12-Hz component of physiologic tremor is invariably associated with modulation of motor unit activity, even when the 8- to 12-Hz tremor is much smaller than the mechanical-reflex oscillation. Participating motor units are entrained at 8–12 Hz, regardless of their mean frequency of discharge. The 8- to 12-Hz and mechanical-reflex oscillations are easily distinguished by examining the effect of inertial and elastic loads on their frequencies. The frequency of the 8- to 12-Hz tremor exhibits little or no change when inertial or spring loads are attached to the limb. By contrast, the frequency of mechanical-reflex oscillation is inversely proportional to the square root of the inertial load and is directly proportional to the square root of the added stiffness. Similarly, the frequency of the 8- to 12-Hz tremor is independent of the length of the stretch-reflex arc. For these reasons, the 8- to 12-Hz tremor probably originates from an oscillating neuronal network within the central nervous system. A prominent 8- to 12-Hz tremor is seen in a minority of normal adults. Hence, entrainment or synchronization of motor units is not a prominent feature of muscular contraction in most people under ordinary circumstances. By contrast, prominent motor unit entrainment is characteristic of all forms of pathologic tremor.

The underlying pathophysiology of virtually all pathologic tremors is poorly understood. In general, tremor could originate from neurons with inherent rhythmical properties (pacemaker neurons) or could emerge from the interaction of neurons within neuronal networks. These two mechanisms of neuronal oscillation are not mutually exclusive. Neuronal networks in the sensorimotor cortex, the thalamus, the inferior olive, and the spinal cord are only a few of many documented sources of physiologic neuronal oscillation that if pathologically altered could produce an abnormal tremor. These specific sources of oscillation have network properties that promote entrainment and contain neurons with membrane properties that produce oscillation through inhibition-rebound excitation, utilizing a calcium-dependent potassium-mediated membrane hyperpolarization and a low-threshold calcium spike. Pathologic tremor could result from increased entrainment of the motor pathways by the oscillator, increased strength of oscillation within the oscillator, or both. GABA, acetylcholine, norepinephrine, glutamate (NMDA receptors), and probably other neurotransmitters modulate these sources of oscillation, and alterations in these and other neurotransmitters could produce increased oscillation within a neuronal network or increased entrainment throughout the motor system. Current investigations into the sites and mechanisms of physiologic neuronal oscillators are crucial to the understanding of pathologic tremors.

The most common form of pathologic tremor is *essential tremor*. Essential tremor occurs in people of all ages but is most common in older people, having a prevalence of at least 0.4% in people over the age of 40. Approximately half of all cases are dominantly inherited. Essential tremor most commonly affects the hands but also occurs in the head, voice, face, trunk, and lower extremities. The *sine qua non* of essential tremor is a rhythmic 4- to 12-Hz entrainment of motor unit discharge that forces the affected bodily part into oscillation. Mechanical loads have little effect on the frequency of essential tremor. The frequency of essential tremor has a negative logarithmic relationship with tremor amplitude, such that mild essential tremor is qualitatively similar to the 8- to 12-Hz physiologic tremor. Both tremors may therefore have a common origin. Essential tremor is a postural tremor with a variable kinetic component. Tremor in repose is observed only in the most advanced cases and differs from the complex pill-rolling hand tremor of Parkinson's disease. Most patients with essential tremor have no other abnormal neurological signs. However, tremor resembling essential tremor is seen in many other neurological disorders, including Parkinson's disease and dystonia. Because essential tremor has no unique diagnostic features, the relationship of essential tremor to similar tremors in other disorders is unknown. Ethanol, primidone, beta-adrenergic blockers, and thalamotomy in or near ventralis intermedius suppress essential tremor. The origin of oscillation is unknown, but the inferior olive is a popular candidate. The thalamus and other parts of the motor system may become secondarily entrained, thus explaining why lesions in so many areas of the motor system suppress essential tremor.

Rest tremor is the best known and most specific feature of Parkinson's disease. Parkinsonian rest tremor is usually seen in one or both hands but also occurs in the feet. The fingers of the hand exhibit complex movements as though pills or other small objects were being rolled in the hand, hence the name "pill-rolling tremor." These rhythmic 3- to 5-Hz finger movements are superimposed on rhythmic extension-flexion of the wrist and pronation-supination of the forearm. Voluntary muscle contraction typically suppresses parkinsonian rest tremor, although advanced rest tremor may persist during posture and movement. The ventrolateral thalamus has long been suspected as being the origin of parkinsonian tremor, even though the principal pathology is loss of dopaminergic cells in the substantia nigra. The loss of nigrostriatal dopaminergic input possibly disinhibits the globus pallidus interna, and the resulting increase in GABA-ergic input to the ventrolateral thalamus may enhance the oscillatory properties of thalamic neurons. Levodopa, dopaminergic agonists, anticholinergics, and thalamotomy in or near ventralis intermedius suppress parkinsonian tremor.

Damage to the dentate and interpositus cerebellar nuclei produces a dramatic 3- to 5-Hz kinetic tremor of the upper extremities. The frequency of *cerebellar tremor* varies predictably with changes in limb inertia and stiffness. Studies in laboratory primates have revealed oscillation in the motor cortex, somatosensory cortex, interpositus nucleus, and somatosensory afferents. Therefore, cerebellar tremor involves the abnormal oscillation of transcortical and transcerebellar sensorimotor feedback loops. The underlying complexity of these oscillating loops is probably the principal reason why effective treatment is not available for most patients.

Many other less common forms of tremor deserve mention to emphasize the broad clinical scope and complexity of tremor pathophysiology. *Rubral* or *midbrain tremor* is an unusual combination of 2- to 5-Hz rest, postural, and kinetic tremor of an upper extremity that results from damage to the red nucleus and neighboring cerebellothalamic and nigrostriatal nerve fiber tracts. Peripheral neuropathies occasionally cause an abnormal mechanical-reflex tremor. Task-specific tremors such as primary writing tremor may in some patients be a variant of essential tremor and may in other patients be a symptom of focal dystonia. Finally, a variety of drugs produce parkinsonian rest tremor (neuroleptics), postural tremor (beta-adrenergic agonists, valproic acid, thyroxin, and methylxanthines), kinetic tremor (lithium), and combinations thereof (lithium, amiodarone, and valproic acid).

In conclusion, the motor system contains central and peripheral feedback loops, oscillating neuronal networks, and underdamped body mechanics. The complex integration of these sources of oscillation has made the identification of the principal origins of tremor exceedingly difficult. No form of tremor is completely understood. This deceptively simple involuntary movement is a tantalizing riddle, the clues to which are contained throughout the broad subject of motor control.

Further reading

Elble RJ, Koller WC (1990): *Tremor.* Baltimore: The Johns Hopkins University Press

Llinás R (1984): Rebound excitation as the physiological basis for tremor: A biophysical study of the oscillating properties of mammalian central neurons. In: *Movement Disorders: Tremor,* Findley LJ, Capildeo R, eds. London: Macmillan, pp 339–351

Paré D, Curro'Dossi R, Steriade M (1990): Neuronal basis of the parkinsonian resting tremor: A hypothesis and its implications for treatment. *Neurosci* 35:217–226

Stein RB, Lee RG (1981): Tremor and clonus. In: *Handbook of Physiology: The Nervous System, Motor Control,* Brooks V, ed. Baltimore: Williams & Wilkins, pp 325–343

V

Vertigo

Joseph G. Feghali

Vertigo is the false perception of motion of oneself or of one's environment. As a first step in the workup of a patient with vertigo, it is essential to determine the exact nature of the symptoms. To some, "vertigo" is a term used to describe all types of dizziness. These include a variety of symptoms described as lightheadedness, giddiness, a spinning sensation, blurry vision, numbness, faintness, unsteadiness, etc. An early determination of the exact nature of the symptoms can allow the physician to establish a diagnosis and initiate rational treatment.

The technical meaning of vertigo is that it is a feeling of rotational motion of the body and/or the surroundings. Vertigo is associated with lesions of the vestibular portion of the labyrinth or the vestibular end organs, and less commonly, it is seen in lesions of the statoacoustic nerve and/or the brainstem. The vestibular end organs consist of three semicircular canals, the utricle, and the saccule. Different parts of the vestibular end organs respond to angular or linear acceleration as well as static gravitational forces. The output of the end organs is transmitted to the vestibular nuclei in the brainstem via the vestibular portion of the statoacoustic nerve (cranial nerve VIII). The vestibular nuclei have multiple neural connections including to the spinal cord, the cerebellum, the cerebral cortex, and the oculomotor system.

The vestibular system works in conjunction with the visual and the proprioceptive systems to maintain spatial orientation and posture. Both physiologic and pathologic stimulation of any one or more of the three systems can result in vertigo. Vertigo can also have other etiologies. Lesions of the brainstem and the vertebrobasilar circulatory system can result in vertigo that is usually associated with other neurologic signs and symptoms.

Vestibular vertigo

Vertigo of vestibular origin is common and can present with dramatic symptomatology. It is characterized by a whirling sensation of the body and/or the surroundings and may be associated with nausea and vomiting. It may also be associated with ataxia and postural unsteadiness. Examination usually reveals an associated nystagmus, a repetitive beating movement of the eyes that results from the stimulation of the oculomotor system via the vestibulo-oculomotor connections.

The following is a description of a few common vestibular syndromes associated with vertigo.

Ménière's disease or syndrome. Prosper Ménière was the first to describe the association between vestibular symptoms and the inner ear. Classical Ménière's syndrome describes a symptom complex of episodic paroxysmal vertigo associated with hearing loss and tinnitus. The vertigo is usually severe and lasts for hours, not weeks, and is typically associated with nausea and vomiting. The hearing loss is of the sensorineural or perceptive type. It usually affects the lower frequencies more than the higher frequencies. Over time, the hearing loss is typically fluctuant but relentlessly progressive. It usually affects one ear but it can be bilateral in about 10% of cases. Many patients with Ménière's syndrome have associated symptoms of a feeling of pressure in the involved ear and many have symptoms of intolerance to moderately loud noises, or recruitment. Recruitment makes it difficult to fit such patients with hearing aids. All too many conditions presenting with vestibular symptoms are erroneously labeled as cases of Ménière's syndrome.

Benign positional paroxysmal vertigo (BPPV). BPPV is a characteristic type of vertigo that occurs when the patient's head is positioned in a specific critical position (e.g., patient supine with head turned to one side). The vertigo occurs after a short latency and is fatigable. These characteristics differentiate it from central positional vertigo resulting from lesions of the central nervous system. BPPV is thought to be due to an irritation of the posterior semicircular canal of the dependent inner ear. It is typically a self-limiting disease. In some rare cases, when the patient is incapacitated for prolonged periods, surgical therapy may be indicated. Different types of surgical therapies have been described. These include severing the nerve to the posterior semicircular canal, obliteration of the posterior semicircular canal, and microvascular decompression of the statoacoustic nerve in the posterior fossa.

Acute unilateral labyrinthine dysfunction. This condition is also known as viral labyrinthitis, vestibular neuronitis, and acute labyrinthitis. This is a syndrome characterized by vertigo without a hearing loss. The symptoms may start suddenly or may have an insidious onset. The vertigo can become severe over 1 to 3 days. It then starts to subside slowly. While some patients are cured in about 10 days, some others continue to have symptoms for several weeks. The cause of this syndrome is unknown. It has been known to follow acute febrile viral syndromes. Treatment is typically supportive. The condition may recur in some patients.

Vestibular schwannoma. This condition is also known as "acoustic neuroma." It is a benign tumor that arises on the vestibular portion of the statoacoustic nerve. It typically causes a progressive unilateral hearing loss and tinnitus. Its slow growth allows for concomitant vestibular compensation to take place. Rarely, it can cause acute vertigo. This is possibly due to the occlusion of the blood supply to the vestibular labyrinth. Vestibular schwannoma should be suspected in patients with associated unilateral change in hearing and/or tinnitus.

Evaluation of the patient with vertigo

A complete history and physical examination remain the most valuable tools in the diagnosis of the vertiginous patient. Special attention to associated conditions, including decreased vision, peripheral neuropathy, and atherosclerotic vascular disease, may indicate to the physician the possible causes of the vertigo.

An audiogram is important to detect and quantify an associated hearing loss. Electronystagmography (ENG) remains the standard test in the evaluation of the vestibular system. It provides an objective record of the nystagmus that is elicited in response either to a physiologic or a pathologic stimulus. ENG is an electronic recording of the change in the difference in potential between the cornea and the retina of the eye, the corneoretinal potential. The corneoretinal potential allows the eye to act as a dipole. Movements of the eyes such as during nystagmus cause a shift of the dipole. This results in a change in potential that can be recorded and quantified. A complete ENG may include a positioning test, a fistula test, and the caloric test. During the caloric test, the patient is placed in the supine position with the head elevated 30 degrees. The external auditory canal is irrigated with water that is 7°C colder or warmer than body temperature. Primarily, this produces a thermal stimulation of the lateral semicircular canal and results in nystagmus that is recorded. Since each ear is examined separately, the caloric test is important in the localization of unilateral vestibular lesions.

Several other vestibular tests have been described. Dynamic posturography is of interest because it can also help in determining rehabilitation strategies in the vertiginous patient. Posturography is a dynamic test of sensory (visual, proprioceptive, vestibular) integration with motor responses. It also gives information about coordination under various conditions that can cause unsteadiness and require a recovery reaction from the patient.

The aim of vestibular testing is to detect the presence of a lesion and to localize it to the central or to the peripheral vestibular systems.

Vestibular test results may be sufficient for a diagnosis or may necessitate further testing, including blood tests and specialized radiologic imaging. Blood tests may include a hemogram, blood chemistry, a screening test for syphilis, immune and endocrine profile, as well as allergy testing. Radiologic imaging includes computerized tomography and/or magnetic resonance imaging.

Treatment

The treatment of vertigo is variable and depends on the associated condition. When the associated condition is known to be self-limiting, no treatment may be necessary. In other cases, a variety of treatment strategies are available.

Medical therapy is symptomatic but should not be a substitute for an adequate effort to establish a diagnosis.

Surgical therapy is highly successful in selected cases of peripheral unilateral vestibular lesions. Surgery is recommended in patients who are incapacitated by their symptoms. Surgical procedures are performed on the inner ear and/or the statoacoustic nerve. Some surgical procedures preserve hearing and vestibular function (e.g., endolymphatic sac procedures) while others preserve hearing but ablate vestibular function in the involved ear (e.g., selective vestibular nerve section). In patients with poor hearing in the involved ear, a labyrinthectomy may be indicated. A labyrinthectomy is destructive to hearing and vestibular functions. The diagnosis and the age of the patient are two of the factors that influence the outcome of surgery. For example, the vertigo of unilateral Ménière's syndrome is controlled in 90% of patients who undergo a vestibular nerve section. Following surgery, younger patients compensate better for the loss of vestibular function.

Vestibular rehabilitation incorporates vestibular, proprioceptive, and visual interactive exercises. In patients who have not compensated for lost vestibular function, vestibular exercises promote the substitution of alternative techniques for the maintaining of posture and spatial orientation. Vestibular rehabilitation may be used in conjunction with other types of therapy.

Summary

Vertigo is the sensation of motion that has not occurred. Usually, it is due to a peripheral vestibular lesion but it may be the result of a variety of other disorders. An etiology of the vertigo should be pursued and established in order to initiate appropriate therapy.

Further reading

Baloh RW (1984): *Dizziness, Hearing Loss and Tinnitus. The Essentials of Neurotology.* Philadelphia: FA Davis

Barber HO, Sharpe JA, eds (1988): *Vestibular Disorders.* Chicago: Yearbook Medical Publishing

Dix MR, Hood JD, eds (1984): *Vertigo.* New York: Wiley 1984

Viscerosensory Functions

György Ádám

The autonomic nervous system, like the somatic, contains special receptors, afferent pathways, and central sensory projections in addition to its important efferent components. For decades, particularly following the work of Gaskell (1916) and Langley (1921), it was believed that the autonomic (visceral or vegetative) nervous system was identical and equivalent to an effector system. This opinion persisted even though Cyon and Ludwig (1866) described more than a century ago the anatomy and the physiology of the depressor nerve originating in the aortic arch, and soon thereafter Hering and Breuer (1868) reported the discovery of the pulmonary receptors initiating respiratory reflexes. Following a long series of morphological, physiological, and even psychophysiological data, it became clear that the visceral afferent system is essential to the function of the autonomic nervous apparatus. It plays an important role in the maintenance of homeostasis, in viscerosomatic regulations, and in influencing animal and human behavior. Langley (1900) initially regarded visceral afferents as an important constituent of autonomic function, dealing with "afferent sympathetic fibers." But it was only in the epoch of new, modern techniques when great emphasis was attached to this sensory apparatus. Electrophysiological multi- and single-fiber recordings started by Adrian (1933) described different kinds of vagal and sympathetic afferent activities; Light- and electron-microscopic examination of autonomic nerves revealed a prominence of afferent fibers in these bundles, and the application of horseradish peroxidase and other tracer substances elucidated the pathways, relays, and projections of the visceral sensory system. In parallel with these morphological and physiological techniques, psychophysiological and behavioral methods had established some important features of nonconscious and conscious perceptive phenomena including signal-detection learning, discrimination, and pain perception.

Visceral receptor structures

The first morphological description of sensory endings in the viscera originates from Dogiel (1878). For the time being three types of endings can be distinguished: (1) unencapsulated simple terminations with poor or no ramifications, (2) unencapsulated arborized endings forming glomeruli, knots, etc., and (3) encapsulated endings of various forms, like the Pacinian corpuscles of the mesentery. Most respiratory and gastrointestinal visceroceptors belong to the first category, most cardiovascular endings to the second and third classes. It is worth mentioning the apparent contradiction between the simplicity of the histological structure and the complexity of the physiological role they play.

Physiologically, visceral receptors are in general classified into three major and two minor groups according to their adequate stimuli. Mechanoreceptors, chemoreceptors, and thermoreceptors constitute the major categories, their lowest threshold (i.e., the highest sensitivity to mechanical, chemical and thermal stimuli) is well established. Osmoreceptors and volume receptors constitute two additional groups, often cited in the literature as special functional endings, but their separation from chemo- or mechanoreception is still a debated issue. It should be mentioned that an increasingly popular conception emphasizes the multimodal character of some visceral receptors; for example, mechanoreceptors of the stomach are also sensitive to chemical substances.

Visceral mechanoreceptors have been divided into slowly and rapidly adapting types. Slowly adapting mechanical receptors have been described in the cardiovascular system, in the respiratory apparatus, in the urinary bladder, and in the kidney. Some of them are called "tension" receptors, most of them being located in the muscular layers of hollow viscera signaling distension and/or contraction. Those located in the serous membrane detect stretching or dislocation of the entire organ. A special class of these receptors are located in the adventitia of big vessels (aorta, carotid artery, etc.) and are extensively branched. They are stimulated by the distension of the arteries, being often called "baroreceptors." Rapidly adapting mechanoreceptors are also distributed in several internal organs; most of them are found in the serosa and in the mucosal layer. Whereas slowly adapting receptors have both myelinated and unmyelinated axons, the rapidly adapting ones include the largest diameter of viscerosensory fibers, for example, the A-beta type of Pacinian corpuscles. However, these rapidly adapting structures were found less frequently in several viscera than the slowly adapting ones. They usually respond to well-defined rapid dislocations of the mucosa of the digestive or respiratory tracts. Seemingly, the diversity of mechanically sensitive visceral receptors is much more abundant than the preceding classification: the blood vessels of the lungs contain J-receptors sensitive to blood-flow changes, "movement" receptors are found in the intestines, and "flow" receptors in the carotid wall.

Visceral chemoreceptors are found classically in several well-known areas, namely near the main arteries (the carotid and the aortic regions), as well as in the digestive tract, in the liver, and in the kidney. The arterial chemoreceptors have been studied the most intensively; they are excited by a decrease in PO_2 and by an increase in PCO_2. Gastrointestinal chemical receptors are widespread and of different modalities; glucoreceptors, alkali and acid receptors, aminoacid receptors, etc. They are located in the mucosa or the submucous layer, responding selectively to their appropriate chemical agent, adapting slowly to steady concentration of the given substance. Liver glucoreceptors, on the contrary, are rapidly adapting structures. Kidney chemoreceptors are sensitive to renal ischemia and most probably to different components of the urine.

Visceral thermoreceptors were first indirectly demonstrated by observing behavioral reactions due to thermal gastrointestinal stimulation (Simanovski, 1881; Neumann, 1906). Later both visceroceptive conditioning and electrophysiological recordings reinforced the early data: fibers are found in both splanchnic and vagus nerves starting from slowly adapting cold and warmth receptors of the digestive tract. They seem to be exclusively unmyelinated axons.

Visceral osmotic receptors in all probability exist in the liver and possibly in the gastrointestinal region. The idea of osmotically sensitive receptor structures was first suggested by Verney (1946) in the hypothalamic area secreting antidiuretic hormones. It is still not clear whether osmosensitive endings constitute a distinct entity of receptors or if they can be included among chemoreceptors.

Volume receptors are presumed to exist owing to the fact that the volume of the body fluids is held constant by a sensitive regulatory system in which receptors of the atrium of the heart may play a prominent role. The existence of such receptors in the skull is also not excluded, but their analogy with blood vessel and interstitial mechanoreceptors cannot be excluded either.

In addition to the aforementioned specific receptor structures, a considerable category of visceral afferent endings consists of nonspecific unmyelinated terminals apparently responding indiscriminately to different classes of stimuli. It is not clear whether this multimodal type of free nerve endings is identical with nociceptors or whether two distinct classes of nonspecific visceral receptors exist: namely nociceptors and nonnociceptive multimodal terminals. Many authors emphasize the fact that the very exposure of a viscus for recording of afferent discharges is per se a noxious stimulus, since normally these internal tissues are protected from mechanical manipulation, dryness, and other concomitants of experimentation. Thus, the dilemma can be solved only by the closer examination of visceral pain.

Visceral afferent pathways and central projections

The primary afferent fibers of viscerosensory pathways are anatomically similar to somatic sensory fibers, having their cell bodies in the spinal (dorsal root) and cranial ganglia. These visceral afferent fibers are usually components of sympathetic and parasympathetic nerves, thus the terms "sympathetic afferent fibers" and "parasympathetic afferent fibers" is common, although some argue that such terms are misleading, since visceral afferents are not truly "sympathetic" or "parasympathetic" but rather parallel to those of other primary afferent systems. The nonselective use of the term "axon" has been extended to both the peripheral and the central ramifications of the bipolar visceral afferent neurons. The centrally directed axon usually has a smaller diameter and is shorter than the peripherally oriented one, regardless of whether the axon is myelinated or nonmyelinated. The somas of these primary afferent neurons are grouped according to the distributions of their axonal bundles. In spinal ganglia the cells are located in peripheral clusters, whereas in the nodose ganglion of the vagus, the sensory neuronal groups are situated more centrally, displaying separate cardiovascular, pulmonary, and gastrointestinal regions. This somatotopic organization seems to be general, that is, it can be followed in the gasserian ganglion too.

Visceroafferent fibers emerge from the major thoracic, abdominal, and pelvic organs (Fig. 1). Some of these afferents join the main autonomic nerves, for example the splanchnic, the vagus, or the pelvic nerves. Recent research has demonstrated a predominance of afferent fibers in parasympathetic nerves and a considerable proportion in sympathetic ones. More than 80% of all fibers in the vagus and 50% of all fibers in the pelvic nerve of the cat are afferent, (corresponding to some 40,000 vagal, and 7500 pelvic, afferents). As far as afferents traveling in sympathetic nerves are concerned, they constitute only about 20% of the total number of fibers (16,000 fibers in the cat). As mentioned previously, the magnitude of the sensory component of autonomic nerves has been considerably underestimated, in part owing to the fact that 90% of them are unmyelinated or lose their myelin sheet in the vicinity of their organ of origin, and are hard to detect by light microscopy.

Visceroafferent fibers associated with the sympathetic system and the lumbosacral parasympathetic system reach

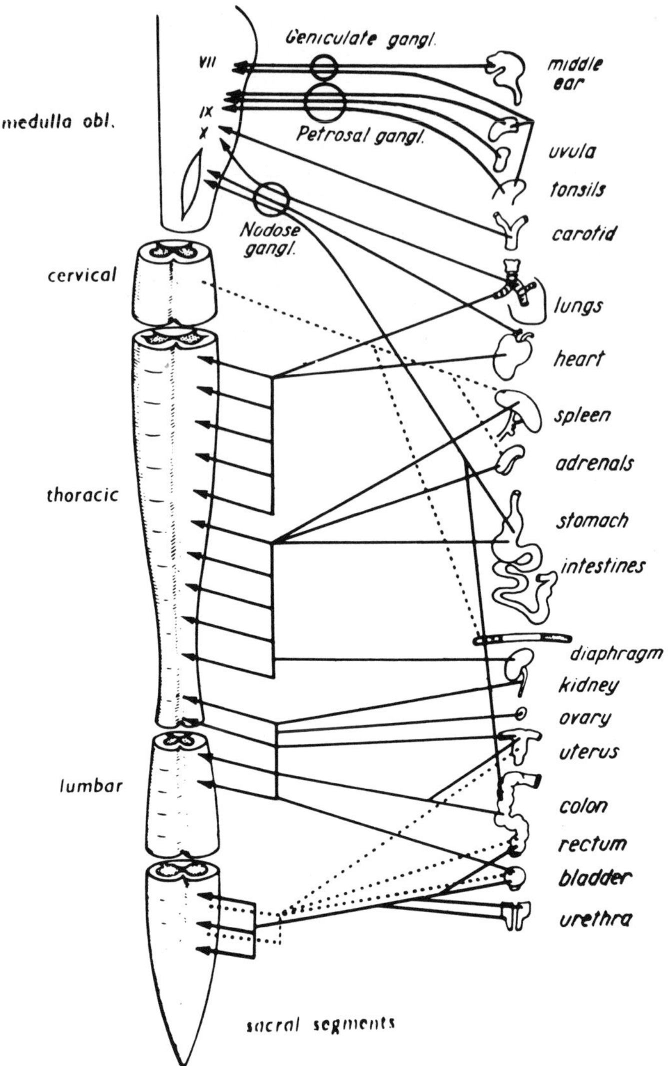

Figure 1. Schematic representation of the entrance of visceral afferent pathways into the medulla and spinal cord. Solid lines represent parasympathetic (input to the medulla and to the sacral segments) and sympathetic (input to the thoracic and lumbar segments) afferent pathways. Dotted lines indicate nerves belonging to the somatic afferent system. (After Ádám, 1967.)

the spinal cord through the dorsal roots; they enter Lissauer's tract, where they may ascend or descend one or two spinal segments. They terminate in lamina I and V of the dorsal horn (but not in the lamina II: substantia gelatinosa). Secondary sensory neurons of these lamina send their axons to the brainstem through the anterolateral tract of the spinal cord, although a minority of these cells may send their fibers through the dorsal (i.e., lemniscal) tract. In parasympathetic cranial nerves afferent fibers run within the same nerve until the region near the entry into the medulla, where they form separate bundles. In the vagus nerve an almost complete separation of sensory fibers can be observed already at the nodose ganglion. However, there are several exceptions to the foregoing general morphological rule. For example, some visceral afferent fibers reach the spinal cord through ventral roots, and others do so through somatic sensory bundles.

The anterolateral system carries most of the viscerosensory messages mainly through its two divisions linking the spinal neurons with different nuclei of the thalamus. The lateral spinothalamic tract ascends to the ventral posterior lateral regions of the thalamus and mediates discriminative aspects of noxious and painless stimuli. The medial spinothalamic tract leads similarly to the same thalamic relay neurons, but, in addition, it ramifies to the midbrain reticular formation and to the intralaminar and medial nuclear complexes of the thalamus. This latter is believed to be involved mainly in the mediation of long-term information about the state of the viscera and of emotional components of viscerosensory impact, including pain.

The central projections of visceral afferent pathways are organized in the same hierarchy along the neuraxis and the hemispheres as the somatosensory representations. The spinal relay neurons have been mentioned previously. As far as the medullar and pontine representations are concerned, they include the nuclei of the facial and the glossopharyngeal nerves, as well as one part of the vagal bulbar projections. All three cranial nerves (VII, IX, and X) carry visceral "parasympathetic" sensory information. The nucleus tractus solitarius (NTS) of the medulla is regarded as the main structure processing viscerosensory input of the lower brainstem. Diencephalic projections include the posterior lateral and the medial nuclei of the thalamus mentioned previously as well as lateral and medial hypothalamic structures. These latter nuclei are regarded as sites of the integration of visceral and somatic behavioral reactions in which viscerosensory input may play a well-defined role. The cerebellar representation of the visceral afferent system has been described too. Projection areas of splanchanic, vagal, and pelvic nerves have been found in the anterior lobe of the cerebellum; however, nothing is known about their functional significance.

The cortical projections of different visceral afferent nerves have been repeatedly confirmed by recording evoked potentials following stimulation of the vagus, the splanchnic, and the pelvic nerves and by removing areas of the cortex using behavioral tests subsequently. Cortical viscerosensory representations include large areas of the sensory motor cortex overlapping with the well-defined primary and secondary somatic projections S1 and S2. Vagal and splanchnic zones have been identified on the orbital, as well as on the cingular, gyri of the limbic system. The former projections may play some role in somatovisceral integration, whereas the latter have a role in the emotional reactions triggered by visceral input.

Role in physiological regulations. The best known viscerosensory activity so far is its role in the maintenance of the constancy of the internal environment of the organism, that is, in serving homeostasis. This many-sided function consists of mechanisms controlling (a) gastrointestinal activity, (b) respiration, (c) cardiovascular stability, (d) endocrine secretion, (e) body temperature, and (f) water and salt balance. Most of the reflexes triggered by visceral afferent messages involve merely the organ(s) from which the sensory signals originate, for example, changes in the frequency and in the amplitude of respiration following the excitation of pulmonary mechanoreceptors traveling to medullar centers through the vagus nerve. Other reflexes of viscerosensory origin affect a whole complexity of organs in addition to the one from which they were elicited. Thus, in addition to single-organ reflexes, visceral afferent impulses evoke multiple-organ reflexes too. For example, the chemoreceptors of the carotid or aortic body excited by hypoxia of the blood transmit messages to the medulla, which initiates a series of reactions of the heart, of the arterioles, of the respiratory muscles, etc. In general, the stability of homeostasis cannot be conceived without the steady signaling activity of the sensitive and well-distributed network of the visceral sensory apparatus.

Viscerosomatic interactions play an important role in the aforementioned homeostatic functions. The behavioral integrated motor activity of the organism serving the constancy of internal environment (e.g., food intake, defecation, vomiting, and voiding) is always triggered partly by visceral afferent signals. Even more general behavioral states, considered extrahomeostatic, like the sleep-wakefulness cycle, sexual behavior or emotional states, are deeply influenced by viscerosensory input.

Visceral pain. The free nerve endings in different sheets of viscera are identified by most authors as nociceptive receptors; however, the existence of free axon terminals detecting innocuous stimuli cannot be excluded. It is not yet decided whether these endings are specifically sensitive to noxious stimuli, or whether they act along an intensity scale, that is, intensive stimuli of any modality initiate responses that are transmitted as pain messages. As a third possibility, summation of weak stimuli cannot be rejected either. The main dilemma is the same in every domain of pain research: it must be decided what is noxious and what is nonnoxious and this decision, based on firm data, must be subsequently applied to visceral function. For the time being, this crucial problem cannot be solved, since visceral nociception displays such diverse phenomena that cannot be incorporated at present in the framework of a consistent theory of pain. For example, occlusion of a coronary artery of the heart is thought to result in referred pain of the chest, sternal region, and/or left arm. Injection of bradykinin into the left atrium imitates these phenomena. It activates receptors of the heart directly; consequently 75% of the spinothalamic cells of the given spinal segment increase their activity in about 10–15 seconds after the injection. It is believed that the spinothalamic neuronal pool is the site of convergence of visceral and somatic inputs, thus constituting the basis for the referred pain that occurs with angina pectoris. But, on the other hand, many coronary occlusions, and even marked myocardial infarctions, pass without any sensation of referred pain.

It was long believed that afferent fibers in the sympathetic pathways were those that were mainly associated with nociception and sensations of pain. Actually, noxious stimuli of the gastrointestinal tract are transmitted mainly by splanchnic afferent fibers. Many visceral afferent signals of pain associated with gastroduodenal ulcers are carried by vagal pathways. Thus, the sympathetic system does not have a monopoly on nociceptive messages, all the more since pain reactions originating from the respiratory apparatus and from the pelvic organs are triggered by afferent fibers traveling with the parasympathetic nerves.

Visceral perception. The term "visceral perception" conventionally signifies the detection and discrimination of nonpainful, innocuous visceral messages. The first description of internal feelings in the framework of our cultural background can be found in Greece. We attribute to Aristotle the concept of "sensorium commune" or common sensations specifying feelings coming from inside the body. It is most likely that oriental cultures (e.g., the Indian) had discovered internal sensations even earlier, judged by the ancient practice of the Yogis, who apparently managed to both perceive and eliminate internal signals. With the advent of modern scientific research it was Ivan Sechenov (1866), who emphasized the "dim feelings," discussing influences coming to the brain from inside the body, and Herbert Spencer (1872), who drew attention to the human internal perceptual environment. Charles Sherrington was the first to demonstrate a generalized extra-homeostatic effect of visceral afferent stimulation (1899), and Konstantin Bykov, an associate of Ivan Pavlov, described the first viscerosensory learned reflexes (1924).

Innocuous visceral perception must be studied by rather sophisticated psychophysical techniques such as conditioning, signal detection approaches, and methods of limits, since except for internal phenomena coupled with states of urgency like hunger, dyspnea, thirst, need to void, or defecation, most visceral events remain unreported or nonverbalized. This latter feature renders a peculiar double-faced nature to visceral sensitivity at least in three aspects: (1) visceral sensitivity serves mainly homeostatic regulatory functions as described previously but at the same time extrahomeostatic traits, influencing (in a covert way) psychological phenomena (e.g., emotions); (2) viscerosensory activities are mostly genetically inborn faculties but, importantly, also demonstrate the plasticity or learning that makes optimal adaptation to the biological and social environment possible in the adult organism; and (3) visceral afferent messages are mostly nonconscious even in states of awareness, but owing to the biological and social needs, a conscious proportion can be observed as well.

The emergence and experimental proof of extrahomeostatic, learned, and conscious viscerosensory phenomena in the recent psychophysiological literature renders an affirmative answer to the substantial question of whether visceral perception can be regarded as a specific class of sensitivity. It is assumed that the infant learns to control micturition and defecation in early childhood by conditioning afferent impulses coming from the mechanoreceptors of the bladder and the rectum following their distension. Apparently, the reinforcement of the social environment pressures the infant to detect and identify the signals arriving from these hollow viscera. The abundance of experimental data on detection and discrimination of gastrointestinal and urogenital nonpainful stimuli via signal detection (forced-choice) paradigms or classical and operant conditioning seem to confirm the possibility of perceiving, articulating, and identifying visceral innocuous phenomena (Fig. 2). Information coming from the internal milieu generally remains below the level of consciousness; the data from the internal milieu have little chance in the normal life of the individual to traverse the long process of recognizing, discriminating, and labeling visceral events. Apparently there is little need for it, except in the few situations outlined previously. But in a relevant situation people can be taught to discriminate and identify these covert events, for example, similarly to cats, which, although poor color detectors, can be conditioned to detect and identify colors.

What theoretical and practical conclusions can be drawn

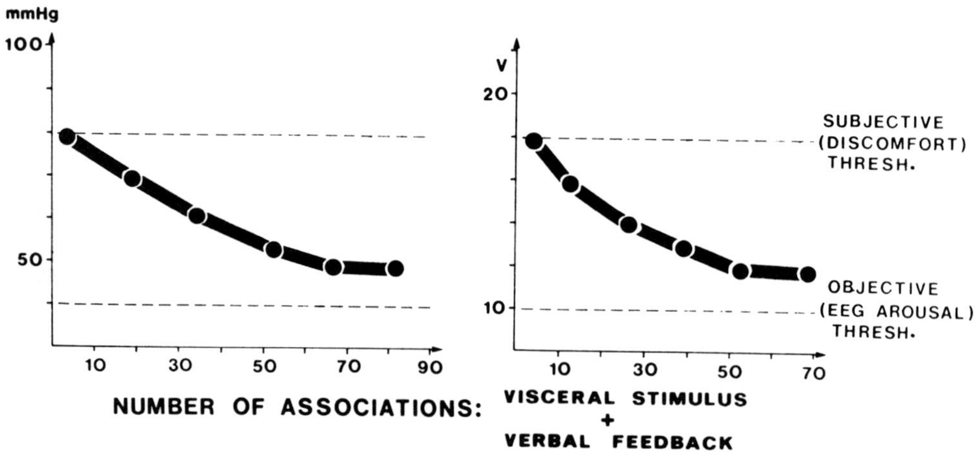

Figure 2. Visceral perception curves in humans. Conscious detection of mild distensions of the duodenum (*left*) and of the weak electrical stimulations of the neck of the uterus (*right*) have been followed in two distinct groups of patients. Learning has been undertaken by using instrumental conditioning paradigms with verbal feedback. The abscissa represents the number of associations (visceral stimulus and verbal reinforcement), the ordinate indicates the intensities of the visceral stimulus: the pression of the duodenal balloon (*left*) and the intensity of the uteral electrical current in volts (*right*). Twenty subjects have been followed up in both one-session experiments. Each subject had to press a button in case of some subjective feeling of pressure or discomfort. Each black dot represents average values of a group of reported intensities. The session started with stimulation at the level of marked feeling of discomfort (subjective threshold). The intensities of visceral stimuli were consequently lowered step by step, associated with a verbal warning of the moment of stimulation. These consecutive verbal reinforcements resulted in a marked decrease of the reported sensations with subjective visceral stimulus intensities causing desynchronization of the EEG (objective threshold), but not detectable by the subjects prior to the conditioning procedure.

from the preceding data? Should experimental psychology and neurophysiology aim at a special "visceral education" to teach people to perceive more and more internal phenomena? Research to elucidate the physiological background and psychological laws of visceral perception must be undertaken to answer such questions.

Further reading

Ádám G (1967): *Interoception and Behaviour.* Budapest: Akadémiai Kiadó

Cervero F, Foreman RD (1990): Sensory innervation of the viscera. In: *Central Regulation of Autonomic Functions,* Loewy AD, Spyer KM, eds. New York, Oxford: Oxford University Press

Cervero F, Morrison JFB, eds (1986): *Visceral Sensation.* Progress in Brain Research, vol. 67. Amsterdam, New York, Oxford: Elsevier

Mei N (1983): Sensory structures in the viscera. In: *Progress in Sensory Physiology,* vol. 4, D Ottoson, ed. Berlin, Heidelberg, New York, Tokyo: Springer-Verlag

Volume (Extrasynaptic) Transmission

Paul Bach-y-Rita

"Volume transmission" is a term proposed by Fuxe and Agnati for electrical and chemical communication in the brain via the extracellular fluid. Although primarily extrasynaptic, it can include activity induced by substances diffusing into synaptic clefts from the extracellular fluid. Volume transmission is a complementary mechanism to classical electrochemical synaptic transmission. In volume transmission, neurotransmitters and other neuroactive substances are released at points that may be remote from the target cells, which they reach by diffusion through the extracellular fluid to exert desynchronized and sustained influences on vast neuronal ensembles.

A neurohumoral transmission mechanism by local diffusion had been proposed by Bach-y-Rita following experimental studies in 1964, as a mechanism of the sustained activity recorded by him in highly convergent (responding to stimulation of several modalities from various parts of the body) brain cells following sensory stimulation. However, further elucidation of the concept could not occur until receptor identification methods were developed. Early studies on regional distribution of enkephalin in the brain pointed out discrepancies, or "mismatches," between opiate peptide distributions and opiate receptors, such as those between dense enkephalin-containing terminals and sparse opiate receptors. These were initially considered to be exceptional instances. However, the evidence recently reviewed by Herkenham suggested to him that, in the brain, mismatches are the rule rather than the exception. In the presence of mismatches, reflecting transmitter release at some point distant from the receptors, volume transmission must be considered as a mechanism of neuronal communication.

Neurons may chemically intercommunicate by the mediation not only of the classical neurotransmitters but also by peptides, and by other neuroactive substances. Herkenham has reviewed the evidence for mismatches for a number of these in the mammalian brain, including peptides, monoamines, and amino acids. Descarries has recently confirmed his earlier findings that the monoamine systems in the brain are highly nonsynaptic. The most nonsynaptic is noradrenaline. Noradrenaline-immunostained profiles from the upper layers of the frontal, parietal, and occipital rat neocortex averaged 0.6 microns in diameter. In contrast to their unlabeled (nonnoradrenaline) counterparts, they rarely showed membrane differentiations characteristic of a synaptic contact (junctional complex). The proportion of cortical noradrenaline varicosities engaged in synaptic contact (synaptic incidence) was 17% or 26%, depending on the use of more or less stringent criteria. The study thus confirmed the initial radioautographic characterization of the cortical noradrenaline innervation as a mostly nonjunctional system.

There is ample intercellular space (approximately 20% of the volume of the living brain) for many kinds of neuroactive substances to diffuse from release points to receptors on or in the neurons of the cerebral cortex; the receptors on the surface of the cell can be contacted by, and bound to, the neurotransmitter and other ligands diffusing in the extracellular fluid. Fuxe and Agnati pointed out the importance of understanding the structure of the extracellular space with its intercellular matrix of polyionic molecules; the structure of the brain cell microenvironment may depend on physical and chemical constraints represented by cellular elements such as astroglia as well as the extracellular matrix. Glial cells constitute a high proportion of the total cell number and have a strategic position with processes reaching both blood vessels and synaptic regions. In primary culture, the astrocytes possess transmitter receptors and receptors for several neuropeptides. The presence of receptor regulation of amino acid neurotransmitter transport over the cell membrane has significant implications, as it suggests that astrocytes may be an integral regulatory component in local control of neuronal activity in the CNS.

Bach-y-Rita has speculated on the role of volume transmission in mediating mass, sustained functions such as sleep and vigilance, and mood. Such a role for volume transmission would be consistent with the parsimonious nature of the brain; the widespread activation of groups of neurons via diffusion should be more energy efficient than synaptic activation of those same neurons. In the example of the noradrenaline system, most of the receptors of this primarily nonjunctional system are located in the distant dendritic tree. In addition to the long time noradrenaline must take to diffuse in the extracellular fluid to the dendritic receptors, the distal location of these ensures that considerable additional time is needed for activation of the soma. Inactivation is also slow in the absence of junctions, with their degradation enzymes and reuptake mechanisms. Thus, massive activation can be produced and sustained over a period of time at a low cost in energy.

The fact that neurotransmitters are found in the extracellular fluid of mammalian brains has led to a series of studies, such as those using microdialysis techniques, to measure changes in concentrations of specific neurotransmitters and amino acids in the extracellular fluid of specific brain regions after certain interventions. Among those found in the extracellular fluid are dopamine, choline, adenosine, noradrenaline, serotonin, and acetylcholine. Iversen in 1979 suggested that the term "nonsynaptic neurotransmitter" is useful: it would be an expansion of the concept of "neurotransmitter," which is usually reserved by synaptic transmission.

When a muscle fiber is denervated, there follows a massive up-regulation of receptors on the entire cell membrane, resulting in hypersensitivity to low concentrations of acetylcholine. Bach-y-Rita has suggested that if up-regulation of specific receptors has occurred following brain damage resulting in partial or total denervation (comparable to the massive up-regulation on the cellular membrane of denervated muscle), those cells may similarly become hypersensitive to specific neuroactive substances, and thus respond to the low concentrations of those substances in the extracellular fluid. Other examples of receptor up-regulation and hypersensitivity, possibly including drug addiction and kindling, may be studied in the context of volume transmission in the future.

An ingenious approach to demonstrating volume transmission has been taken by Arshavsky et al., who impaled single command neurons of the isolated pedal ganglion of a mollusc and stimulated intracellularly through the microelectrode, producing bursts of activity in locomotor cells. They then lifted the command neuron free of the ganglion by elevating the impaling microelectrode. Any synapses present were thus disrupted. They then replaced the impaled neuron in its original site, and again stimulated it intracellularly. In 25%

of the cases (a high percentage, since many cells must be damaged due to the aggressive nature of the procedure), the intracellular stimulation again drove the locomotor neurons, exciting those that had previously been excited, and inhibiting those that had previously been inhibited. Since no direct anatomical connections survived the physical separation of the cell from the ganglion, the information transmission to the locomotor cells had to be by volume transmission. They interpret their findings as supporting the idea that neurotransmitters are released into the extracellular space in sufficient quantity to produce changes in target cells in the absence of synaptic contacts.

The Arshavsky et al. studies demonstrated that volume transmission is a neurotransmission mechanism in invertebrates. Other evidence for extrasynaptic (volume) transmission from molluscs includes studies of the effects produced by the stimulation of neuroendocrine bag cells in mollusc abdominal ganglia: upon excitation, the bag cells released several peptides, including egg-laying hormones, which diffused into the ganglia and evoked various effects on a number of abdominal neurons. A crustacean study has shown that the somatogastric ganglion can generate a variety of motor patterns depending on the exposure to various neuromodulators. The gastric central pattern generator (CPG) driving chewing movements can operate in different modes; switching between these modes is under the control of the pentapeptide proctoline, which can induce endogenous bursting and plateauing in specific target neurons, and can also switch the activity of the motoneurons to either operate synchronously with the rhythm of the gastric CPG or the pyloric CPG. In insects, Hoyle showed that complex electrophysiologic effects are induced by octopamine released nonsynaptically by identified locust neurons.

In the sympathetic and parasympathetic nervous systems, the axon terminals that release norepinephrine or acetylcholine rarely if ever make conventional synaptic contacts with effector tissues; transmitters diffuse for distances of several millimeters to reach their targets, such as smooth muscle. In a series of studies starting in 1969, Fuxe and his collaborators used monoamine fluorescence histochemistry to show that monoaminergic drugs can lead to the appearance of extraneuronal fluorescence surrounding the monoamine nerve cells, probably reflecting diffusion of monoamines in extracellular fluid.

Evidence for volume transmission in the brain has been provided with a transplant model, using intrastriatal adenopituitary transplants. In that model, the prolactin secretion from the adenopituitary transplant could be modulated by endogenous dopamine concentrations in the extracellular fluid space. The D2 receptors of the prolactin cells in the adenopituitary transplant were reached by endogenous dopamine diffusing from the surrounding neuropil into the transplant. However, this evidence for volume transmission in a brain transplant model does not directly demonstrate that it is a mechanism for neuronal communication in the normal brain.

The synaptic basis of information flow has been the major focus of neuroscience research for most of this century. In view of the evidence for volume transmission, interneuronal communication based on transmitter-receptor interactions not requiring preservation of impulse frequency coding enabled by tight physical coupling at the synaptic junction must now be considered to play a role in neuronal communication. Present evidence suggests that volume transmission is a mechanism of electrochemical communication that is slower in both activation and deactivation than synaptically driven communication. While convincing evidence for volume transmission as a normal mechanism of information transmission exists in invertebrates, to date the evidence in mammalian brains is indirect and suggestive. It may be an efficient mechanism for mass, sustained functions, but there is insufficient direct evidence to draw conclusions at present.

Acknowledgment. The helpful comments of an anonymous reviewer are greatly appreciated. She or he also noted that the concept of extrasynaptic or volume transmission is revolutionary and if proven will require major shifts in thinking and research.

Further reading

Bach-y-Rita P (1990): Receptor plasticity and volume transmission in the brain: Emerging concepts with relevance to neurologic rehabilitation. *J Neurol Rehabil* 4:121–128

Dismukes RK (1979): New concepts of molecular communication among neurons. *Behav Br Sci* 2:409–448

Fuxe K, Agnati LF, eds (1991): *Volume Transmission in the Brain*. New York: Raven

Schmitt FO (1984): Molecular regulators of brain function: A new view. *Neurosci* 13:991–1001

Wilson's Disease

Dominique Muller and Joseph A. Ghika

In 1912, Kinnier Wilson described a disease characterized by various neurological manifestations, such as tremor, rigidity, spasmodic contractions, emotionalism, and mental symptoms, which he concluded were related to lesions of the lenticular nucleus. Four years later, Bramwell reported the case of seven children, four of whom died of liver failure, and proposed that the neurological manifestations of the disease could be preceded by liver dysfunctions. The disease is now referred to as Wilson's disease or hepatolenticular degeneration, and it is known to be a disorder of copper metabolism. It is inherited as an autosomal recessive trait and results from the expression of a locus on chromosome 13. The prevalence is low, about 1 case in over 200,000, and only individuals homozygous for the abnormal gene suffer from clinical manifestations.

Pathogenesis

One of the major alterations observed in this disease is an accumulation of copper in the body tissues.

Copper, which is a component of an ordinary diet (about 2 mg/day), is found essentially in muscles, bones, and liver, where it is used as a cofactor of different mitochondrial, cytoplasmic, or nuclear enzymes (cytochrome oxydase, superoxide dismutase, tyrosinase). In the brain, the highest concentrations of copper are found in noradrenergic neurons of the locus coeruleus, where copper is associated to dopamine-beta-hydroxylase, the enzyme that catalyzes the conversion of dopamine to noradrenaline. Over 90% of copper is transported in the blood associated to ceruloplasmin, a 151 Kd protein, that binds up to eight atoms per molecule. For its excretion, copper is removed from the blood by hepatic uptake and then partially excreted in the bile.

In Wilson's disease, hepatic copper accumulation starts very early during the first years of life. This process begins first asymptomatically until hepatic copper binding sites are saturated. Then hepatic insufficiency and hepatocellular necrosis may develop and copper accumulates in other tissues. The most affected organs are the brain, erythrocytes, and kidneys. Copper deposition also usually occurs in the Descemet's membrane of the cornea, forming a golden-brown pigment ring known as the Kayser-Fleischer ring. This constitutes one of the most pathognomonic signs of the disease.

The clinical manifestations of Wilson's disease seem to be directly related to a toxic effect of copper in the different regions where it accumulates. If copper is removed, this results in a recovery from both the hepatic and neurological signs. The mechanisms of copper toxicity are, however, not yet clearly understood. In the brain, lesions predominate in the basal ganglia, although they may also be seen in the subthalamic nucleus, thalamus, claustrum, myelinated fiber bundles, or even brain stem and cerebral cortex. An enlargement of ventricles is usual as well as an atrophy of cerebral cortex. Histological findings include an increase in the size and number of protoplasmic astrocytes, many of which become multinucleated, and a degeneration of neurons and glial cells, a process which may lead to cavitations. Cell death is not restricted to a specific type of neurons, but it is more important in basal ganglia than in cerebral cortex. Vascular changes may also be seen, consisting of capillary swelling and proliferation. These lesions are not specific for Wilson's disease and may also be seen in cases of hepatic encephalopathy. Thus, they probably reflect general suffering of neurons, and their distribution might be related to regional differences in vulnerability. One possibility that has been proposed to account for copper toxicity is that copper might inhibit enzymes involved in glycolysis. At the concentrations found in the brain of patients suffering from Wilson's disease, copper significantly reduces the production of lactate from glucose and inhibits hexokinase and pyruvate kinase.

Another aspect for which there is yet no definitive answer concerns the reasons for copper accumulation in patients suffering from Wilson's disease. A usual finding in these patients is a reduced level in serum ceruloplasmin. It has thus been proposed that the genetic defect might involve a difficulty in synthesizing this protein. It should be noted, however, that some homozygous patients do have normal levels of ceruloplasmin, whereas asymptomatic obligate carriers may exhibit reduced levels of the protein. Another proposition has thus been that copper accumulation might result from the synthesis of an abnormal protein implicated in the metabolism of copper by hepatocytes. This protein might have an increased affinity for ionic copper and this might lead to a reduced excretion of copper in the bile.

Clinical picture

Wilson's disease generally starts during the second decade (between 6 and 20 years) as a liver disease (50–60%) or with neurological symptoms (30%) and sometimes both of them. Exceptionally, patients may present with other systemic manifestations such as osteoarticular (10%; osteoarthropathy, osteomalacia), cardiac (1–2%; cardiomyopathy), ophthalmological (2%), or hematological (2%; hemolytic anemia) symptoms. Isolated cases starting with endocrinological (hypoparathyroidism, pancreatitis) or renal (Fanconi syndrome) symptoms have been reported.

The liver dysfunction in patients with Wilson's disease is either an asymptomatic high level of transaminases or a chronic, recurrent hepatitis with icterus, nausea, vomiting, and asthenia, which may rapidly progress to cirrhosis with hepatic failure and portal hypertension. At this stage of the disease, copper deposition will occur in extrahepatic organs.

Neurological symptoms are essentially an atypical corticobulbar syndrome with important dysarthria, potentially dangerous dysphagia, hoarseness of the voice, drooling, emotional lability, and a strange "vacuous smile." Other symptoms include slowness of distal motility, corticospinal signs, parkinsonism (bradykinesia, rigidity, posture, and gait abnormalities), tremor, or involuntary movements (choreoathetosis, dystonia, or exceptionally myoclonus). Personality changes are very frequent—in 10–15% of the cases they precede the neurological symptoms—and are essentially personality or behavioral changes (aggressivity, asocial behavior, depression, or psychosis). Dementia is exceptional. Seizures or supranuclear oculomotor disturbances (slow saccades, gaze distractibility, apraxia of lid opening, recurrent ophthalmoplegia, accommodation defects) are sometimes found. Brown-yellow Kayser-Fleischer rings at the corneal limbus are present in almost 100% of the cases with neurological symptoms. Sometimes, however, they are only unilateral or can only be seen with a slit lamp.

The diagnosis is confirmed by decreased serum ceruloplasmin (<20 mg/dL) and copper levels (<80 μg/dL), high 24-hour copper urine excretion (>100 μg/24 h), and high transaminases levels. Glucosuria, aminoaciduria, phosphaturia, and bicarbonaturia (Fanconi syndrome) may also be present. High levels of copper in hepatic biopsy can be found. Hypodensities or low signals in the basal ganglia are shown on, respectively, computed tomography (CT) and magnetic resonance imaging (MRI) scans.

Treatment and perspective

The major objective of the treatment is to reduce the accumulation of copper in liver and other tissues. This may be achieved in two different ways: by using chelating agents, such as D-penicillamine or trientine, which remove copper from tissues; or by inhibiting the absorption of copper from the intestinal tract using potassium sulfide, zinc salts, tetrathiomolybdate, or low copper diets (no liver, cacao, chocolate, nuts, mushrooms, shellfish). The first method is the most commonly used. Treatment usually results in a significant improvement of both hepatic and neurological manifestations. Occasionally recovery of organic lesions may be observed. Kayser-Fleisher rings as well as hypodensities or hyposignals in the basal ganglia will eventually disappear on CT and MRI scans. Cases of successful pregnancies under treatment have been reported. Techniques such as magnetic resonance imaging, positron emission tomography, or recording of sensory-evoked potentials are very helpful not only for the diagnosis of the disease but also to assess the extent of structural alterations of the brain and the efficacy of treatment. Another interesting aspect concerns results obtained with positron emission tomography using 2-deoxy-2-fluoro-D-glucose or F-6-fluorodopa, which may show metabolic dysfunctions before lesions are seen on CT or MRI scans. This will certainly help clarify the sometimes poor correlations observed between clinical pictures and structural alterations.

The major source of progress in the coming years, however, is expected to come from experiments in molecular biology. Precise identification of the gene responsible for the disease (the esterase locus of chromosome 13) and of the protein that it codes for will obviously constitute a decisive step in opening new perspectives for the understanding and treatment of this disorder.

Further reading

Grimm G, Prayer L, Oder W, Ferenci P, Madl C, Knoflach P, Schneider B, Imhof H, Gangl A (1991): Comparison of functional and structural brain disturbances in Wilson's disease. *Neurol* 41:272–276

Hawkins RA, Mazziotta JC, Phelps ME (1987): Wilson's disease studied with FDG and positron emission tomography. *Neurol* 37:1707–1710

Patten BM (1988): Wilson's disease. In: *Parkinson's Disease and Movement Disorders,* Jankovic J, Tolosa E, eds. Baltimore and Munich: Urban and Schwartzenberg

Walshe JM (1988): Wilson's disease: Yesterday, today, and tomorrow. *Movement Disorders* 3:10–29

Zebrins: Compartment Markers in the Cerebellum

Richard Hawkes

The simple, uniform cytology of the mammalian cerebellar cortex conceals an elaborate underlying complexity. Rather than being homogeneous, the cerebellar cortex consists of a family of parasagittal tissue compartments, each with specific patterns of afferent and efferent connections. The fundamental parasagittal organization is reflected in the pattern of expression of numerous molecules, including various antigens, enzymes, receptors, and neurotransmitters. A partial list is shown in Table 1. Furthermore, several genetic probes have been described, such as those for the *L7* and *11p15* genes, that reveal sagittal bands of expression in young animals, and finally, the pattern of cell elimination in certain murine neurological mutants such as *nervous* also appears to be organized according to a system of parasagittal compartments.

The clearest molecular markers of compartmentation are two polypeptide antigens, *zebrin I* (apparent molecular weight 120 kD) and *zebrin II* (apparent molecular weight 36 kD), identified in rodents by means of monoclonal antibodies. Both zebrins are restricted to the same subset of Purkinje cells. In both cases, the *anti-zebrin* immunoreactivity is intracellular, and spread throughout the cell, including the somata, dendrites, axons and axon collaterals. The function of neither antigen is known.

Zebrin$^+$ Purkinje cells are clustered to form an array of parasagittal bands, separated by similar zebrin$^-$ bands (Fig. 1). The characteristic striped pattern is responsible for the name "zebrins." This can be seen more easily in Figure 2, which shows a drawing of the posterior lobe of the cerebellum. A more schematic view of the compartmentation, based upon the unfolded cerebellar cortex, is provided in Figure 3.

In the vermis, there are three zebrin$^+$ compartments, P1$^+$, P2$^+$ and P3$^+$, on each side of the midline (the single P1$^+$ midline compartment is actually two abutting bands with no intervening zebrin$^-$ cells). At the interface of vermis and hemisphere there is the P4$^+$. The only exceptions to this arrangement are in parts of lobules VI/VII and lobule X, where all Purkinje cells express zebrin and compartment interfaces cannot be distinguished. In the hemisphere, the arrangement is more complicated: by one schema, there are two minor zebrin$^+$ bands, P4b$^+$ and P5a$^+$, and three more lateral larger bands, P5b$^+$, P6$^+$ and P7$^+$; an alternative view is that rather than parasagittal bands, there is a mosaic of patches (e.g., see Fig. 2). Whatever the case, the cerebellar pattern is highly reproducible between individuals in the number of compartments, in their positions, and in their size.

Cerebellar compartmentation is not a parochial feature of rodents. By using *anti-zebrin II* immunoreactivity as a probe, a similar Purkinje cell compartment set has been identified in a large range of species, including monkey, opossum, cat, shrew, bat, chicken, and pigeon. Thus, it appears that the cerebellum enlarges during evolution by increasing the number of Purkinje cells per compartment, rather than by the addition of new compartments. The human cerebellum contains both zebrin$^+$ and zebrin$^-$ Purkinje cells, but their distribution has not yet been analyzed. Curiously, in lower vertebrates zebrin II has never been detected in amphibians or reptiles, but is present in many fish, and in several species of weakly electric teleosts both zebrin II$^+$ and zebrin II$^-$ compartments are present (although not as alternating bands).

Given the numerous molecular markers of cerebellar com-

Table 1. A Partial List of Markers of Chemical Compartmentation in the Cerebellar Cortex

Marker	Species	Relation to Zebrin I	Cell Type
5′-nucleotidase	Mouse	Corresponds[a]	Bergmann glia
Cytochrome oxidase	Rat, primate	Corresponds (rat) Opposite[b] (primate)	? All cells
CRF	Opossum, primate	?	Climbing fibers
Enkephalin	Opossum	?	Mossy fibers +
Acetylcholinesterase	Cat, rat	Corresponds	? Mossy fibers +
Zebrin I	Rodent, cat, primate		Purkinje cells
Zebrin II	Many	Corresponds	Purkinje cells
Motilin	Rat, mouse	?	Purkinje cells
Antigen B30	Mouse	Corresponds	Purkinje cells
NGF receptor	Rat	Corresponds	Purkinje cells
Antigen B1	Rat, cat, primate	Corresponds	Purkinje cells, etc.
N-CAM (embryonic)	Rat, mouse	Corresponds	Purkinje cells, etc.
CSADCase	Rat, etc.	?	Various neurons
Taurine	Rat	?	Purkinje cells
Synaptophysin	Mouse	?	Various neurons
P-Path	Mouse, rat	Opposite/overlapping	? Purkinje cells

[a]"Corresponds" to zebrin I means that codistribution at the compartment level has been demonstrated experimentally.
[b]"Opposite" means expressed preferentially in the zebrin$^-$ compartments.
Abbreviations: CRF, corticotropin-releasing factor; NGF, nerve growth factor; N-CAM, neuronal cell adhesion molecule; CSADCase, the taurine-synthesizing enzyme cysteine sulfinic acid decarboxylase.

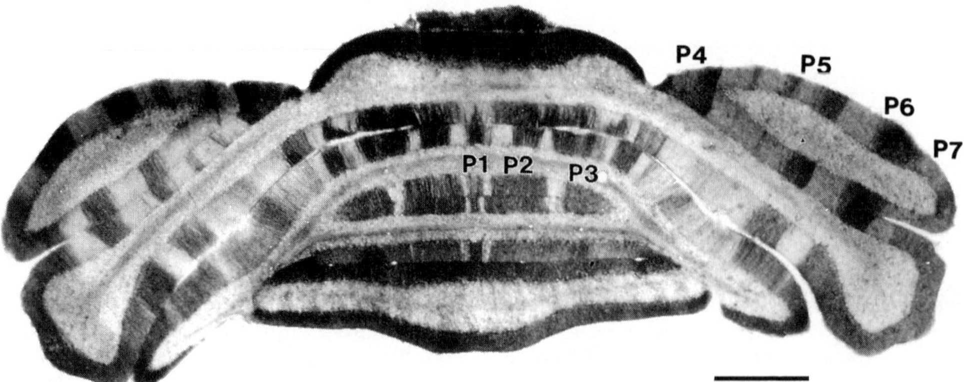

Figure 1. A 50 μm frontal section through the posterior lobe of the adult mouse cerebellar cortex immunoperoxidase stained for zebrin II, to reveal the intrinsic compartmentation. Only a subset of Purkinje cells is immunoreactive. The P1$^+$–P7$^+$ compartments are labeled. Scale bar = 1 mm.

partmentation, it is imperative to know whether each has its own peculiar distribution or whether many follow a common ground plan. When comparisons have been made by using zebrin-immunoreactivity as a reproducible reference frame, many markers have been found to codistribute. Thus, zebrin I$^+$ compartments also express high levels of zebrin II, 5′-nucleotidase, the antigens B1 and B30, and the nerve growth factor receptor. Positive markers of the zebrin$^-$ Purkinje cells are much rarer. Patches of higher cytochrome oxidase activity coincide with the zebrin$^-$ compartments in primates, but in rats the opposite is the case and zebrin I and cytochrome oxidases are coexpressed. The only clear example of a *positive* marker of the P$^-$ bands is the monoclonal antibody *P-Path,* which recognizes a family of gangliosides expressed in all zebrin$^-$ compartments. However, here the situation is also complicated by the observation that a few bands—for example, the P3$^+$—are double labeled by both *anti-zebrin* and *P-Path*. This finding of overlapping topographies may presage a new level of complexity in cerebellar organization.

The presence of so many, apparently independent, markers of compartmentation in the cerebellar cortex suggests that a similar ground plan may also apply to the topography of cerebellar connections. Many studies have revealed a general parasagittal organization to cerebellar connections. For example, when anterograde tract tracing studies have been combined with zebrin immunocytochemistry, a clear correlation has emerged between the organization of cerebellar afferent terminal fields and zebrin compartments. The mammalian cerebellum has two primary classes of afferent input, climbing fibers from the inferior olivary complex, and mossy fibers from multiple sources. When small tracer injections are placed in the inferior olive, a banded pattern of climbing fiber terminal fields can be revealed in the contralateral cerebellar cortex. In many cases, these bands of terminals align precisely with the boundaries between Purkinje cell compartments. Thus, there seems to be a straightforward relationship between cerebellar compartments and climbing fiber innervation. A similar relationship can also be demonstrated for some mossy fiber pathways—for example, the spinocerebellar tract—but the details are rather more complex, with some zebrin compartment boundaries clearly respected, others apparently ignored, and still other boundaries respected by the mossy fiber terminals that have no

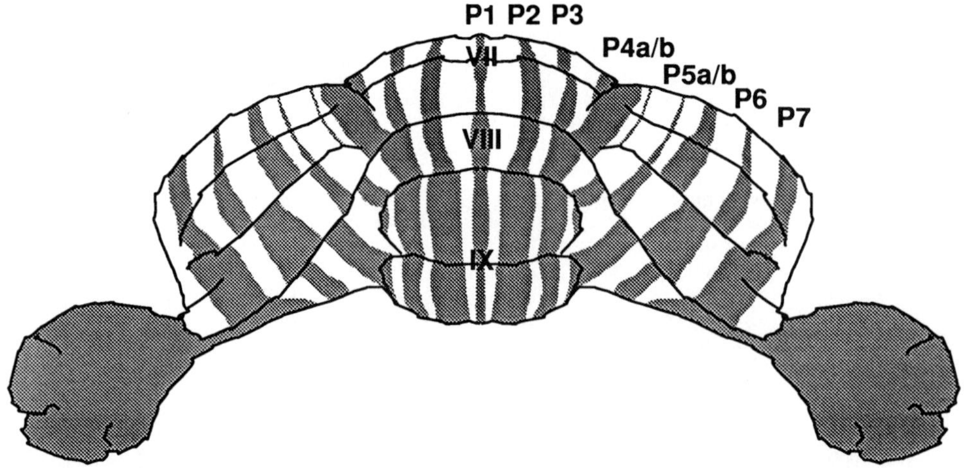

Figure 2. A drawing of the zebrin compartments in the posterior lobe of the adult rat cerebellum as seen in whole mount. The zebrin-immunoreactive compartments P1$^+$–P7$^+$ are labeled, as are lobules VII, VIII, and IX. (Based upon Hawkes and Leclerc (1987): *J. Comp. Neurol.* 256:29–41.)

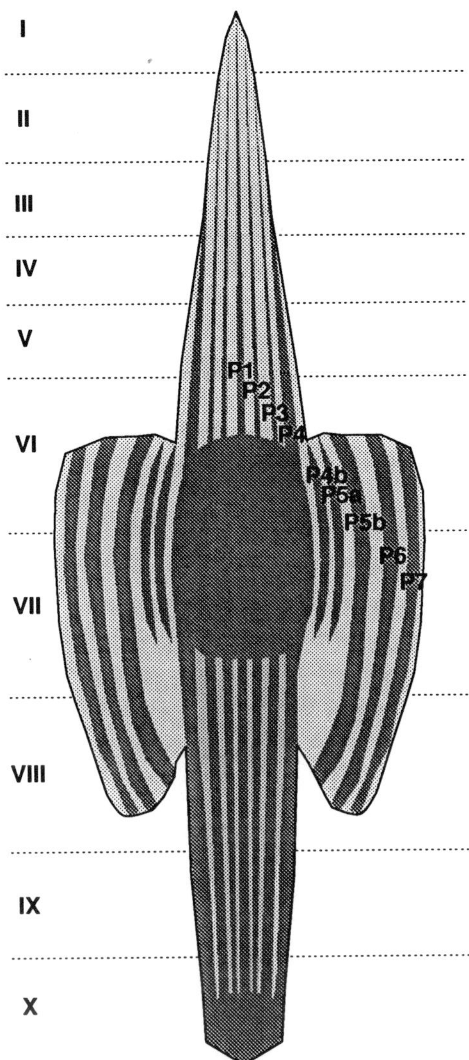

Figure 3. A generalized schematic view of the unfolded adult mammalian cerebellar cortex to illustrate the parasagittal band organization. This is based upon the mouse cerebellum but would apply to many other mammals with only minor modifications. The zebrin-immunoreactive compartments P1$^+$–P7$^+$ are labeled, and more darkly shaded. The P$^-$ compartments are named according to the immediately lateral P$^+$. The position of each lobule is indicated roughly at the left.

The most studied species is the rat, although similar events occur in mice, opossums, and certain fish, and so are probably quite general. Up until 5 days postnatal (P5) there is no zebrin immunoreactivity in the cerebellum. This is important because it immediately precludes any role for zebrins in the *establishment* of compartmentation; for example, early segregation of Purkinje cells into compartments can be revealed as early as embryonic day E15, and afferent axon compartments are already present at birth. At P5–P6, zebrins first appear in a small population of Purkinje cells of the posterior lobe vermis. There is a sharp developmental boundary along the dorsal face of lobule VIII, such that all Purkinje cells located anteriorly are zebrin$^-$ and all located posteriorly are zebrin$^+$. At this stage, some differential expression of zebrin I and II can be seen within lobules VIII–X, but its relationship to the adult band display is obscure. At P7, zebrin expression spreads first throughout the rest of the vermis, and then to include the hemispheres. By P12–P15 all the Purkinje cells are immunoreactive, including those destined to be zebrin$^-$ in the adult. The mature pattern of alternating zebrin$^+$/zebrin$^-$ compartments is sculpted from P15 onwards by the suppression of zebrin expression in the Purkinje cells of the P$^-$ compartments, and the cerebellar cortex appears mature for zebrin expression from about P25 onward.

Even though differential zebrin expression does not develop until a month after the terminal Purkinje cell division, a number of experiments have clearly shown that the adult zebrin phenotype is a cell-autonomous property, fixed at the time of cell birth and not determined later by extracerebellar factors such as the pattern of afferent input. For example, if the cerebellar *anlage* is removed from an embryo at E12–E15 (i.e., during the period when Purkinje cells are generated, and prior to their migration and contact by afferents), grafted ectopically into an adult host, and allowed to mature, Purkinje cells of both zebrin phenotypes are generated. Similarly, ectopic zebrin$^+$ Purkinje cells are occasionally seen stranded in zebrin$^-$ compartments, again suggesting that the phenotype is not a function of the cell's environment. Finally, a wide range of afferent lesioning experiments, both in the adult rat and the newborn, have conspicuously failed to alter the mature pattern of zebrin expression.

What is the significance of compartmentation? One reason might be embryological—to achieve the precise correlation of afferent and efferent projections. By having alternating clusters of different chemospecific cells the target presents a set of sharply differentiated internal boundaries to the afferent growth cones. For example, once pathway guidance and axon fasciculation have brought the climbing fiber to approximately the correct position, precise terminal fields are identified thanks to the steplike borders.

In the adult, one consequence of the projection of complex afferent terminal fields onto intrinsic Purkinje cell subsets is that the parasagittal bands may become functionally subdivided into a complex mosaic of hundreds of cerebellar patches—each averaging around 300 Purkinje cells—with different combinations of afferent inputs, different chemical signatures, and different efferent projections. The subdivision of the cerebellum into hundreds of parallel pathways may serve a functional role in the encoding and planning of motor controls. Multiple simultaneous representations of afferent information in different combinations is a prerequisite for the parallel processing of motor information, where the timing constraints are so severe that to evaluate and integrate afferent input serially would be unrealistic. In addition, this kind of modular segregation of afferents also ensures

zebrin equivalent. Furthermore, shifts in the mossy fiber terminal fields with respect to Purkinje cell compartments may lead to the subdivision of parasagittal bands into multiple independent patches. Despite the complexity, the general conclusion for both climbing and mossy fibers is that the afferent organization is reproducibly maintained in register with the intrinsic compartmentation of the cerebellar cortex. This is supported by electrophysiological studies of tactile receptive fields in the rat cerebellum, both in crus II of the hemisphere and in lobule IX of the vermis, where a strong correlation between functional patches and zebrin compartments has been demonstrated.

One of the most interesting features of cerebellar compartmentation is the way in which zebrin expression is regulated during development. Zebrin I and II have identical developmental timetables and may be considered together.

that minor inputs to the cerebellum can be independently evaluated. In a homogeneous structure an input that represents, say, only 1% of the total would be swamped by the other inputs. By segregating this information to a discrete module, it becomes possible to evaluate its contribution separately. Likewise, by segregating modules of afferent inputs and associated interneurons, the possibility is opened of a higher degree of local structural and functional specialization than could occur in a larger, more homogeneous structure. For example, epigenetic adaptations within a module, such as the modulation of interneuron connectivity, could act to optimize the processing of specific combinations of inputs.

Further reading

Hawkes R, Gravel C (1991): The modular cerebellum. *Prog Neurobiol* 36:309–327

Hawkes R, Brochu G, Doré L, Gravel C, Leclerc N (1992): Zebrins: Molecular markers of compartmentation in the cerebellum. In: Sotelo C, Llinás R, eds. *The Cerebellum Revisited*. NY: Springer-Verlag

Zoster and Postherpetic Neuralgia

Anne A. Gershon

Varicella-zoster virus (VZV) is one of the seven recognized herpesviruses that affect humans. The herpesviruses are well known for their ability to cause latent infection with the potential for subsequent reactivation, following primary illness. VZV has a poorly understood predilection for infection of both the skin and the nervous system. VZV infection is first manifested as varicella (chicken pox), a common childhood disease usually characterized by a generalized vesicular rash and mild to moderate constitutional symptoms. In the United States, virtually every adult over the age of 50 can be expected to have had varicella, and most cases occur before the age of 10 years. The virus gains access to sensory nerve endings owing to its presence in the skin during varicella; latency in the nervous system is then established after the virus reaches sensory (dorsal root) ganglia, presumably by retrograde axonal transport. During latency only a few genes of VZV are demonstrable in ganglia; if reactivation of latent virus occurs, the whole repertoire of VZV genes is expressed and infectious VZV is again produced, resulting in the clinical illness, zoster (shingles). The intricate interrelationship between VZV and the human host is probably the result of a delicate balance between the immune system of the host and the virus. Changes in the immune system, whether due to age, disease, chemotherapy, or radiotherapy potentially hamper the balance to create a situation in which the latent virus reactivates and again causes disease.

Zoster

Clinically, zoster is an illness characterized by a unilateral vesicular skin rash with or without associated constitutional symptoms. The neuropathology of zoster is poorly understood, but presumably, VZV reactivates and travels down the sensory nerve by anterograde transport to reach the skin. About 10–15% of the population may be expected to develop zoster during their lifetime. The thoracic and lumbar areas are most commonly involved, but infection of the ophthalmic branch of the trigeminal nerve is also common and can result in ophthalmologic complications. Commonly, the first symptom of zoster is pain, presumably due to sensory neural infection. The pain is usually followed within a few days by development of the striking vesicular rash on one side of the body. The acute pain of zoster may be the most troublesome aspect of this disease. It is also not unusual to develop a few skin lesions outside the localized area of rash; immunocompromised patients are predisposed to develop greater numbers of such lesions, resulting in a syndrome termed disseminated zoster.

Persons with zoster are contagious to others in that individuals who have not previously had varicella may develop chicken pox 2 to 3 weeks after exposure to a zoster patient. A newly described clinical phenomenon is that of chronic zoster, seen in immunocompromised persons, especially those with human immunodeficiency virus infection, in which scattered new vesicular skin lesions continue to develop over long periods of time, some of which may persist for months and become hyperkeratotic, resembling warts.

It is known that involvement of the central nervous system is common in zoster, although there may be no symptoms possibly other than headache; the cerebrospinal fluid is often temporarily abnormal. Rarely, the internal carotid artery may become inflamed, with the development of paralysis on the side opposite the skin rash. This complication of zoster is termed granulomatous angitis, and whether it represents viral invasion or an inflammatory reaction to VZV or both is unknown.

Postherpetic neuralgia. In addition to the pain associated with acute zoster, from 25 to 50% of persons over the age of 50 years may develop protracted pain, termed postherpetic neuralgia, about 1 month after the onset of rash. This pain may persist for months to years after the skin vesicles have ceased to appear and may be so severe as to interfere with daily life and function. It is often associated with scarring of the skin although vesicles are no longer present. There are elements of both anesthesia and hyperesthesia in the affected skin, and some patients may have difficulty wearing clothing covering the affected area. Two types of pain, one steady and aching or boring, and another described as paroxysmal jabbing, have been described and may coexist. Pathologic findings consist of varying degrees of nerve degeneration, but the cause of postherpetic neuralgia is unknown. Possible hypotheses to explain the phenomenon are continuing low-level multiplication of virus in the ganglion or that nerve damage with pain are the result of attempts at repair. Both the incidence and duration of postherpetic neuralgia are directly related to age. In one study, 50% of persons with zoster over age 60 and 75% of those with zoster over age 75 developed postherpetic neuralgia. Eventually, two thirds of patients will experience significant reduction of pain.

Immunity to varicella-zoster virus

Immunity to VZV is very complex and may be incomplete in perhaps 15% of the population, based on the rate of zoster and rare second attacks of varicella. Antibodies play a role in prevention of second attacks of varicella, acting in concert with various forms of cell-mediated immunity (CMI). That CMI is of great importance in recovery from VZV-induced disease is illustrated by the observation that children with agammaglobulinemia recover successfully from varicella, while those deficient in CMI are at risk for fatal infection. Antibodies are not sufficient for prevention of zoster, as zoster occurs in patients who have antibodies to VZV. Second attacks of varicella and zoster are unusual but may occur. CMI against VZV is composed of a variety of reactions mediated by various white blood cells, including cytotoxic lymphocytes and antibody-dependent cellular cytotoxicity. A specific decline in CMI to VZV in aging has been demonstrated, and probably plays a role in the increasing incidence of zoster with age. Cytokine production, such as the T cell growth factor interleukin-2, and interferon, and decreases in macrophage function have also been reported in the elderly. Presumably these play a role in development of zoster in the elderly; why the incidence of postherpetic neuralgia is increased in the elderly, however, is not understood.

Treatment of VZV infection

Both varicella and zoster are usually self-limited in the normal host, and both varicella and zoster may result in severe morbidity and even mortality in immunocompromised individuals. Passive immunization can be accomplished in immunocompromised patients who have not previously had chicken pox with varicella-zoster immune globulin (VZIG), given within 3 days of a close exposure to a person with varicella or zoster. VZIG is of no use to prevent zoster, nor is it useful to treat either varicella or zoster.

The most frequently used specific antiviral therapy for VZV infection, used mainly in immunocompromised patients, is the drug acyclovir (ACV), which selectively inhibits formation of viral DNA. ACV has minimal side effects, and it may be administered by mouth as well as intravenously. It is, however, poorly absorbed orally and it must be administered in large doses by this route. Orally administered ACV has a small beneficial effect on the course of varicella in healthy children when administered within 24 hours of onset of rash. Similarly, if administered promptly to patients with zoster, there is more rapid recovery from this disease. Unfortunately, however, there is no clear effect on the incidence of postherpetic neuralgia, although the pain of acute zoster responds favorably to ACV. In general, ACV is administered intravenously to immunocompromised patients for potentially life-threatening infections and orally to those patients who are less ill and in whom a decrease in morbidity rather than mortality is sought. Of some concern is that widespread, indiscriminate use of ACV may result in increased resistance of VZV to this drug.

A variety of treatments have been proposed for postherpetic neuralgia, all of which remain at least somewhat controversial. These include antidepressants, neuroleptics, anticonvulsants, steroids, various forms of nerve block, and neurosurgical procedures. Clearly there is much to be learned about the pathogenesis and treatment of this disease.

Varicella vaccine

Active immunization can be accomplished with live attenuated varicella vaccine, now licensed in Japan, Korea, and some European countries, for use in healthy and immunocompromised children. In the United States, the vaccine is not licensed, although the Food and Drug Administration now permits its use on a compassionate basis for selected children. Although varicella vaccine appears to have been developed mainly to protect immunocompromised children against severe varicella, it has become more and more accepted that healthy children have much to gain from immunization against chicken pox. The best responses to the vaccine have been observed in healthy children, with a seroconversion rate of about 95%, after one dose. Adverse effects, such as rash, are mild and rare in healthy children.

Varicella vaccine is of special interest because it appears that vaccinated individuals have a lower chance of developing zoster than those who have had natural varicella. For example, zoster occurs in about 15% of leukemic children who have had natural varicella, but in only 2% of vaccinated leukemic children. Leukemic vaccines may be viewed as a sentinel population in whom the incidence of zoster after immunization of healthy populations can be predicted. Whether the incidence of zoster will also be lower in elderly individuals who were vaccinated against chicken pox as children will only be known many years from now, but it seems likely that what is true for leukemic children at high risk to develop zoster will be true for healthy children (many of whom will become elderly adults) as well. It is hypothesized that the vaccine has less propensity to cause zoster not only because the vaccine virus is attenuated and less virulent than the natural (wild type) virus, but also because there is less opportunity for the vaccine virus to reach the ganglia via the skin after vaccination than after the natural chicken pox. Thus, varicella vaccine may eventually play a role in controlling development of zoster and therefore postherpetic neuralgia as well.

Further reading

Gershon AA, LaRussa P, Steinberg S (1991): Live attenuated varicella vaccine: Current status and future uses. *Semin Ped Infect Dis* 2:171–178

Mahalingham R, Wellish M, Wolf W, Dueland AN, Cohrs R, Vafai A, Gilden D (1990): Latent varicella-zoster viral DNA in human trigeminal and thoracic ganglia. *N Engl J Med* 323:627–631

Watson PN, Evans RJ (1986): Postherpetic neuralgia: A review. *Arch Neurol* 43:836–840

Weller TH (1983): Varicella and herpes zoster: Changing concepts of the natural history, control, and importance of a not-so-benign virus. *N Engl J Med* 309:1362–1368, 1434–1440

Index

RELATED TITLES

Encyclopedia of Neuroscience, Volumes 1 and 2
George Adelman
1987
ISBN   0-8176-3335-9
ISBN   3-7643-3335-9

Neuroscience Year
Supplement 1 to the Encyclopedia of Neuroscience
George Adelman
1989
ISBN   0-8176-3383-9
ISBN   3-7643-3383-9

Neuroscience Year
Supplement 2 to the Encyclopedia of Neuroscience
Barry Smith and George Adelman
1992
ISBN   0-8176-3507-6
ISBN   3-7643-3507-6

Neuroscience Year
Supplement 3 to the Encyclopedia of Neuroscience
Barry Smith and George Adelman
1993
ISBN   0-8176-3592-0
ISBN   3-7643-3592-0